Innovative Production Machines and Systems

Sixth I*PROMS Virtual Conference

15-26th November 2010

Edited by

D.T. Pham

E.E. Eldukhri and

A.J. Soroka

Organised by: FP6 I*PROMS Network of Excellence

Sponsored by: The European Commission

ISBN 978-1-4710-5026-8

Contents

Preface

The 2010 Virtual International Conference on Innovative Production Machines and Systems (IPROMS) took place on the Internet between 15 and 26 November 2010. During the Conference, several thousand registered delegates and guests from around the world participated in the Conference, making it a truly global phenomenon.

This book contains the Proceedings of IPROMS 2010 with the peer-reviewed technical papers presented at the Conference have been grouped into eleven sections:

- Advanced Manufacturing Technology
- E-manufacturing, E-business and Virtual Enterprises
- Fit Manufacturing
- Innovative Design Technologies
- Intelligent and Competitive Manufacturing
- Intelligent Optimisation
- Intelligent Planning and Scheduling
- Production Automation and Control
- Production Organisation Management
- Sustainable Manufacturing Systems

Many of the IPROMS 2010 papers were written by partners and associate partners in the I*PROMS EU-funded FP6 Network of Excellence for Innovative Production Machines and Systems, but there were also large numbers of authors external to the Network. In total, IPROMS 2010 authors were from some 17 countries across five continents.

Numerous people and organisations have helped make IPROMS 2010 a reality. We are most grateful to the IPROMS 2010 sponsors, I*PROMS partners, Conference Programme and Organising Committees, Session Organisers, Chairmen, Authors, Referees, and the I*PROMS Coordination Team.

D.T. Pham, E.E. Eldukhri and A.J. Soroka
MEC, Cardiff School of Engineering, Cardiff University

Programme and Organising Committees

D. Pham (Chair), MEC, Cardiff University, UK
E. Eldukhri (Organising Committee Chair), MEC, *ibid*
A. Soroka (Programme Committee Chair), MEC, *ibid*
S. Dimov, MEC, *ibid*
A. Fahmy, MEC, *ibid*
M. Packianather, MEC, *ibid*
P. Pham, MEC, *ibid*
P. Hines, Cardiff University, UK
M. Naim, Cardiff University, UK
R. Setchi, Cardiff University, UK
N. Aspragathos, University of Patras, Greece
W. Beinhauer, Fraunhofer IAO, Germany
K. Cheng, Leeds Metropolitan University, UK
A. Colombo, Schneider Electric, Germany
B. Denkena, IFW, University of Hannover, Germany
P. Dietz, Clausthal University of Technology, Germany
J. Efstathiou, University of Oxford, UK
F. Feenstra, TNO, The Netherlands
B. Grabot, ENIT, France
R. Harrison, Loughborough University, UK
S. Hinduja, University of Manchester, UK
M. Hoepf, Fraunhofer IPA, Germany
E. Hohwieler, Fraunhofer IPK, Germany
A. Kusiak, The University of Iowa, USA.
R. La Brooy, RMIT University, Australia
A. Labib, University of Portsmouth, UK
V. Marik, Czech Technical Univ., Czech Republic
F. Meo, Fidia S.p.A, Italy
A. Nearchou, University of Patras, Greece
G. Putnik, University of Minho, Portugal
E. Oztemel, Sakarya University, Turkey
V. Raja, University of Warwick, UK
Z. Salem, University of Aleppo, Syria
T. Schlegel, University of Stuttgart, Germany
R. Teti, University of Naples Federico II, Italy
A. Thatcher, University of the Witwatersrand, S. Africa
M. Wu, University of Derby, UK
X. Xie, INRIA, France
X. Yu, RMIT University, Australia

Special Session Organisers

R. Teti (Intelligent and Competitive Manufacturing Engineering - ICME)

Session Chairs

Advanced Manufacturing Technology - Eldaw Eldukhri
E-manufacturing, E-business and Virtual Enterprises - Richard Barton
Fit Manufacturing - Andrew Thomas
Innovative Design Technologies - Kok Ng
Intelligent and Competitive Manufacturing - Roberto Teti and Michael Packianather

Intelligent Optimisation - Mei Choo Ang
Intelligent Planning and Scheduling - Ashraf Afify
Production Automation and Control – Ashraf Fahmy
Production Organisation and Management - Oludare Williams
Robotics - Eldaw Eldukhri
Sustainable Manufacturing Systems - Zuhriah Ebrahim

Reviewer Panel

T. Alirezaee,
Z. Ahmed,
M. Alhamar,
F.L. Amorim,
M.C. Ang,
D.B.H. Anh,
N.A. Aspragathos
R.A. Barton,
I. Bassani,
G.R. Belmonte,
L. Benyoucef,
C. Bitondo,
K. Bönisch,
A. Britton,
C.P. Buckley,
C. Burvill,
C.U. Bussa,
F.J. Campa,
G. Campatelli,
F. Cao,
M. Castellani,
S.Y. Coleman
T. Czelusniak,
D. D'Addona,
I. De Iorio,
X-T. Doan,
M. Durante,
S. Dusch,
K.M. Ebrahimi
E.E. Eldukhri
A. Fahmy
A. Favi,
B. Field,
A. Formisano,
A. García-Domínguez,
S. Genna,
N. Giannopoulos,
G. Goli,
T. Goszczynski,
A. Hassan,
L.C. Hieu,
N.V. Hieu,
C.F. Higa,
A. Langella,
M. Leary,
J.Y. Lee,
C. Leone,
L.N. Lopez de Lacalle
V. Lopresto,
M. Marcos-Bárcena,
A. Marzi,
H. Marzi,
E. Marzi,
I. Masood,
M. Mazur,
I. Medina-Bulo,
F. Memola Capece Minutola
N. Müller,
A. Nearchou
K. Ng
M.A. Noronha,
M.I. Okereke,
M.S. Packianather,
N. Pagano,
B. Palumbo,
H.S. Park,
V. Popov,
D.T. Pham,
Z. Pilat,
U. Prisco,
A. Prisco
J. Ramirez,
M.H. Saw,
T. Schlegel,
A. Scippa,
G. Serroni,
R. Setchi,
I.G. Shäffer,
E. Siemann,
K. Singhal,
M. Slowikowski,
A.J. Soroka
A. Squillace,
A. Subic
A. Synodinos,
I. Tacheva
F. Tagliaferri,
R. Teti
S. Thiel,
A.J. Thomas,
D.Q. Tran,
H. Tutsch,
A. Velenosi,
J.L. Victory,
K. Vidackovic,
J. Zielinski,
N. Zlatov,
A. Zouggari,
J. Zulaika

I*PROMS Central Coordination and MEC Teams

C. Rees
C. Matthews

Innovative Production Machines and Systems
D.T. Pham, E.E. Eldukhri and A.J. Soroka (eds)

Analysis of wood cutting mechanics with different grain orientation

G. Campatelli[a], G. Goli[b], A. Scippa[a], A. Velenosi[a]

[a] *Department of Mechanical Engineering and Industrial Technology, University of Firenze, Italy*
[b] *Department of Agricultural and Forest Economics, Engineering, Sciences and Technologies, University of Firenze, Italy*

Abstract

The study of wood cutting mechanics during has been poorly studied if compared to metal cutting because of the easiness of machining operations and the highly variable characteristics of the material that are difficult to model. However with the rising of technological level in the wood machining sector arise the need to improve the efficiency and the productivity of such cutting processes. One of the most critical issue is the strong anisotropy of the material due to the wood grain orientation. The aim of this work is the analysis of the different cutting mechanics that arise for different orientation of the grain respect to the cutting direction. To analyze the cutting mechanics many experimental tests has been carried out, acquiring the cutting force direction and modulus and an high speed video of the process. The analysis of the experimental tests has allowed the identification of four different cutting mechanics for which a brief explanation has been provided in the paper mainly using geometrical reasoning.

Keywords: Wood cutting, Cutting forces prediction, Orthogonal cutting

1. Introduction

The evaluation of the cutting forces for wood cutting is strategic for the advance of manufacturing technologies for non consumable goods. The ever increasing demand for ecological, biodegradable and recyclable materials is improving the perspective to use wood for the production of furniture, houseware and structural part for buildings; house made of wood have good insulation and low cost and are becoming very attractive for the market. The prediction of cutting forces in different conditions of the material would allow the increase of the productivity both enabling the operator of the machine to chose the highest possible speed of cut considering the maximum power of the spindle and to optimize the positioning of the wood on the machine to minimize the power consumption and improve the quality of the resulting surface. The cutting of metal materials has been studied in detail and many authors have developed ad hoc cutting models; on the contrary wood cutting processes are still a nearly explored field. This is due mainly to two reasons: the great variability of characteristics of the wood depending on species and ageing and to the relevant anisotropy of the material [1]. Some partial cutting models has been developed [2,3] and they take into account the different characteristic of the wood species considering factors such as the strength, the hardness, the resilience and the humidity of the wood. The cutting models developed in such way are able to predict the cutting force depending only on the depth of cut and front rake angle. This angle is responsible for the production of three chip types associated to different cut mechanics, as presented in figure 1.

Fig. 1. Three different type of wood chip.

In order to take into account also the anisotropy of the material is necessary to consider also the fibers orientation. The general idea of this work is to develop a cutting model that could link the modulus and direction of the cutting force to the fiber orientation. Most of the authors have been studied the influence of the cutting speed on the mean cutting force but neglected this aspect of the cutting process, strongly connected with the anisotropy of the material.

Franz [1] developed a first study on the subject evaluating the influence of cutting parameters, such as speed and depth of cut, on the cutting forces. Many other authors focused on specific issue of the cut, such as Thibaut [4] that observed a decrease of friction coefficient when the cutting speed was increased, or Marchal [5] that evaluated the increase of cutting forces values on rake and clearance faces with increasing cutting speeds.

2. Experimental setup

The experimental campaign has been conducted at high speed cutting condition (~ 8 ms^{-1}) on the Douglas-fir wood specie (*Pseudotsuga Menziesii* Franco Var. Menziesii [6]), using different grain orientation. The cutting technique was "free orthogonal cutting", with different tangential orientation of the grain. The cutting device was built by the Unversity of Reading and the tests performed in cooperation with the university of Firenze whom a complete description is reported in [8]

The samples (14 mm width, 20 mm long and a minimum 35 mm deep) were cut from straight grain tangential boards and in close proximity (see Fig. 2), in order to minimize the effects of the wood variability, and equilibrated at nearly the same moisture content (about 11%).

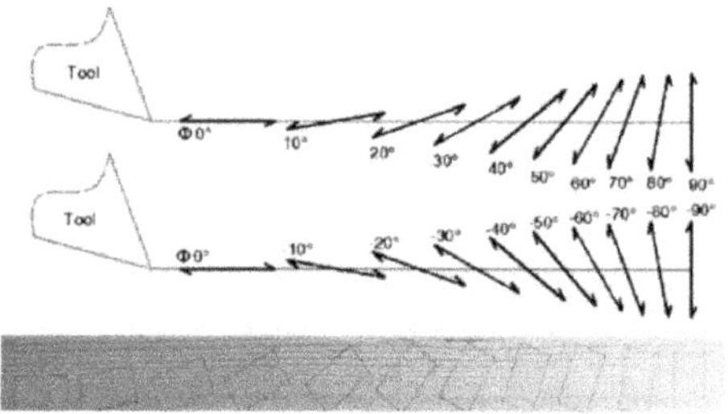

Fig. 2. A piece of Douglas-fir with the specimens to be cut and grain orientation description.

In order to perform high cutting speeds a Hopkinson bar system was adapted to incorporate a specially designed cutting rig (see Fig. 3). The rig consisted of a base that housed a brass guide ring and a sliding bar to which a cutting blade was fixed. The blade was a commercial insert for wood made of sintered tungsten carbide (rake γ: 20°; wedge β: 55°; clearance α: 15°).

Fig. 3. The experimental setup.

For each test has been acquired the cutting forces (normal and tangent components) and a high-speed video to identify qualitatively the cutting mechanics and chip shape.

The sample holder was fixed to an orthogonal dynamometer (see Fig. 4), built and calibrated in order to measure the cutting forces: the adopted ring was that proposed by M. C. Shaw [7], which allowed readings of two cutting force components (parallel and normal)

Fig. 4. The load cell with sample holder.

The cutting process was filmed using a Phantom

7 high-speed camera (13 kfps).

The analysis of the experimental data has shown a "dynamically weak" load cell's supporting structure with many resonances in the range of the measured cutting forces, acquired at 100 kHz.

Because the same resonance frequencies are present in all the cutting configurations, despite the very different cutting mechanics, it makes sense to consider these frequencies as normal modes of the supporting structure.

For this reason only the mean cutting forces have been considered in this work: the normal component (F_n) and the parallel one (F_p) of the cutting forces are evaluated according the following standard (Fig. 5).

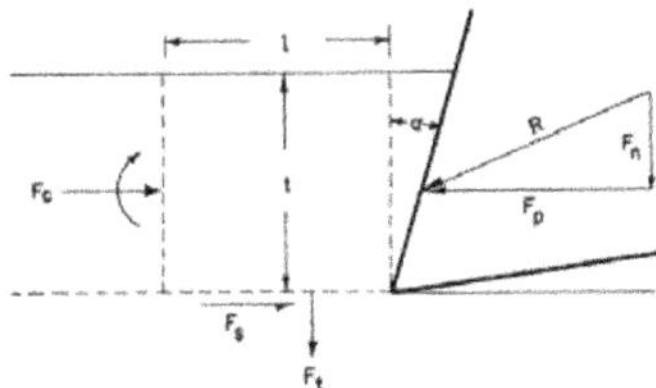

Fig. 5. Standard for the sign of the normal and parallel cutting force components.

The results of the experimental measures are summarized in the following graph in which is reported the relationship between the orientation of the grain and the cutting forces (see Fig. 6) for a depth of cut of 0,2 mm.

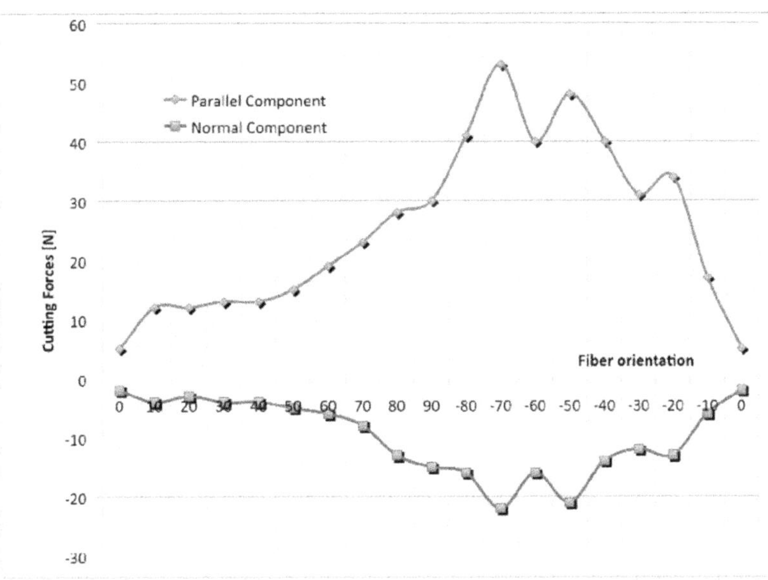

Fig. 6. Mean cutting forces vs. grain's orientation.

In this experimental campaign the attention has been focused on the grain orientation angle only, not considering the effect of the rake angle on cutting force. This choice has been made to analyze more specifically a factors that few models have till now considered. Moreover the effect of the grain orientation is more relevant respect to the effect of the rake angle. Geometrically the cutting force mechanics depend by the combination of these two angles but, where the rake angle have only a reduced variation, the grain orientation angle could vary from 0° to 360°. For the experimental campaign the value of the rake angle has been considered fixed to 20°, the recommended angle for the material chosen.

For the same reason to reduce the focus of the work on the effect of grain orientation the cutting speed has been fixed to a high value, compatible with the thermal stability of material-cutting tool. Some tests has been carried out also with lower speed, also reported in the following figures in order to have a more clear picture of the cutting phenomena, but the study of the cutting forces has included only the test at the highest speed, the only ones that has an industrial interest.

Finally, analyzing the experimental relation between the mean cutting forces and the orientation of the fibres (see Fig. 6) could be observed four range of grain orientation where chip formation offers different cutting mechanics.

3. Analysis of the chip formation and model of geometric interpretation

Analyzing the experimental relation between the mean cutting forces and the orientation of the fibres (see Fig. 6) could be observed four range of grain orientation where chip formation offers different cutting mechanics.

1) Cutting with positive grain angle orientation θ: cutting mechanics is described by a steady shear plane given by the grain orientation (Fig. 7). Fibres are transversally cut and slide along the grain.

2) Cutting approximately perpendicular to the grain orientation: fibres are initially bent so presents positive angle orientation that allows fibres to slide along themselves such as the previous case (Fig. 8).

3) Cutting against the grain orientation: fractures starts from the cutting edge and propagates along the fibres, that are cyclically raised up on the rake face of the tool and sheared by the tip of the tool (Fig. 9). The surface obtained is the most irregular among the different cases previously described.

4) Cutting along the grain: fractures propagate cyclically and chip is formed by flexural rupture at the end of fracture (Fig. 10).

In order to predict cutting forces with different grain orientation in wood cutting, a model to evaluate chip formation has to be developed considering each of the four cutting mechanism. The model proposed in this paper is articulated in four 2-D sub-model, considering the orthogonal cutting condition.

Fig. 7. Regular chip formation sliding along the shear plane. Low speed cutting on the left, high speed cutting on the right.

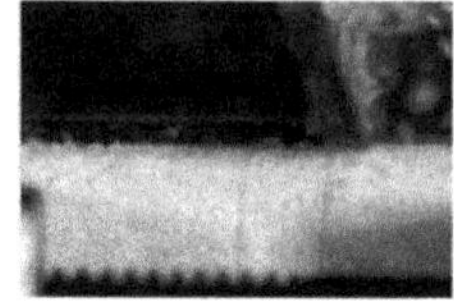

Fig. 8. Chip formation sliding along the shear plane after grain orientation bending. Low speed cutting on the left, high speed cutting on the right.

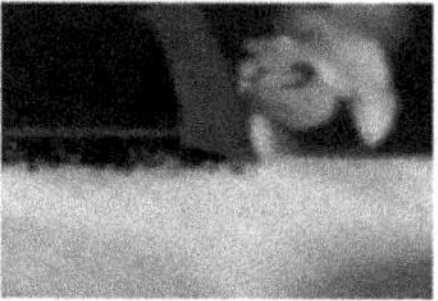

Fig. 9. Fractures propagated in the specimens and fibers sheared by the tool. Low speed cutting on the left, high speed cutting on the right.

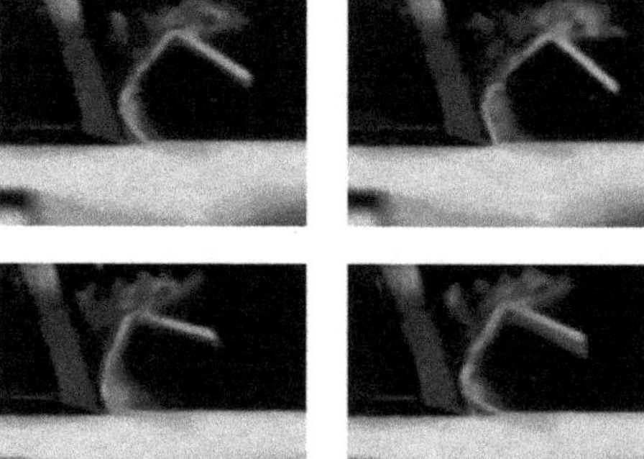

Fig. 10. Cyclic fracture and flexural rupture in high speed cutting along the grain.

3.1. Cutting with positive grain angle orientation

In this configuration with positive grain angle orientation θ, from 10° to 60°, cutting forces can be correlated to tangential yield stress along the grain.

Moreover, tangential yield stress can be correlated to the rate of sliding along grain direction, V_s. Considering V_c as the chip flow rate, V_s could be obtained as the resulting vector between the grain angle orientation and speed cutting V_t (see Fig. 11). In this case a fit model for the forces is the following, where τ is the tangential strength of the wood.

$$R = \frac{\tau_s \cdot A}{\cos(\vartheta - \gamma)}$$

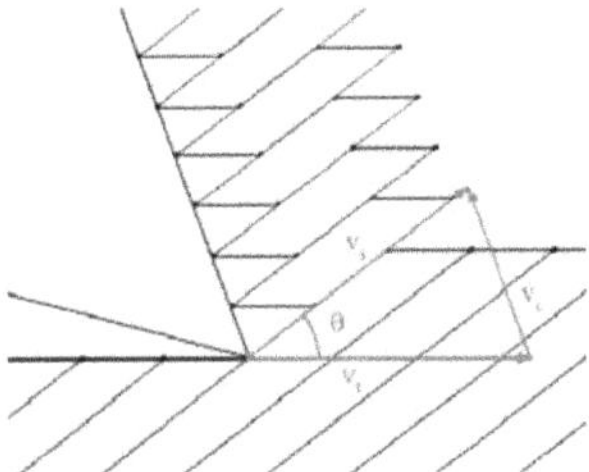

Fig. 11. Geometric interpretation of chip formation with positive grain angle orientation.

3.2. Cutting approximately perpendicular to the grain orientation

This cutting mechanic could be associated to the grain orientation from 70° to -50°. In this case could be observed the presence of a bending plane in front of the cutting edge. The fibers that cross this plane are bend till a positive grain orientation that makes the mechanics of cutting similar to the previous case, with the sliding of the grains.

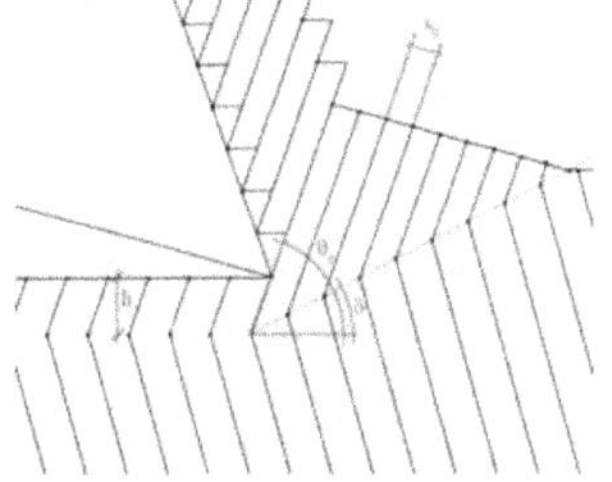

Fig. 12. Geometric interpretation of chip formation processing approximately perpendicular to the grain orientation. The *bending plane* in red.

This model could be partially validated by the observation of the final positive bending of the fibers near the machined surface (see Fig. 13).

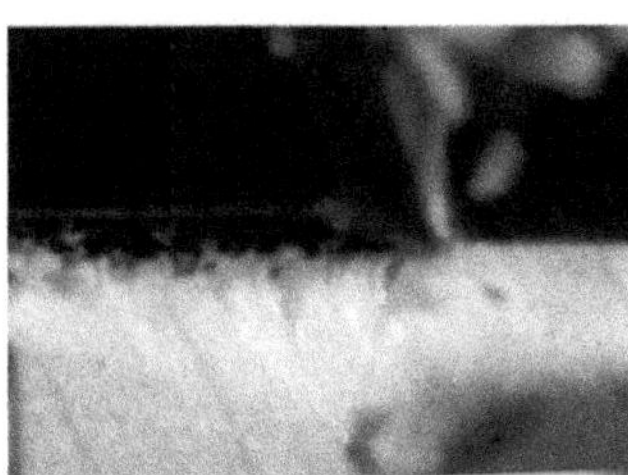

Fig. 13. Image of the bent fibres for nearly perpendicular cutting.

In this region the cutting forces are higher than the previous one because to the force necessary to slide the fibers along themselves is added the bending force that act on the bending plane. A fit model could be the following, where δ is the angle of the banding plane and S_0/S the compression rate of the fiber.

$$\delta = \vartheta_0 + \arccos\left[\frac{s_0}{s} \cdot \cos\left(\vartheta - \delta - \frac{\pi}{2}\right)\right] - \frac{\pi}{2}$$

3.3. Cutting against the grain orientation

Processing against the grain orientation, from -40° to -10°, the third geometric sub-model considers the cyclic lift and the elastic flexional behavior followed by a blanking of the fibers. Fig. 14 shows how the original fibers in the triangular area ADB are curved on an arc and a fracture propagates along the grain. In this case the fibers do not have a sliding but only a blanking when the cutting edge reach the base of the fiber.

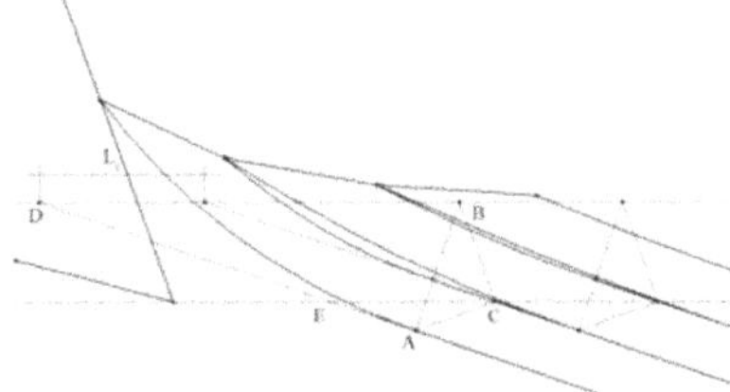

Fig. 14. Geometric interpretation of chip formation processing against the grain orientation.

In this case, if it is considered that the highest contribution to the cutting force is given by the bending of the fibers and their sliding the cutting force could be modeled by the following equation where F is the force orthogonal to the front of the tool.

$$R = \frac{F_{\perp\theta}}{\cos\left(\vartheta - \gamma - \frac{\pi}{2}\right)}$$

3.4. Cutting along the grain

The cutting forces for a configuration with a grain orientation angle very low can be evaluated considering the wood elastic behavior and the fracture toughness of material. In this case, while the blade goes forward, the specimens is elastically deformed storing potential energy that is released when a fracture starts along the grain (Fig. 15). In this case the cutting force is proportional to the fracture toughness of the material multiplied for the width of the cutting zone.

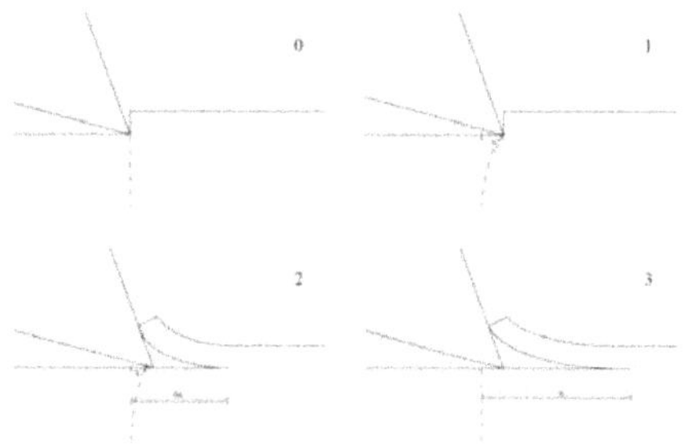

Fig. 15. Geometric interpretation of chip formation processing along the grain orientation.

The geometric interpretation proposed for cutting along the grain orientation suggests that the cutting force F_t is not constant while the tool is going forward and it will show a cyclic trend that oscillated between the fracture creation and fracture propagation zone.

3.5 Summary of the models

To provide a schematic representation of the four cutting zones the most relevant features of each zone is reported in Table 1. The mean forces are calculated suing the experimental test data and not the interpolating curve.

Table 1. Cutting zone and mechanics table

Zone	1	2	3	4
Angle range (θ)	10° to 60°	70° to -50°	-40° to -10°	0°
Cutting mechanics	Sliding of fiber	Bending and sliding	Bending and fracture	Fracture
Mean cutting forces (N)	14,0	37,6	30,5	5,0

Conclusions

The analysis of the cutting forces for wood with different grain orientation respect has allowed to identify that the cutting mechanics of wood is very complex. The authors has evaluated at least four different zone, respect to the grain orientation angle, that are characterized by different cutting mechanics. For each zone a brief description of the possible model has been provided, using a geometrical approach. This research could be the base for the development of a more detailed analytical model for wood, and other anisotropic materials, cutting.

Acknowledgements

The authors want to thank the University of Reading for the implementation and availability of the test equipment.

References

[1] Giordano G., 1985, Tecnologia del Legno, v. III, Parte prima, 2 ed., Torino, Editrice Torinese.

[2] Franz N.C., 1958, An Analysis of the Wood-Cutting Process, University of Michigan Press, Engineering Research Institute Publications Reports.

[3] McKenzie W.M., 1967, The Basic Wood Cutting Process, Proc. Of the 2nd IWMS, pp.3-8.

[4] Thibaut B., 1988, Le processus de coupe du bois par déroulage. Thèse del'Université des Sciences et Techniques du Languedoc.

[5] Marchal R., 1989, Valorisation par tranchage et déroulage des bois de chênes méditéranéens. Thèse de l'Institut National Polytechnique de Lorraine/ Université des Sciences et Techniques du Languedoc.

[6] AA.VV., 1986, Le Douglas, Cahier du Centre Technique du Bois et de l'Ameublement, CTBA.

[7] Shaw M C., 2004, Metal Cutting Principles. Oxford Series on Advanced Manufacturing.

[8] Wyeth, D. J., Goli, G., Atkins, A. G., 2009, Fracture toughness, chip types and the mechanics of cutting wood. A review COST Action E35 2004–2008: Wood machining – micromechanics and fracture. Holzforschung, 63(2), 168-180. doi: 10.1515/HF.2009.017.

Innovative Production Machines and Systems
D.T. Pham, E.E. Eldukhri and A.J. Soroka (eds)

Micromechanical modelling of finite deformation of thermoplastic matrix composites

M.I. Okereke[a], C.P. Buckley[b], L.C.Hieu[a]

[a] *School of Engineering, University of Greenwich, Kent ME4 4TB, UK*
[b] *Department of Engineering Science, University of Oxford, Oxford OX1 3PJ, UK*

Abstract

The prediction of the constitutive behavior of thermoplastic matrix composites from quasi-static up to impact rates demands a detailed understanding of the behavior of the polymeric constituents of these materials; this is due to the pronounced rate dependence of the polymeric matrix. This paper is an attempt at approaching the prediction of finite deformation of thermoplastic matrix composites, using a multi-scale approach in which the fibre and the matrix are separately modelled and combined within a finite element scheme to determine the constitutive response of the test composite. A micromechanical model comprising a finite element implementation of constitutive laws for the fibre and matrix constituents are discussed. The robust formulation for predicting the behavior of the semicrystalline polymer was successfully developed, including the techniques of generating the 3D representative volume element (RVE) of composites as well as prescribing the periodic boundary conditions on the 3D RVE. Finally, the validation studies for predicting the elastic properties of the composite using the Finite Element (FE) methods and the effect of spatial arrangement of the fibre inclusions within the matrix at finite strains are illustrated.

Keywords: Finite element method, thermoplastic matrix composite, micromechanical modelling

1. Introduction

Continuous fibre, thermoplastic matrix, composite materials are attractive for the high volume production, because they combine the good features of manufacturing economics with some of the stiffness, strength and density advantages of the more widely used thermoset matrix composites. They have the potential for industrial and advanced engineering applications, including manufacturing of the components of light-weight cars of the future. However, these materials offer new challenges about prediction of their properties in-use, arising from the pronounced viscoelasticity and plasticity of the matrix polymer, and its sensitivity to the thermal and mechanical history during processing. This study is an attempt at approaching the prediction of the finite deformation of thermoplastic matrix composites in which the modelling approach involves a multi-scale modelling of the composite by tracking deformation from very small strains where linear viscoelasticity conditions apply to finite strains which are dominated by nonlinear viscoelasticity effects. The modelling strategy is divided into microscale and mesoscale levels of analyses. The former deals with a microscale representation of the constitutive models for semicrystalline polymers and the fibre. The later considers the lamina level representation and modelling of the composite.

In this paper, a finite element implementation of the proposed micromechanical model, at the lamina-level, is emphasised. The test composite used in the study is Plytron, a glass fibre polypropylene matrix composite. The following topics are presented in the next sections: (a) development of the robust matrix model for semicrystalline polymers; (b) design of a novel method of generating the 3D representative volume element (RVE) of continuous fibre composites; (c) implementation of the periodic boundary conditions (PBCs) and application of single load cases to the 3D RVE; (d) and FE implementation of the chosen homogenization strategy at the lamina-level to predict a nonlinear finite deformation of the composite.

2. Development of the robust matrix model

A robust physically-based constitutive model was developed for modelling the experimentally observed constitutive response for polypropylene; and the test matrix is polypropylene, a semicrystalline polymer. The 1D mechanical analogue for the model is shown in Fig. 1, while Fig. 2 shows the comparison between the model and experimental data for the compression tests on polypropylene. The model prediction is thought to capture accurately the observed experimental response. The matrix modelling principle is an extension to two-process viscoelastic relaxation of a single-mode glass-rubber constitutive model for amorphous polymers[1-3].

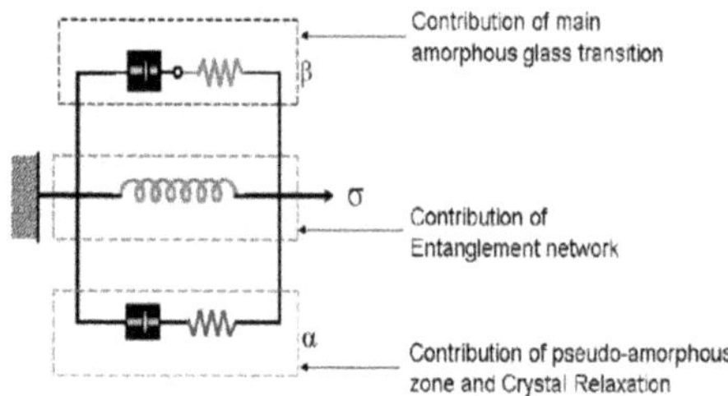

Fig. 1. 1D Mechanical Analogue for modelling of semicrystalline.

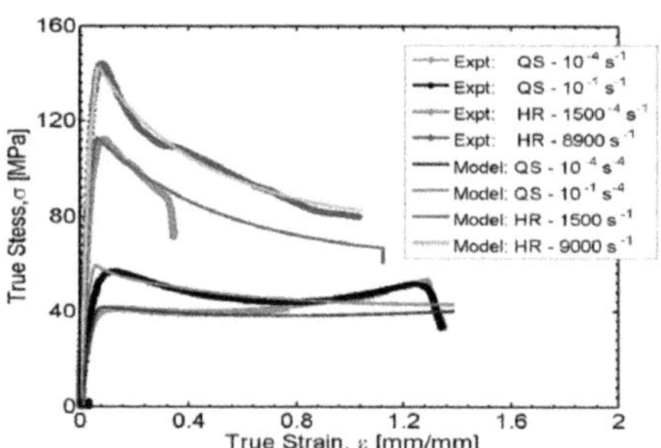

Fig. 2. Comparison of experiments with model prediction for compression tests on polypropylene.

3. A new algorithm for generating 3D RVEs

Having developed a robust matrix model, the next stage of the micromechanical modelling approach is the development of a RVE for the test material. A MATLAB algorithm was developed based on the Monte Carlo Method or Hard Core model[4-6] in which a defined 2D RVE window is populated randomly until a defined volume fraction is achieved. An extra constraint of *periodicity of material* was applied on the generated RVE. For every fibre inclusion that is cut by a boundary wall, the corresponding half of the inclusion is replicated at a corresponding opposite and parallel wall.

Typical RVEs generated using the above approach is shown in Fig. 3 while Fig. 4 shows the strategy for creating a 3D RVE for use in the micromechanical modelling.

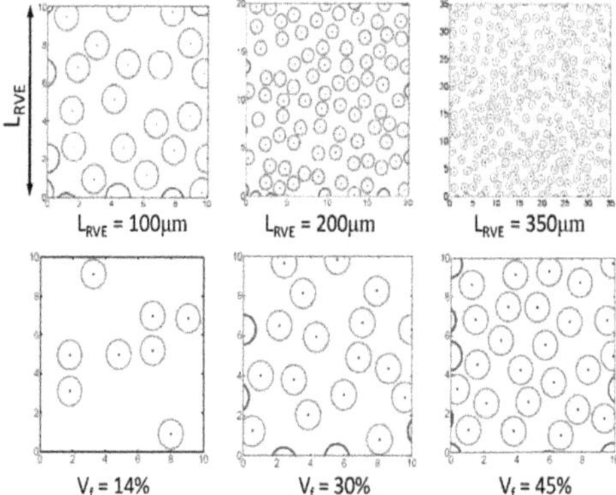

Fig. 3. Typical 2D RVEs generated using the Monte Carlo Algorithm for the different RVE window sizes (LRVE) and different volume fractions (Vf).

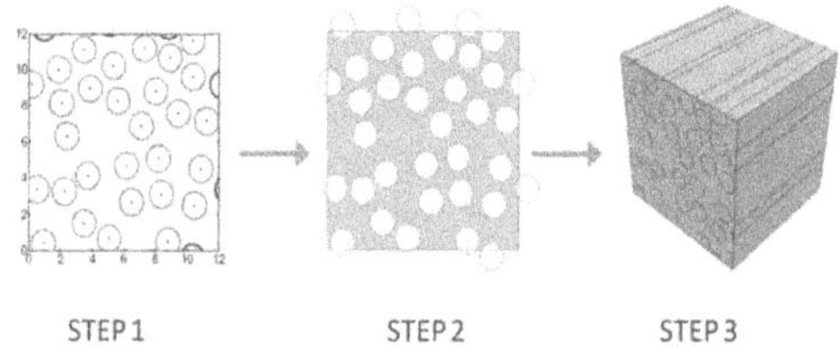

Fig. 4. A three step implementation for creating 3D RVEs. Step 1: 2D RVE generated using Monte Carlo Algorithm. Step 2: Use of Python script within ABAQUS to covert the 2D RVE to the ABAQUS assembly model where white circles are the fibres and a green region is the matrix. Step 3: Extrusion of the 2D model to create a 3D RVE.

4. Implementation of the PBCs on 3D RVEs

Traditionally, the periodic boundary conditions are generally applied to 2D RVEs such that the homogeneous deformation is enforced on the boundary nodes of a given 2D RVE as shown in Fig. 5. This work serves as the first instance where the PBCs are applied to 3D RVEs with the random spatial arrangement of inclusions. This implies applying the homogeneous deformation equations shown in Fig. 6, to all six surface nodes, 8 corner nodes and 12 edge nodes of a 3D RVE.

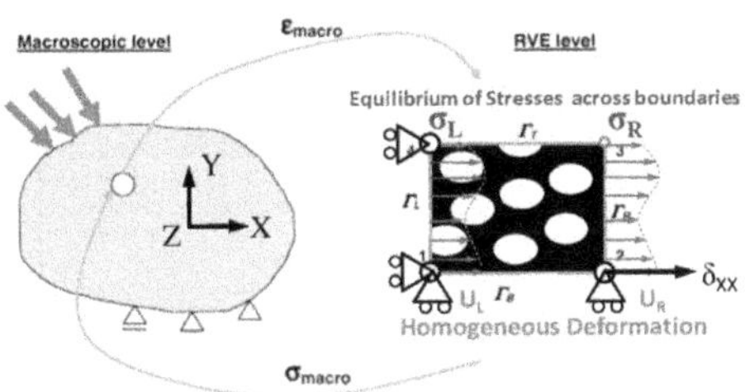

$$U_T - U_B = U_4 - U_1 \quad \text{and} \quad U_R - U_L = U_2 - U_1$$

Retained Nodes (1,2 & 4)

Fig. 5. A strategy for applying the Periodic Boundary Conditions (PBCs) on 2D RVEs.

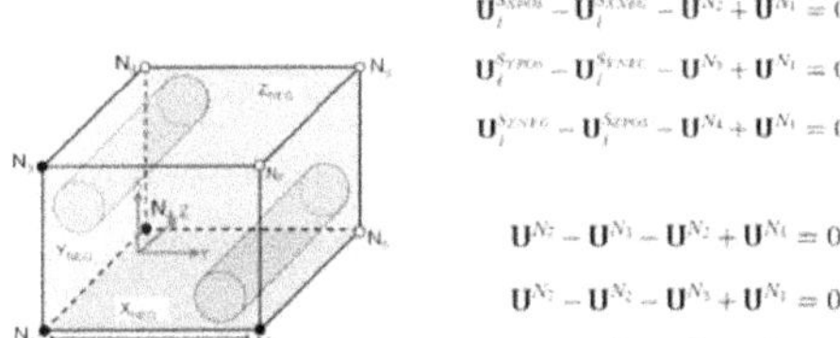

Fig. 6. Strategy for applying Periodic Boundary Conditions (PBCs) on 3D RVEs and list of applicable homogeneous deformation equations for the given 3D RVE domain.

The following shows examples of the simulations based on 3D RVEs of the polypropylene-glass fibre composite where the z-axis corresponds with the fibre direction. Figure 7 presents the logarithmic strain for the compression tests along the x- and y-axes. Figure 8 shows the von mises stress in the z-axis and out-of-plane shear deformation (xy). The logarithmic strain for the out-of-plane (xz) and in-plane (yz) shear deformations is shown in Fig. 9.

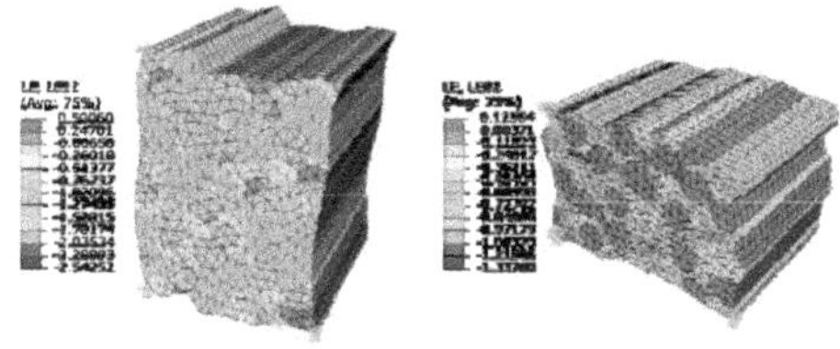

Fig. 7. The logarithmic strain for the compression tests along the x- and y-axes.

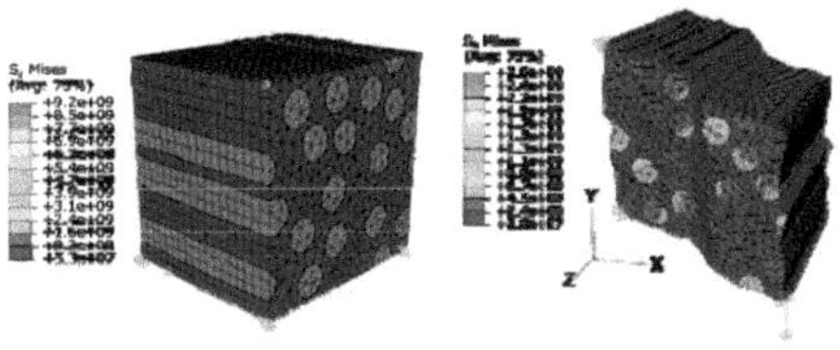

Fig. 8. The von mises stress in the z-axis and out-of-plane shear deformation (xy).

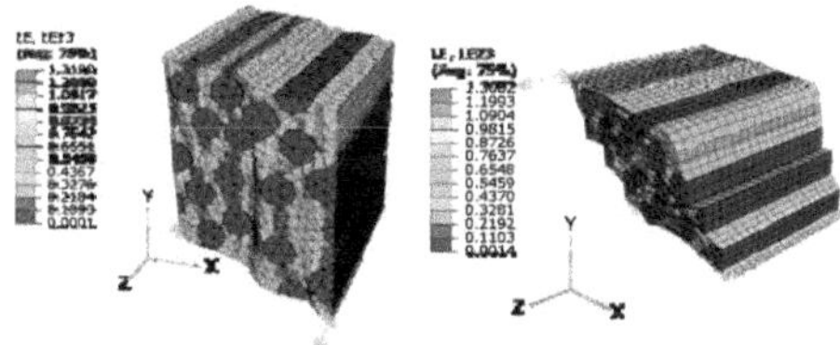

Fig. 9. The logarithmic strain for the out-of-plane (xz) and in-plane (yz) shear deformations

5. Homogenization strategy

In order to derive the constitutive properties of the test composites, the generated 3D RVE implemented with the PBCs has to be used to determine homogenized properties. A homogenization strategy based on the Direct macro-micro relationship[7-11] was adopted in this work. The previous works adopted a 2D RVE where three retained nodes (for the RVE) are used to prescribe any desired load case. This work extended the homogenization strategy above for a 3D RVE such that four retained nodes are used to prescribe the 3D homogeneous deformation for the given RVE.

Consider a typical 3D RVE domain (Ω_{RVE}) such that there exist four retained nodes (N_1, N_2, N_3 and N_4). The coordinate positions for these nodes become: $\mathbf{x_1}$, $\mathbf{x_2}$, $\mathbf{x_3}$, and $\mathbf{x_4}$). The corresponding reactions forces (in 3D) for the four nodes include: f_{N1}, f_{N2}, f_{N3}, and f_{N4}.

The formulations for determining the overall stresses and strains at macroscale based on the reaction forces; and coordinate positions of chosen retained nodes are shown below:

$$\sigma_{macro} = \langle \sigma \rangle = \frac{1}{V}\left\{ \mathbf{x}_1 \otimes \mathbf{f}_{N1} + \mathbf{x}_2 \otimes \mathbf{f}_{N2} + \mathbf{x}_3 \otimes \mathbf{f}_{N3} + \mathbf{x}_4 \otimes \mathbf{f}_{N4} \right\}$$

$$\mathbf{u}_2 = \epsilon_{macro}(\mathbf{x}_2 - \mathbf{x}_1), \quad \mathbf{u}_3 = \epsilon_{macro}(\mathbf{x}_3 - \mathbf{x}_1), \quad \text{and} \quad \mathbf{u}_4 = \epsilon_{macro}(\mathbf{x}_4 - \mathbf{x}_1),$$

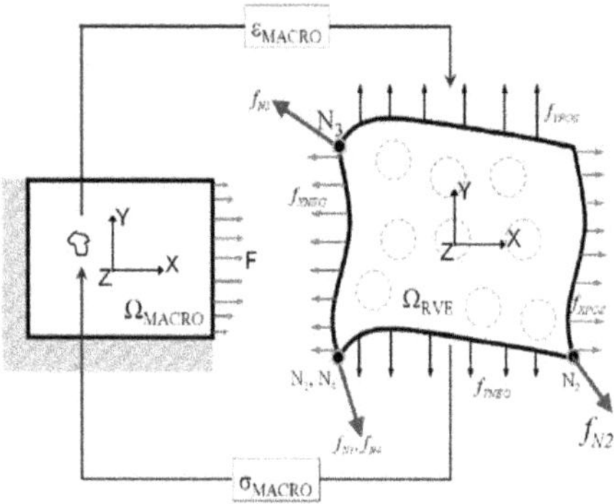

Fig.10. Macro-micro links for an RVE subjected to the periodic boundary conditions.

6. Model predictions

In order to validate the modelling strategy, a boron-aluminum composite of volume fraction 47% was simulated using the above strategy. The experimental data on tests carried out on the boron-aluminum composite [12] and predictions from several prediction approaches were compared with predictions based on this work. One of the other approaches include that attributed to Sun and Vaidya [13] which is an FE method approach using a single-fibre square fibre array 3D RVE. Other approaches are Hashin-Rosen analytical approach based on energy variational principles [14-16], as well as semi-empirical classical laminate theory [17]. This work used two 3D RVEs consisting of (a) one fibre (FEM Small) and size 30μm^2 and (b) 27 fibres (FEM Big) and size 100 μm^2. Table 1 shows the results of comparison between the two approaches.

In order to determine the elastic properties using the above approach, an optimal RVE window size needs to be determined for the test composite. This is the RVE window size at which there is a convergence of all elastic properties for the given RVE window. Figure 11 shows the graph of elastic properties against the RVE window size for Young Modulus. Also, model predictions of rate-dependent transverse compression for polypropylene-glass fibre composite are shown in Figure 12 using the above homogenization approach. Again, the effect of spatial fibre arrangement at nonlinear finite deformation is illustrated in Figure 13. This shows the transverse strain (ε_{22}) contour plots for six different realizations of a 90x90 μm^2 RVE window tested at 25^{0}C.

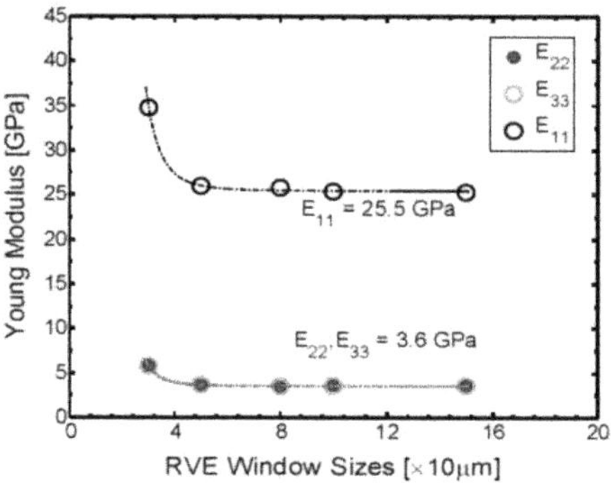

Fig. 11. Variation of predictions of Young Modulus with RVE Window sizes.

Table 1
Comparison of elastic properties of the boron-aluminium composite (v_f=47%) based on different approaches. The fibre axis is along 1-axis and the transverse directions are 2- and 3-axes. Unit: GPa.

Properties	Expt	FEM Big	FEM Small	FEM Sun	Analyt.	Empi.
E_{11}	216	208.7	213.4	214	215	214
E_{22}	140	134.3	141.0	135	139.1 {131.4}	156
E_{33}	-	135.2	141.0	-	-	-
G_{12}	52	52.9	51.9	51.1	53.9	62.6
G_{13}	-	52.8	52.0	-	-	-
G_{23}	-	49.4	45.0	-	54.6 {50.0}	43.6
ν_{12}	0.29	0.299	0.300	0.19	0.195	0.20
ν_{13}	-	0.299	0.300	-	-	-
ν_{23}	-	0.193	0.198	-	0.31 {0.280}	0.31

7. Conclusions

A FE micromechanical model for prediction of the finite deformation of a polypropylene-glass fibre composite is presented. The modelling overview includes (i) development of the robust matrix model, generation of 3D RVEs, (ii) implementation of PBCs on 3D RVEs, (iii) definition of micro & macro homogenization relationships, and (iv) the model predictions for elastic and finite deformations. The effect of spatial arrangement at the finite deformation suggests that in order to obtain the homogenized responses at such large strains, a large RVE window is required. This presents opportunities for further work.

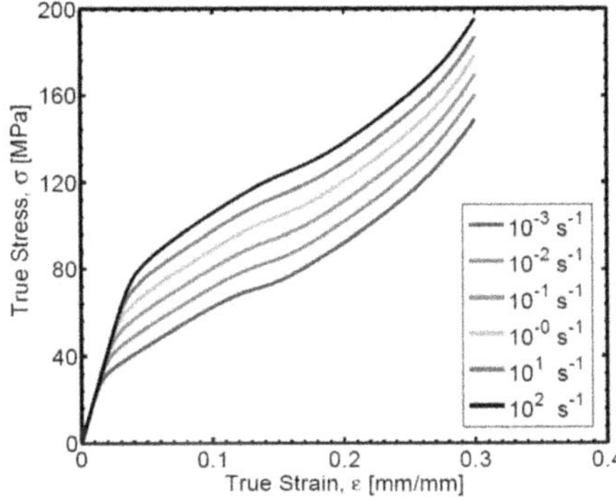

Fig.12. Prediction of rate-dependent transverse compression of test composite.

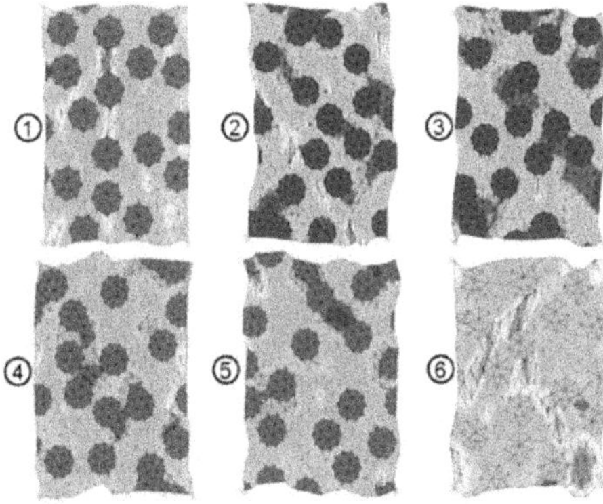

Fig.13. Transverse compression (ε_{22}) contour plots for 6 different spatial arrangements of the test composite.

References

[1] Buckley CP, *et al.*. Deformation of thermosetting resins at impact rates of strain. Part 2: constitutive model with rejuvenation. Journal of the Mechanics and Physics of Solids, 2004, 52(10), pp 2355.

[2] Buckley CP and Jones DC. Glass-rubber constitutive model for amorphous polymers near the glass transition. Polymer, 1995, 36(17), pp 3301.

[3] Buckley CP, Jones DC, and Jones DP. Hot-drawing of poly(ethylene terephthalate) under biaxial stress: Application of a three-dimensional glass-rubber constitutive model. Polymer, 1996, 37(12), pp. 2403.

[4] Buryachenko VA, *et al.* . Quantitative description and numerical simulation of random microstructures of composites and their effective elastic moduli. International Journal of Solids and Structures, 2003, 40(1): pp 47.

[5] Melro AR, Camanho PP, and Pinho S. Generation of random distribution of fibres in long-fibre reinforced composites. Composites Science and Technology, 2008, 68(9): pp 2092.

[6] Wongsto A and Li S. Micromechanical FE analysis of UD fibre-reinforced composites with fibres distributed at random over the transverse cross-section. Composites Part A: Applied Science and Manufacturing, 2005, 36(9), pp 1246.

[7] Van der Sluis O. *et al.*. Overall behaviour of heterogeneous elastoviscoplastic materials: effect of microstructural modelling. Mechanics of Materials, 2000, 32(8): pp 449.

[8] Van der Sluis, O., P.J.G. Schreurs, and H.E.H. Meijer, Homogenisation of structured elastoviscoplastic solids at finite strains. Mechanics of Materials, 2001. 33(9): p. 499.

[9] Van der Sluis O, Schreurs PJG, and Meijer HEH. Effective properties of a viscoplastic constitutive model obtained by homogenisation. Mechanics of Materials, 1999, 31(11), pp743.

[10] Kouznetsova VG, Brekelmans WAM, and Baaijens FPT. Approach to micro-macro modeling of heterogeneous materials. Computational Mechanics, 2001, 27(1), pp 37.

[11]Kouznetsova VG, Geers MGD, and Brekelmans WAM. Multi-scale second-order computational homogenization of multi-phase materials: a nested finite element solution strategy. Advances in Computational Plasticity, 2004, 193(48-51), pp5525.

[12] Kenaga D, Doyle JF, and Sun CT. The Characterization of Boron/Aluminum Composite in the Nonlinear Range as an Orthotropic Elastic-Plastic Material. Journal of Composite Materials, 1987, 21(6), pp 516.

[13] Sun CT and Vaidya RS. Prediction of composite properties from a representative volume element. Composites Science and Technology, 1996, 56(2). pp 171.

[14] Hashin Z. Analysis of Composite Materials - A Survey, in Journal of Applied Mechanics, Transactions ASME, 1983, pp 481.

[15] Hashin Z, and Shtrikman S. A variational approach to the theory of the elastic behaviour of multiphase materials. Journal of the Mechanics and Physics of Solids, 1963, 11(2): pp 127.

[16] Hashin Z. Theory of Fibre Reinforced Materials, in NASA Contractor Reports, 1972.

[17] Chamis CC. Simplified Composite Micromechanics Equation for Hygral, Thermal, and Mechanical Properties. Sampe Quarterly, 1984, 15(3): pp 14 -23.

Innovative Production Machines and Systems
D.T. Pham, E.E. Eldukhri and A.J. Soroka (eds)

Some results on the performance of EDM electrodes produced by SLS technique

F.L. Amorim[b], N. Müller[a], I. Bassani[b], I., G. Schäffer[a], E. Siemann[a], T. Czelusniak[b], M.A. Noronha[b], C.F. Higa[b], K. Bönisch[a], G.R. Belmonte[b]

[a] *Institute for Mechanical Engineering - IMW, Technical University of Clausthal, Robert-Koch-Str.32, 38678 Clausthal-Zellerfel. Germany*
[b] *Laboratory for Research on Machining Processes - (LAUS), Pontifical Catholic University of Paraná, Rua Imaculada Conceição, 1155 – Prado Velho, 80215-901 Curitiba, PR. Brazil*

Abstract

Electrical Discharge Machining (EDM) is a non-conventional machining process widely applied to manufacture of complex, small and hard material components which are not easily machined by conventional processes. The production of complex EDM electrodes is difficult, time consuming and it can account for about 50 per cent of the total process costs. Selective laser sintering can be an alternative technique to produce EDM electrodes in a faster way. This work has carried out an experimental study on the performance of EDM electrodes made by SLS using pure copper, bronze-nickel alloy, copper/bronze-nickel alloy and steel alloy powders. Important EDM performance measures such as material removal rate Vw and volumetric relative wear ϑ were investigated. The results showed that all the electrodes made by SLS promoted a poor EDM performance when compared to solid copper electrode.

Keywords: SLS technique, EDM electrodes, process performance.

1. Introduction

The unique feature of machining any electrical conductive material regardless of hardness makes the electro discharge machining process (EDM) one of the most extensively used non-conventional material removal process. Examples of applications include the precision machining of hardened steels, carbides, ceramics and any material that offers an electrical conductivity higher than 0,01 S/cm [1]. Also, geometrically complex parts that would be extremely difficult to machine using conventional methods can be manufactured with this process [2], therefore ranking EDM as one of the major processes applied in precision mechanic industry to generate complex three dimensional components [3].

In the EDM process, the tool and workpiece electrodes are separated by a gap filled with a dielectric liquid, where pulsed arc discharges occur [4]. Although there is no complete and definite model describing all the processes that occur in a single discharge [5], the best theory to the explanation of the EDM process is the thermoelectric phenomenon, researched by authors such as Van Dijck et al. [6] and DiBitonto et al. [7]. The theory states that the material removal in EDM is related to the erosive effect produced when discrete discharges occur between two electrical conductive materials. Four phases can be distinguished during a single discharge (Fig. 1). The first is the ignition phase, represented by the ignition delay time t_d (s), which corresponds to the interval time between the occurrence of the breakdown of the high open voltage

u_i (V) and the low discharge voltage u_e (V). Right after the first stage, the current rapidly increases to the specified peak current i_e (A), creating the plasma channel. The third phase is the discharge phase, where the energy and pressure of the plasma is sustained for a period of time t_e (s) causing the melting and evaporation of a small amount of material in both electrodes. The forth and last phase is the plasma channel collapse, when the electric energy is turned off, causing the molten material to be violently ejected. This period of time is named pulse interval time t_o (s).

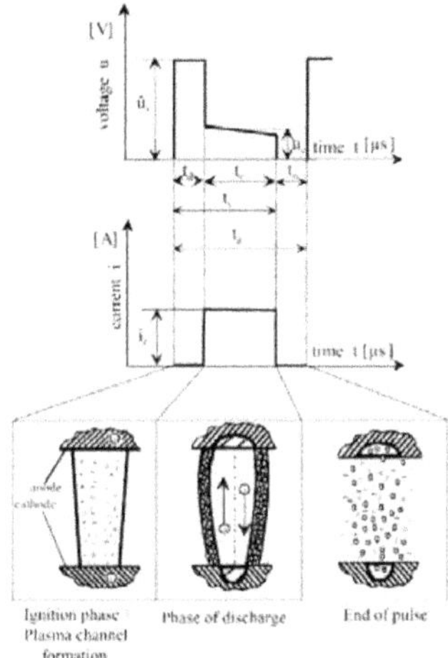

Fig. 1. The phases of an electric discharge in EDM [8].

In the EDM process, the electrode shape is copied in the workpiece. Therefore is essential that the electrode shape replicates the desired workpiece geometry. EDM electrodes are commonly manufactured by conventional machining, e.g. milling. The more complicated the geometry of the desired part, the more difficult, time consuming and expensive is to produce the corresponding electrode by conventional machining. In the mould making industry for example, where EDM is frequently applied, the mould complexity is an industrial demand and this results in an even more time consuming and expensive electrode manufacturing. Hence reducing the electrode manufacturing time and costs will also reduce the component machining time and costs.

Also, as the role of EDM is constantly being redefined, due to new machining technologies, such as high speed milling, it is important to develop new technologies to maintain and improve the use of EDM in precision mechanic industry. One aspect related to this is the production of electrodes in a faster way. An efficient technology for producing EDM electrodes can be selective laser sintering (SLS), one of the Layer Manufacturing Techniques, which can be classified into Rapid Tooling techniques, where the electrodes can be made directly.

The SLS process can build parts with complex shapes directly from a CAD design without any tools. The parts are built from powder materials directly by the layer by layer technique. The materials for this process are different sorts of powder which are processed by the laser beam. This beam is deflected and guided by mirrors and transmits the energy input to the current layer. This causes the geometric data from the slice layer process into the construction process. In this case the powder is melted and there are links between the molten powder particles and the already solidified layer. Shortly after the construction platform is lowered by one layer thickness and the wiper mechanism apply new material at the building platform. Then the construction process repeats.

Fig. 2 illustrates some influencing variables of the building process. These can be divided in the main groups of laser, materials, environment, exposure and machine [9].

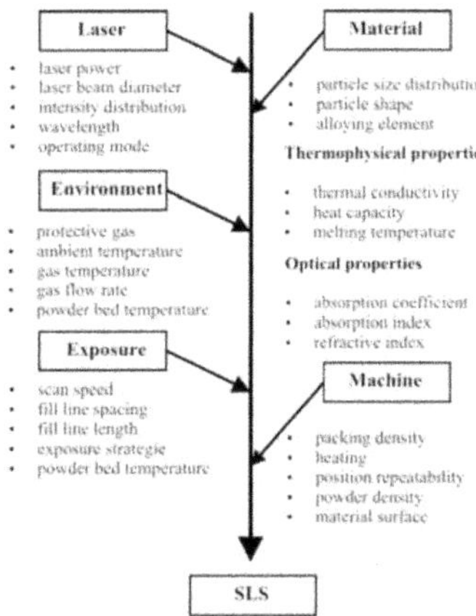

Fig. 2. Ishikawa diagram for SLS/DMLS [9].

The main influence variable in the building process is the energy density, which can be expressed by Eq. 1

$$E_V = \frac{P_L \cdot t_{ww}}{V} = \frac{P_L}{h_s \cdot D_s \cdot v_s} \tag{1}$$

Here, the energy density E_V (J/mm³) is a function (see Eq. 1) of the laser power P_L (W), the exposure time t_{ww}, the scan speed v_S (mm/s), the degree of overlap by hatch distance h_S (mm) and also the layer thickness D_S (mm). If the energy density is too low, the SLS electrode will be porous and have a bad surface quality. On the other side, if the energy density is too high, the electrode can have defects, as well as residual stresses and a poor surface quality also hence a bad EDM performance.

The present work has investigated the EDM performance of electrodes made by the SLS process. This performance was measured in terms of the material removal rate V_w, which represents the volume of material removed from the workpiece per unit of time (mm³/s), and volumetric relative wear ϑ, which represents the ratio electrode wear rate to material removal rate, expressed in percentages.

2. Experimental Procedure

The experimental procedure was divided in two parts, one for the electrodes manufacturing and the other for the EDM tests made. These will be presented below.

2.1. Electrodes manufacturing

The electrodes were manufactured at the Institut für Maschinenwesen – IMW of TU Clausthal, Clausthal-Zellerfeld/Germany. The following materials, equipment and methods were applied:

(i) SLS machine: a EOS M 250 machine equipped with a 200 W CO_2 laser, 0,4 mm of spot radius, pre-heating of the building platform of 80 °C and using nitrogen (N_2) as inert atmosphere.

(ii) Powder materials: four different powder materials were used: pure copper with an average particle size of 50 µm; standard EOS steel alloy (DS20) with an average particle size of 20 µm; standard EOS bronze-nickel alloy (DM20) with an average particle size of 20 µm; a mixture composed by 50 vol% of pure copper and 50 vol% of standard EOS bronze-nickel alloy.

(iii) Electrodes geometry: cubes of 10 mm with a support structure of 3 mm were manufactured for all materials cited above. The support structure is made of the same material as that of the electrodes and serves as a connector between the electrodes and the building platform, facilitating the cutting of the samples.

(iv) SLS parameters: The mains SLS parameters used are shown in Table 1. Except for the steel alloy, two different laser speeds were used for each material. The layer thickness was set according to the average particle size of the material, while the hatch distance was fixed for all materials (0,3 mm). As for the laser sintering strategy, the stripes strategy was used, which means that the laser passes along the powder bed performing lines.

2.2. EDM tests

The Electrical Discharge Machining experiments were conducted at the Laboratory for Research on Machining Processes (LAUS) of the Pontifícia Universidade Católica do Paraná (PUCPR), Curitiba/Brazil. The following materials, equipment and methods were applied for all series of tests:

(i) EDM machine: a Charmilles ROBOFORM 30 CNC machine equipped with a isoenergetic generator, which means that it is possible to set the discharge duration t_e and control the ignition delay time t_d as a percentage of t_e.

(ii) Tool electrodes: As described above, 10 mm cubic electrodes were used to perform the EDM tests. All electrode materials had the same geometry and dimensions.

(iii) Workpiece: AISI H13 steel 12 mm cubic samples were selected as workpieces. They had their surfaces grinded to achieve parallel surfaces.

Table 1. Electrodes materials and main SLS parameters.

Electrode Name	Material	Scan Speed [mm/s]	Hatch Distance [mm]	Layer Thickness [µm]	Laser Strategy
Cu1	Pure copper	100	0,3	50	Stripes
Cu2	Pure copper	50	0,3	50	Stripes
DM1	Bronze-nickel alloy	250	0,3	20	Stripes
DM2	Bronze-nickel alloy	150	0,3	20	Stripes
CM1	Copper/Bronze-nickel alloy	100	0,3	50	Stripes
CM2	Copper/Bronze-nickel alloy	50	0,3	50	Stripes
DS1	Steel alloy	250	0,3	20	Stripes

Table 2. EDM experimental parameters settings

Operation Mode	Generator Mode	Open voltage u_i [V]	Discharge current i_e [A]	Discharge duration t_e [μs]	Interval time t_o [μs]	Tool polarity	Duty factor τ [%]
Finishing	Isoenergetic	160	4	25	6,4	(-)	83,5
Semi-finishing	Isoenergetic	120	12	100	12,8	(-)	90,4
Roughing	Isoenergetic	120	32	200	25	(-)	90,6

(iv) Dielectric: a hydrocarbon dielectric fluid with 3 cSt at 400 °C, flash point of 134 °C and 0,01 wt.% of aromatic contents were used for the tests.

(v) Flushing method: a lateral flushing technique was applied. This method was sufficient to evacuate the excess of eroded particles away from the working gap as well as to promote adequate cooling.

(vi) Electrical variables: the EDM tests were carried out under three different standard machining conditions: roughing (i_e=32 A), semi-finishing (i_e=12 A), and finishing (i_e=4 A). The main electrical parameters for each condition are shown in Table 2. The machining time was 40 min for 4 A, 25 min for 12 A and 8 min for 32 A.

(vii) In order to quantify the material removal rate V_w (mm³/s) and volumetric relative wear ϑ (%) the following procedure was adopted: the density of the electrodes was measured using the buoyancy principle; the mass of the electrodes and workpiece were measured with an accurate balance (resolution 0,0001 g) before and after the EDM process and, regarding the electrodes, their mass after the EDM tests was measured only after they were dried in a furnace at 150 °C for 24 hours, because the electrodes are porous and absorb some quantity of the dielectric fluid. Having the density and mass of both tool electrodes and workpieces V_w and V_E could be determined. The surface roughness of the workpieces was measured using an Taylor Hobson measuring instrument. All the performance parameters were compared with the ones achieved by solid copper under the same machining conditions.

3. Results and Discussion

Fig. 3A shows the results of material removal rate V_w for all the electrodes made by the SLS process for EDM finishing conditions. The solid copper electrodes are not shown on the graphs, but the material removal rate achieved by it under the same conditions was V_w= 1,8 mm³/min. Comparing this value with the ones presented in Fig.3A a large difference can be seen. The steel alloy (DS 1) electrode achieved the highest V_w, but still is very low compared to solid copper. SLS made copper electrodes showed the smallest material removal rate for these conditions. Regarding the bronze-nickel electrodes (DM 1 and DM 2), DM1 had a better performance than DM 2, almost twice as better in terms of V_w. That is, the electrode made with higher speed (less energy) was better in this case. This difference couldn't be seen in the electrodes made by a mixture of copper and bronze-nickel alloy (CM 1 and CM 2) where both electrodes had similar results, regardless of the laser speed that were manufactured.

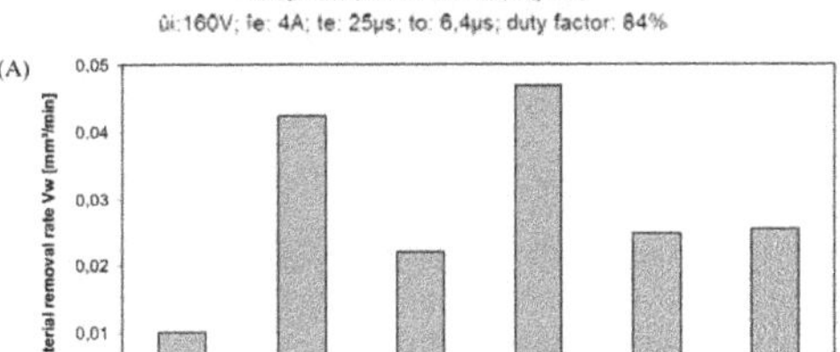

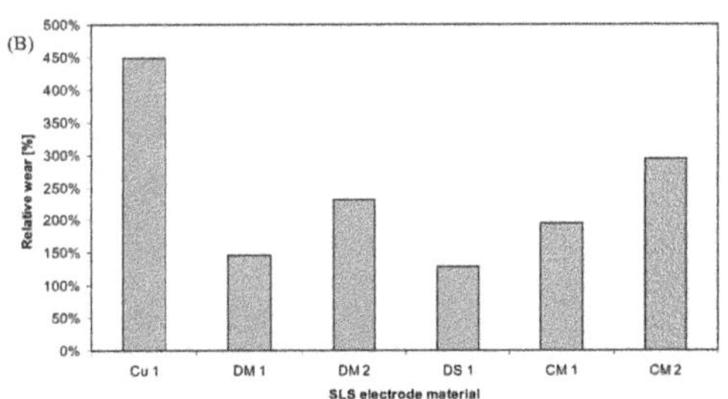

Fig. 3. Material removal rate (A) and relative wear (B) of SLS made electrodes for finishing conditions.

Fig. 3B presents the results of ϑ for all electrodes for EDM finishing conditions. All electrodes had a volumetric relative wear superior than 100%, meaning that more material was removed from the electrode than from the workpiece, while solid copper suffered 2% of volumetric relative wear for the same conditions. DS 1 electrodes had the lowest ϑ among the SLS made materials, while Cu 1 had the highest. Also, the bronze-nickel electrode manufactured with less laser energy (DM 1) had a lower relative wear than DM 2. It is also interesting to notice that while CM 1 and CM 2 had similar results regarding V_w, CM 1 suffered less wear than CM 2.

Results of the electrodes performance for semi-finishing machining conditions are depicted in Fig. 4A and Fig. 4B. Fig. 4A shows results regarding material removal rate. Again great differences can be seen between solid copper electrode performance, that achieved a V_w= 20 mm³/min and the SLS made electrodes performance, where all electrodes presented a material removal rate not greater than 1 mm³/min. Bronze-nickel alloys had the best results for this conditions, with DM 1 being slightly better than DM 2. Similar results were obtained by CM 1 and CM 2, but the differences between both are greater than for DM 1 and DM 2. Pure copper electrodes had the worst V_w, followed by the steel alloy electrode, that had its performance lowered for this conditions, when compared to finishing machining process. Comparing the finishing and semi-finishing conditions, it can be seen that the material removal rate is higher for the last than for the former, regardless of the electrode material. This happens because not only the discharge current is increased, but also the discharge duration gets higher, making possible to remove material for a longer period of time.

In Fig. 4B the volumetric relative wear for the different tested electrodes are presented. Except for copper, the other SLS made electrodes had similar values of relative wear, with DS 1 having slightly higher wear values. Although a increase in the performance of all electrodes could be seen for this EDM conditions, as the same lowered their relative wear when compared to finishing conditions, these results are still poor when compared to pure copper (ϑ =1% for the same conditions).

Results for roughing machining conditions regarding V_w shown in Fig. 5A. For these conditions the steel alloy DS 1 had the worst results, removing almost no material, followed by copper. DM 2 bronze-nickel alloy electrodes had the best material removal rate, together with CM 1, achieving a V_w=18 mm³/min. When compared to solid copper, which had a material removal rate of 45 mm³/min, this difference is less significant than for the other machining conditions. It is worth to mention also that DM 2 removed more material than DM 1, in contrast to the other EDM conditions, where DM 1 was better.

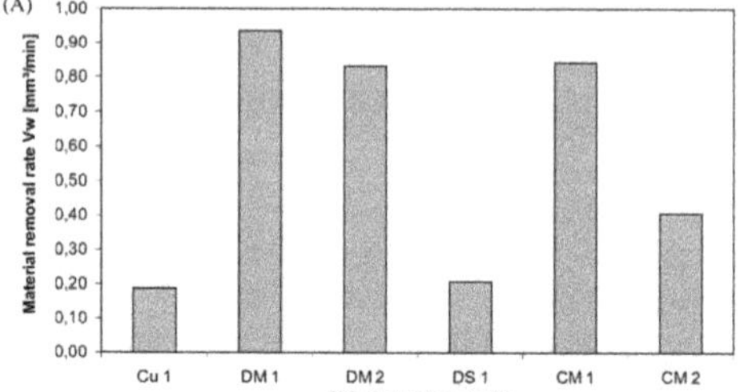

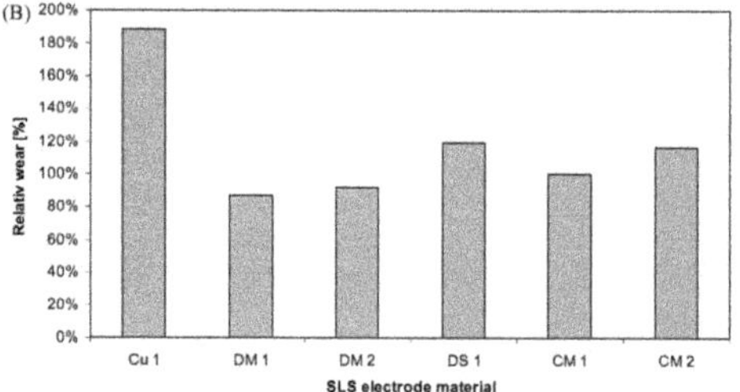

Fig. 4. Material removal rate (A) and relative wear (B) of SLS made electrodes for semi-finishing conditions.

Fig. 5B presents the relative wear of the electrodes. DS 1 had the highest relative wear (more than 200%), followed by Cu 1 (about 100%). The bronze-nickel and copper with bronze-nickel alloys had similar results (between 50 and 60%). Except for DS 1, all the other electrodes improved their performance for roughing condition, having their V_w increased and relative wear decreased.

As it was shown, regarding the EDM performance of the electrodes, the disparity between the ones made by the SLS process and the solid copper electrode is evident. For copper materials, this difference can be credited to the behaviour of the same during the SLS process. Copper powder has a high thermal conductivity and a low absorption coefficient for CO_2 laser [10]. This characteristics lead to a poor SLS performance, as the heat provided by the laser is not well absorbed and is dissipated very quickly, thus resulting in a poor sintering/melting process, a porous part and, consequently, a poor EDM performance.

Regarding the bronze-nickel alloy, this material provides a good sintering/melting behaviour, producing lower porosity parts with good connection between the powder particles. On the other side, the EDM process demands electrode materials that are both good electrical and thermal conductors. Bronze and nickel does not suits this EDM criteria, therefore its poor performance as an EDM electrode. The same idea is applicable to the steel alloy materials, which has good properties for the SLS process but is not suitable as EDM electrode. As for the mixture of copper with bronze-nickel powders, the same analysis can be made. Although in principle the addition of copper may increase the alloy thermal and electrical conductivity, it decreases the SLS behaviour of the same, therefore producing porous electrodes and with thermophysical properties that may be even lower than bronze-nickel materials, leading also to a poor EDM performance.

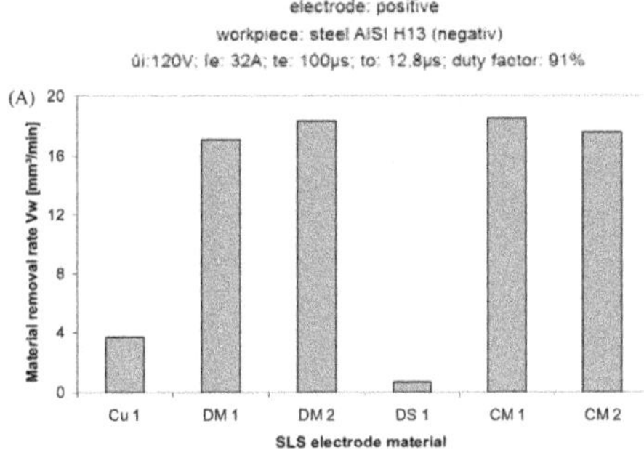

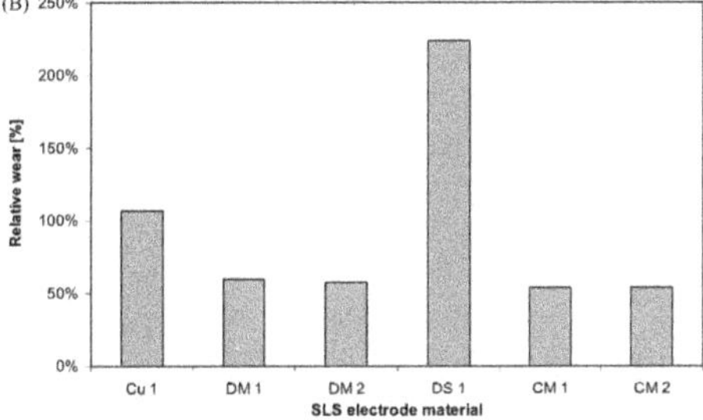

Fig. 5. Material removal rate (A) and relative wear (B) of SLS made electrodes for roughing conditions.

4. Conclusion

In this work a sequence of experiments were performed to evaluate the performance of EDM electrodes made by selective laser sintering (SLS) process. Electrodes of four different material alloys were manufactured and tested under roughing, semi finishing and finishing EDM regimes, being investigated with reference to the material removal rate *Vw* and volumetric relative wear ϑ. From the experimental investigations the following conclusions can be drawn:

(a) The EDM performance of all electrodes was poor for all regimes, achieving low material removal rate and high volumetric relative wear when compared to solid copper electrodes.

(b) Except for steel alloy materials, all the other electrodes improved their performance as the machining condition varied from finishing to roughing regime. For steel alloy electrodes, the opposite behaviour was observed.

(c) In general copper electrodes showed the worst results, removing little material from the workpiece and having a high volumetric relative wear for all the analyzed EDM conditions.

(d) Bronze-nickel electrodes (DM 1 and DM 2) presented the best EDM performance among the investigated electrodes, followed by the mixture of copper and bronze-nickel electrodes (CM 1 and CM 2). For roughing conditions, both materials had similar results, while for finishing and semi-finishing, the bronze-nickel alloy proved to be better.

From what was discussed in this work, further research is needed in order to produce electrodes feasible to be used in EDM applications using SLS technology. The direction of this research has to involve the following topics: selection of new materials that are both suitable for SLS and EDM process demands; optimization of SLS parameters, reducing the electrodes porosity and increasing the bonding between the powder particles; optimization of EDM parameters.

5. Acknowledgements

The authors wish to thank the following founding agencies: from the brazilian side - CAPES (Coordenação de Aperfeiçoamento de Pessoal de Nível Superior), FINEP (Financiadora de Estudos e Projetos), CNPq (Conselho Nacional de Desenvolvimento Científico e Tecnológico); from the german side - DFG (Deutsche Forschungsgemeinschaft.

This work is part of the BRAGECRIM program (Brazilian-German Collaborative Research Initiative on Manufacturing Technology), which aims to intensify the cooperation between Brazil and Germany in the field of manufacturing engineering. In this research project are involved the Pontifical Catholic University of Paraná - PUCPR, at the Laboratory for Research on Machining Processes - LAUS, and the University TU Clausthal with the Institute for Mechanical Engineering - IMW.

References

[1] König, W., Klocke, F., Lenzen, R., 1996. The Electrical machining Processes – What Demands will they face in the future? Proceedings of the International Journal of Electrical Machining, (1), 3-7.

[2] Ho, K.H., Newman, S.T., 2003. State of the art electrical discharge machining (EDM), International Journal of Machine Tools & Manufacturing, 43, 1287-1300.

[3] Amorim, F. L., Weingaertner, W. L., 2004. Die-Sinking Electrical Discharge Machining of a High-Strength Copper- Based Alloy for Injection Molds, Journal of the Brazilian Society of Mechanical Sciences and Engineering, 26 (2), 137-144.

[4] Kunieda, M., Lauwers, B.,Rajurkar, K.P, Schumacher, B.M, 2005. Advancing EDM through Fundamental Insight into the Process. Annals of the CIRP – Manufacturing Technology, 54 (2), 64-87.

[5] Amorim, F. L., Weingaertner, W. L., 2007. The Behavior of Graphite and Copper Electrodes on the Finish Die-Sinking Electrical Discharge Machining (EDM) of AISI P20 Tool Steel, Journal of the Brazilian Society of Mechanical Sciences and Engineering, 24 (4), 366-371.

[6] Van Dijck, F., Dutré, W.L. 1974, Heat conduction model for the calculation of the volume of molten metal in electric discharges, Journal of Applied Physics, 7, 899-910.

[7] DiBitonto, D.D, Eubank, P.T., Patel, M.R, Barrufet, M.A, 1989. Theoretical Models of the Electrical Discharge Machining Process I. A Simple Cathode Erosion Model, Journal of Applied Physics, 66 (9), 4095-4103.

[8] König, W. and Klocke, F., 1997. Fertigungsverfahren - 3: Abtragen und Generieren, Berlin: Springer, Vol 3.

[9] Song, Y.A, 1996. Selektives Lasersintern metallischer Prototypen, Aachen: Shaker Verlag.

[10] Tolochko, N.K., Laoui, T., Khlopkov, Y.V., Mozzharov, S.E., Titov, V.I., Ignatiev, M.B., 2000. Absorptance of powder materials suitable for laser sintering , Rapid Prototyping Journal, 6 (3), 155-160.

Innovative Production Machines and Systems
D.T. Pham, E.E. Eldukhri and A.J. Soroka (eds)

E-business - product and customer support for SME

T. Goszczynski, Z. Pilat, M. Slowikowski, J. Zielinski

Industrial Research Institute for Automation and Measurements – PIAP, Al. Jerozolimskie 202, 02-486, Warsaw, Poland

Abstract

Nowadays SMEs need to extend their offer with different product & customer support services, such as: condition based maintenance, problem solving, equipment reconfiguration services etc., and to be able to cost–effectively provide these services to customers distributed worldwide. For this purpose users need Information and Communication Technologies (ICT) solutions, which will allow smooth transition from the current (mostly locally oriented) service provision, to provision of services at global market, partly in 'virtual world'. The paper presents current state of application for ICT based supplier-customer collaboration, with the special attention for the new tools and possibilities which are available within the Web 2.0 concept and approaches to define objectives actually possible to be fulfilled by ICT research.

Keywords: Web 2.0, SME, CRM, Maintenance, Diagnostics, ICT

1. Introduction

A cost-effective and prompt provision of product & customer support services for geographically distributed customers is a key challenge for globally acting SMEs. It is especially important for suppliers of the components for manufacturing installations. Such efficient services may allow them to provide new business models: instead of classical selling of equipment (e.g. control and measurement systems, automation systems, electrical equipment), equipment manufacturer SMEs may overtake guarantee for optimal use of equipment. This in turn will ask for innovative ways to support selection and specification of equipment, monitoring of the equipment usage conditions & functions, including diagnostics and maintenance. This way SMEs will sell not only products, but also their knowledge on optimal selection and use of their products, and/or overtake full responsibility for the execution of processes where their equipment is installed. Therefore, such SMEs need to extend their products with different product & customer support services, such as: condition based maintenance, problem solving services, reconfiguration of equipment services, etc. The Application problems are introduced in Section 2, Collaborative work in SME in section 3, Reconfigurable Manufacturing Systems (RMS) in section 4, Customer Relationship Management (CRM) in section 5, Maintenance and Diagnostics in section 6 and Conclusions in the last section.

2. Application problems

An essential aspect of such Product Extension

Services (PES) is that they require means for efficient collaboration of different actors within a supply chain on one side and customers (end-users of products) on the other side, within a geographically distributed Extended Enterprise (EE) context. Large companies are able to attack the global market by building their own networks worldwide (e.g. subsidiaries, service centres providing local presence in different regions). SMEs, although often being part of different networks, cannot afford to build highly competent service centres world-wide, having sufficient expertise to efficiently support high variety and permanent evolution of product models and variants. Currently, many SMEs use ICT tools for PES (e.g. for diagnostics, maintenance) which require local presence of SMEs' experts. Therefore time-zone differences, in business models for product extensions present an important challenge to globally acting SMEs. The modern ICT based Collaborative Working Environment (CWE) and Knowledge Management (KM) technology, offer possibilities to realise product extensions and cost effectively provide product & customer support services to geographically distributed customers by supporting work in 'virtual world' (i.e. over Internet). However, again large companies are the ones that have the means to develop and use complex CWE solutions for collaboration, leaving SMEs behind due to the lack of resources and knowledge they can apply to employ and learn such systems.

Building of PES using CWE and KM technology often requires high investments and ICT specialists. Such services have often to be adapted/reconfigured to meet specific and evolving requirements concerning specific products/customers, which implies high, unaffordable costs for SMEs. Therefore, user SMEs refrain from investment in such ICT solutions. This in turn implies a problem for ICT vendor SMEs since most of their potential clients are SMEs.

The Web 2.0 principles, aiming to enable users to build/compose software (SW) services by themselves, may offer a promising approach to solve these problems of both equipment users and ICT vendors, i.e. offer possibilities to cost effectively build PES for specific SMEs' needs. New software should allow for an efficient 'virtual' collaboration within an SME driven EE, taking into account wide spectrum of dynamically changing needs regarding services, collaboration patterns and different technical backgrounds of the involved actors.

The key problem is to enable non-IT experts in SMEs to effectively create/manage services to meet such dynamically changing needs.

The effective solution is to apply advanced Web 2.0 principles and Service Oriented Architecture (SOA) paradigm for CWE in real industrial practice, within an innovative ICT solution affordable for SMEs (low cost, easy-to-integrate). Such CWE solution has to enable cost-effective product & customer support independently of geographical locations of customers and manufacturers. The key Research & Technology Development (RTD) challenge, however, is to move from the service composition at individual level of Web 2.0 to company and group level (thus moving from 'mySpace' to 'mySME').

The objective is to provide a new ICT solution allowing user SMEs to easily combine application specific SW tools (e.g. 'classical' SW solutions for diagnostics, maintenance) and generic collaborative and KM SW services (addressing application independent functionality to support collaboration, such as discovery of expertise for specific problem, traceability of work, provision of knowledge on products/processes etc.), in order to extend their products with new product & customer support services according to their specific, dynamically changing needs.

Such ICT solution should effectively support collaborative work and knowledge sharing among different actors within an EE. It will be open for different information middleware, allowing for remote access over Internet to the equipment/automation system for which PES are provided.

3. The collaborative work concept

The collaborative work in international manufacturing SME networks in general, and especially for product & customer support, requires solving several critical work, synchronization and persistence of workspaces, knowledge activation etc. Especially collaboration among e.g. shop-floor teams using equipment and maintenance/design teams at equipment manufactures (geographically dislocated) in product support processes is constrained by problems concerning distribution of existing tools and do not support dynamically changing collaboration patterns. Advanced platforms for effective creation of collaborative services for industrial applications by non-IT experts are not available in general, and particularly not for SMEs. Many of the addressed problems are common for collaboration work in many different domains. However, there are several specific issues related to CWE in manufacturing SMEs which impose the needs for RTD activities specifically focused upon manufacturing SMEs. Such issues are: high differences in working environments of the collaborative teams (e.g. shop-floor, logistics area, office area for design teams, etc.), teams in modern and flexible manufacturing industry require often different collaboration patterns (e.g. combination of synchronous and asynchronous collaboration during product support processes, taking into account time zones, cultural aspects etc.), different technical background of teams collaborating on common problems (e.g. shop-floor workers with practical experience regarding product use but often low ICT background, designers with high technical expertise), specific security requirements and Intellectual Property Rights (IPR) issues. Product and Customer support by Web 2.0 solutions will be provided by an innovative comprehensive software solution, following the Web 2.0 principles, extending products of automation systems and electrical equipment manufacturer SMEs acting on the global market. Such solution will allow:

- ICT and service engineering vendor SMEs to provide a new SW system highly competitive at the ICT market;
- Equipment manufacturer SMEs - first users of such SW system to strengthen their competitiveness at the global market by providing new product & customer support services and enabling new business models.

User SMEs need to extend their products with different product & customer support services, such as: condition based maintenance, problem solving, equipment reconfiguration services etc., and to be able to cost-effectively provide these services to customers distributed worldwide. An essential aspect of such product extension services is that they require a means for efficient collaboration of different actors in a supply chain and customers, within a geographically distributed extended enterprise context. Users need ICT solutions, which will allow smooth transition from the current (mostly locally oriented) service provision, to provision of services at global market, partly in 'virtual world'. Product and Customer support by Web 2.0 solutions should concern mainly: Reconfiguration of Product, Customer Relationship Management and Maintenance and Diagnostics.

4. Reconfigurable Manufacturing Systems

RMS is one designed at the outset for rapid change in its structure, as well as its hardware and software components, in order to quickly adjust its production capacity and functionality in response to sudden market changes or intrinsic system change. Ideal Reconfigurable Manufacturing Systems possess six core RMS characteristics: Modularity, Integrability, Customized flexibility, Scalability, Convertibility, and Diagnosability. A typical RMS will have several of these characteristics, though not necessarily all. When possessing these characteristics, RMS increases the responsiveness of manufacturing systems to unpredicted events, such as sudden market demand changes or unexpected machine failures. The RMS facilitates a quick production launch of new products, and allows for adjustment of production quantities that might unexpectedly vary. The ideal reconfigurable system provides exactly the functionality and production capacity needed, and can be economically adjusted exactly when needed.

One promising approach, which has the potential to surmount the technical, organizational and financial limitations inherent to the most current approaches, is to consider the production entities as a conglomerate of distributed, autonomous, intelligent and reusable units, which operate as a set of collaborating entities. From a functional point of view, each collaborative unit can, at each time, initiate collaborative actions and dynamically interact with each other in order to achieve both local and global objectives, when they are considered within a cross-layer infrastructure like a manufacturing enterprise [1]. Based on these principles, different approaches have been developed and analysed to cover the requirements of reconfigurable production systems, such as Multi-Agent Systems (MAS) [2] and Holonic Manufacturing Systems (HMS) [3]. Indeed, in spite of some agent and holonic approaches' success, a significant incursion in manufacturing plants in use today is still missing [4].

5. Customer Relationship Management

Customer relationship management is a broadly

recognized, widely-implemented strategy for managing and nurturing company's interactions with customers and sales prospects. It involves using technology to organize, automate, and synchronize business processes - principally sales related activities, but also those for marketing, customer service and technical support [5]. The customer views its interactions with you from its side of the fence. Hence, its interactions must feel part of a logical process and deliver an end-to-end experience; Leveraging the customer data to improve overall performance in product development, process improvement, sales targeting and the operational management. Therefore, excellent analytics have to be core to the business.

Technology is an enabler to offer self-service, and puts the customer in control, for all customer types. This gives the customer the opportunity to take part in the whole process and maintain control. The critical aspect is using it to make the services accessible to customers when they need them. The analytical capabilities can thus drive performance in two main areas: revenue growth by faster identification of better-qualified opportunities and improve operational performance through effective resource utilization. The best businesses have a central analytics function that is dedicated to the task of extracting knowledge from the data. Starting from a hypothesis - a belief that has to be proven or rejected - knowledge is developed through experimentation, in the form of pilot programs in live environments [6]. Because of this focused approach, over a short period of time the root cause activities that impact cost, revenue and the customer experience can be identified. The key to creating a rapid analysis and action loop is an integrated marketing and operations function. Another aspect is the leverage of skilled resources to improve operational performance. A business that understands the customer behaviour drivers through analytics is more able to predict demands on operational resources. With a greater degree of customer self-management in CRM, resources can be freed up and service levels improved. By ensuring that the operations and marketing loop can be taken off front line staff, analytical tools can predict more accurately the expected demand for the services and the contact can be driven to the self-service channels [7].

6. Maintenance and Diagnostics

One of the biggest goals is the ability to integrate information from plant-floor measurement and control systems with production and maintenance management systems. There is a great deal of momentum in the plant automation and condition monitoring industries to provide integrated solutions based on open industry standards that leverage off-the-shelf, commercial computer hardware and software technology such as Ethernet networks, XML and the Internet to provide access to information. OLE for Process Control (OPC) has emerged as the worldwide industry standard, enabling connectivity and interoperability of plant-floor information between disparate fieldbus networks, programmable controllers, distributed control systems, condition monitoring, plant asset management, and production management systems [8]. For maintenance and diagnostics Web 2.0 can give very strong help when used as a collective intelligence of users. Then all problems in maintenance can be shared with other users of the Web as well as diagnostic knowledge [9]. Product extension based on product & customer support services, especially maintenance and diagnostics services, has been subject of intensive RTD work in last few decades. For example, even though the use of advanced automatic diagnostics solutions in manufacturing systems is increasing, diagnosis of run-time problems is still centred on human operators. Thus, methods are being searched which can help the human decision process and (virtual) human collaboration in diagnostics. In modern networked SMEs an effective sharing of knowledge is a critical issue, related to a number of fundamental problems such as human acceptance and motivation issues, ontology problems, correlation of different types of knowledge, treatment of experience-based knowledge etc. Existing, often-powerful methods and ICT tools nevertheless do not satisfy many of these requirements. The objective is to develop generic and widely applicable, modular ICT system, applicable by various user SMEs equipment manufactures but also in other sectors, which can be applied to support product extensions. The platform would provide various PES: services to provide support in solving problems related to customers, services for selection of product, for product use (including support in diagnostics, maintenance and service), etc. System would be open for easy generation of various services to support different products/equipment and to involve different actors (product designers and service providers, maintenance providers, shop-floor operators, end-customers) [10].

7. Web 2.0 perspectives and future work

Web 2.0 principles promote service building and composition on individual level. It is expected that future developments would continue the move from the service composition at individual level of Web 2.0 to company and group level. There is a clear trend to seek applications outside of individuals and small groups to organisations and enterprises. It is also likely that machines will be included as a part of own extended information and collaboration network, and that collaboration will take place on a massive scale and global geographical plan. The emphasis is on services and tools for collaboration design, i.e. tools which either off-line or on-line support design of collaboration: development of core functions exposed as services and their composition, customization, update, maintenance etc. The objective is 'to contribute to blurring the boundary between designing and using, addressing implicitly or explicitly Web 2.0 aspects' [11]. By this the ultimate goal of building collaboration services quicker, cheaper, more flexible, with higher quality, will be achieved. The objective is also to build reference models for CWE that support 'automatic' composition and integration of basic functions exposed as services, taking into account Web 2.0 paradigm. There is a need to provide means for SMEs to effectively build/compose services and trying to solve critical problems related to moving from 'mySpace' to 'mySME', such as composition of services taking into account IPR and privacy issues, sharing of knowledge objects etc.

Service Oriented Architectures (SOA) is a paradigm for combining distributed resources in different domains. Most common SOA technologies are Web services, grid services and peer-to-peer services. With the advent of semantic Web, SOA is seen as an adequate approach to develop agile and rapid complex services. The objective is to obtain a universal, comprehensive information layer. In fact, there is a strategic demand for SOA systems capable of realising the vision of any content, anytime, anywhere, any platform.

Another key innovations are advanced information middleware allowing to have (on-line) access to equipment/processes at geographically distributed customers' plants and a new methodology for practitioners in SMEs on how to combine services supporting collaboration in virtual world and KM technology in order to provide innovative PES, addressing both technological and organisational aspects related to collaboration and KM within EE in global market. Other innovations relevant for the overall RTD and industrial community should be: an application of Web 2.0 principles for industrial community (and especially SME community), aiming to move these principles from the service composition at individual level to company level; a combination of innovative collaborative services with 'classical' SW tools for diagnostics, maintenance, equipment reconfiguration, upgrading of classical tools to become more 'cooperative'; solutions supporting different collaboration patterns & technical background of the collaborating people; solution to security and IPR issues specific for manufacturing SMEs. Other goals are: to apply advanced CWE and KM technology in manufacturing SMEs allowing for product extensions and new business models, and to apply SOA paradigms and Collaborative Reference Architecture [12, 13] to support product extension processes in SMEs. The common business objectives of ICT vendor SMEs, are to develop a new ICT system and industrial user SMEs to extend their products offered at the global market. The future work will provide an advanced system, based on Web 2.0 principles and open SOA approach, enabling SMEs to cost & time effectively create and manage innovative product & customer support services. System, allowing flexible collaborative work of different teams (including mobile ones) and customers within international settings, fully adaptive and reconfigurable to specific, dynamically changing, needs of SMEs.

The system will include:

- set of innovative so-called core collaborative services (CCS) and KM services which can be combined with application specific (partly existing) tools for product & customer support;
- interfaces to automation systems/equipment (information middleware) and other legacy systems providing (on-line) data needed for PES;
- set of service engineering tools, enabling SMEs to effectively build new/maintain existing services for

product extension i.e. adapt them to the user current needs;

- methodology providing an innovative approach for SMEs in applications of collaborative SW services to realise new product & customer support, based on new forms of collaboration between equipment manufactures and their customers and suppliers.

8. Conclusions

Based on the analysis of the requirements of manufacturing SMEs, and on the analysis of state-of-the-art (SotA) the main gaps between the requirements and SotA were identified. This led to the definition of the key RTD problems which have to be addressed in order to satisfy the requirements of industrial SMEs regarding ICT solutions for their product extensions.

Key innovations proposed for the SME community are:

- a new SOA-based system which combines SW services supporting collaborative work with existing SW solutions (e.g. tools for diagnostics and maintenance, customer data-basis etc.) to cost-effectively provide innovative services for product & customer support;
- innovative means for easy generation of different services for product extension allowing users to generate/update PES by themselves, following Web 2.0 principles;
- a set of generic core collaborative SW services, adapted to the needs of SMEs and
- a set of service engineering (building) tools for efficient development of PES and their orchestration.

Acknowledgement

This paper presents the first results of work conducted as a part of the project of the FP7 - Research for SMEs "WEB 2.0 Driven Service-Builder for Product Extension in Globally Acting SMEs" (WEB2SME). Grant Agreement No: 232125.

References

[1] Wooldrige M., An introduction to Multi-Agent Systems. John Wiley & Sons, 2002

[2] Deen S., Agent-based Manufacturing: Advances in the Holonic Approach, Springer Verlag Berlin Heidelberg, 2003.

[3] Leitao P., Restivo F., ADACOR: A Holonic Architecture for Agile and Adaptive Manufacturing Control, Computers in Industry, 2006 pp. 121-130.

[4] Jammes F., Smit H., Service-oriented Paradigms in Industrial Automation. IEEE Transactions on Industrial Informatics, 1(1), 2005, pp. 62–70.

[5] Gummesson E.; Practical Value of Adequate Marketing Management Theory, European Journal of Marketing, 36 (March), pp. 325–49. (2002)

[6] Bendapudi N.and Leone R.; Psychological Implications of Customer Participation CoProduction, Journal of Marketing, 67 2003 (January), pp. 14–28.

[7] Ebner M., Hu A., Levitt D., and McCory J., How to Rescue CRM, McKinsey Quarterly, (Special Edition), pp. 49–57. (2002)

[8] ISO, Geneva, Switzerland, 2002, ISO 13374-1, Condition monitoring and diagnostics of machines – Data processing, communication and presentation - Part 1: General guidelines.

[9] Jaradine A., Lin D., Banjevic D., An review on machinery diagnostics and prognostics implementing condition-based maintennance, Mechanical Systems and Signal Processing Volume 20, Issue 7, October 2006, pp. 1483-1510

[10] Koren, Y. and Katz, R., Reconfigurable Apparatus for Inspection During a Manufacturing Process. US patent # 6,567,162 Issue date: 5/20/03

[11] Expert group, Towards Activity-oriented Collaborative Working Environments – A Research Roadmap 2007 – 2020. Responsible EC officer and editor: Isidro LASO-BALLESTEROS, DG Information Society, European Commission. Rapporteurs: Nikolay Mehandjiev and Dragan Stokic.

[12] Ralli, C., Collaboration Reference Architecture. Workshop on eCollaboration in working environments VUB, Brussels, November, 2005.

[13] Correia A.T., Grama C. Stokic D., et al., An Architecture for CWE in Manufacturing Industry, 16th European Concurrent Engineering Conference, Porto.

Innovative Production Machines and Systems
D.T. Pham, E.E. Eldukhri and A.J. Soroka (eds)

Inference and propagation of performance constraints from abstract to concrete business workflows

A. García-Domínguez[a], I. Medina-Bulo[a], Mariano Marcos-Bárcena[b]

[a] *Department of Computer Languages and Systems, University of Cádiz, C/Chile 1, CP 11003, Cádiz. E-mail: {antonio.garciadominguez,inmaculada.medina}@uca.es*
[b] *Department of Mechanical Engineering and Industrial Design, University of Cádiz, C/Chile 1, CP 11003 Cádiz. E-mail: mariano.marcos@uca.es*

Abstract

Currently, manufacturing organizations of all sizes need to collaborate with others in order to meet their goals. However, achieving an adequate level of communication between their people and their information systems is difficult. Organizations are urged to formalize their processes and implement their information systems flexibly. These tasks can be achieved using workflows and service-oriented architectures, respectively. Abstract business-level workflows can be refined into executable workflows which compose services from the established service-oriented architecture, integrating disparate systems. However, it is hard to ensure that the target performance of the business process is achieved from the concrete software components. In this work, we present two algorithms which infer performance constraints for tasks in a workflow from its global and local restrictions. These tasks can be then further decomposed into nested subgraphs automatically. By repeatedly decomposing the tasks and inferring their constraints, target key performance indicator values can propagate from abstract business processes down to concrete activities.

Keywords: workflow modelling, service-oriented architectures, holonic enterprise, business processes, Web Services.

1. Introduction

In the current rapidly changing market, manufacturing enterprises need to focus on their key value adding processes in order to remain competitive. Non-core processes are outsourced or shared with external organizations. This is the case for both large firms and their collaborative manufacturing meganetworks (CMMNs) [1] and small firms which join together to compete with them [2]. Virtual enterprises pool the firms' resources into specific projects, and extended enterprises take form from trust relationships between the organizations [3].

However, these collaborations introduce new challenges. Business practices and software platforms will usually differ, hampering communications at the business and information system levels.

In order to reach a common understanding, enterprises need to agree on well-defined business processes in order to collaborate. At the same time, the information systems needs to implement these business processes, accommodating their reconfiguration and monitorization and integrating disparate software platforms.

In the business process modelling field, several workflow modelling languages are being developed, such as BPMN 2.0 [4] or WS-BPEL 2.0 [5]. These workflows can describe the business process, run it and monitor its progress, seamlessly aggregating human and computer resources. Workflows can be used at every abstraction level, from the high-level business activities down to the individual activities performed by each participant. At the information system level, the basic building block in these workflows is a *service*: a self-contained piece of software which performs a task and can be reused across the entire organization.

However, the more detailed the model, the harder it is to see the high-level picture and check if the individual activities are meeting the high-level key performance indicators. When the desired level of performance is not obtained, it is important to be able to pinpoint the cause quickly. This could be achieved if the low-level activities were annotated with constraints which could be traced back to the high-level key performance indicators.

In this work, we propose a top-down approach to derive the desired performance of the individual activities performed by each resource from the target values of the key performance indicators of the abstract business process. After introducing some basic concepts and our simplified workflow modelling language, we present two algorithms which infer local performance constraints from global ones, and a transformation which expands the graph so the user can increase the level of detail. Finally, we offer some conclusions and future lines of work.

2. Background

Before presenting the algorithms, we will introduce some basic concepts and related works.

2.1. Workflow modelling

Formalizing the business practices of an organization is a common practice nowadays. A formal business process can guide the collaboration between different organizations or different parts of an organization, and provides a holistic view of all the work required. Aguilar-Savén classifies business process models into 4 categories, depending on their purpose [6]:

a) Learning about the existing process;
b) Improving or developing new processes;
c) Providing decision support during process execution and control; and
d) Supporting the enactment of Information Technology.

In this work, we are most interested in purposes b) and c): we want to model the existing or projected business process and transform it so the information systems can support their execution and control. There are quite a few alternatives for modelling business processes, such as function diagrams (i.e. IDEF0 [7]) or Gantt charts, but not all cover both purposes.

Workflows are defined by the Workflow Management Coalition as "the automation of a business process, in whole or part, during which documents, information or tasks are passed from one participant to another for action, according to a set of procedural rules". As such, they are clearly oriented towards purpose c), and can serve purpose b) depending on their abstraction level. Normally, these workflows are described in a specific language, supported by a Workflow Management System (WFMS). The WFMS communicates with the participants (both humans and software entities) to execute the workflows and monitor them.

There are several workflow modelling languages currently in use. Some of them are designed to be directly run and monitored by a WFMS automatically, such as the Web Service Business Process Execution Language (WS-BPEL) 2.0 [5]. Others work at the business level, describing the process in a more abstract way, such as the Business Process Modeling Notation (BPMN) 2.0 [4]. Usually, modellers start with abstract workflows such as BPMN, and later on derive concrete workflows (i.e. in WS-BPEL 2.0) from them.

2.2. Service-oriented architectures

Previously, we described how workflows can help an organization formalize their practices, assisting the improvement and control of their business processes. However, the information systems need to provide the basic building blocks to make them work.

This requires a change in the way the systems are conceived: instead of closed silos of information which are solely manipulated through a user interface, they must now provide a catalogue of *services* which can be flexibly reused over all workflows in the organization. These new systems are known as *service-oriented architectures* (SOAs).

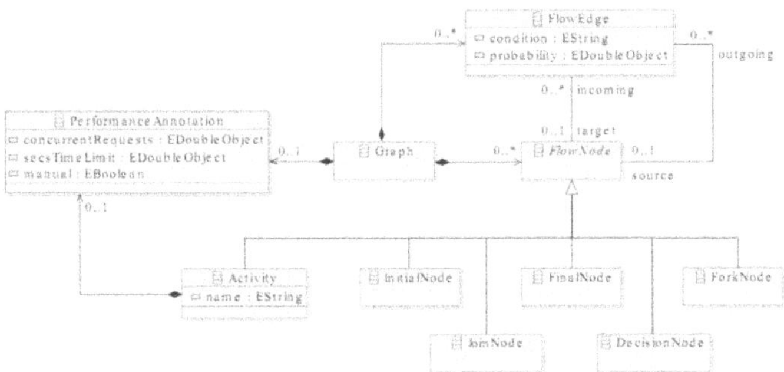

Figure 1. UML class diagram for the elements of our simplified workflow language

According to Erl, services "exist as physically independent software programs with distinct design characteristics that support the attainment of the strategic goals associated with service-oriented computing" [8]. The definition is intentionally vague: any combination of technologies can be used to build a SOA. However, in practice, most SOA development uses Web Services (WS) as defined by the W3C [9]. By using a standardized technology stack, users can interoperate between differing software platforms.

It is interesting to note the interactions between workflows as business processes and SOA: a business process can be modelled as a concrete WS-BPEL 2.0 composition of several existing Web Services. After performing this composition, the WS-BPEL 2.0 workflow is now yet another Web Service which can be reused from anywhere else in the organization. This allows for building more advanced business processes in a bottom-up fashion, and improve the processes more quickly. In a way, this view of the system as a network of collaborating entities which join simpler entities and take part in several others can be seen as a software implementation of the *holonic enterprise* [10].

At the same time, reusing the services to this degree increases our reliance in them: we depend on their performance in order to meet the target values for the key performance indicators of the business processes. However, it is hard to know if we are meeting these high-level indicators from the individual services. For this reason, we propose in this paper several techniques to derive the required performance values for each service, from the high-level abstract workflows down to the low-level, executable workflows.

3. Inference of performance constraints

In the previous sections, we introduced the key concepts required to understand the work presented in this paper, and our motivation. In this section, we will describe the inference algorithms used to annotate the workflow activities with the expected performance, using the global and local information set by the modeller on the workflow.

3.1. Workflow language

We use a simplification of UML activity diagrams to describe our workflows [11]. Figure 1 shows an UML diagram class with each of the element types:

- **Activities** encapsulate some behaviour, described textually.
- Activities can have manual or automatic **performance annotations** on the number of requests they must handle per second and the time limit for each of these requests.
- **Initial nodes** are the starting execution points of the graphs. There can be only one per graph.
- **Final nodes** end the current execution branch. There can be more than one.
- **Decision nodes** select one execution branch among several, depending on whether the condition for their outgoing edge holds or not. Only the outgoing flow edges from a decision node may have a non-empty condition and a probability less than 1.
- **Fork nodes** split the current execution branch into several parallel branches.
- **Join nodes** integrate several branches back into one, whether they started off from a decision node or a fork node. This is a simplification from UML, which uses different elements to integrate each kind of branch: join nodes and

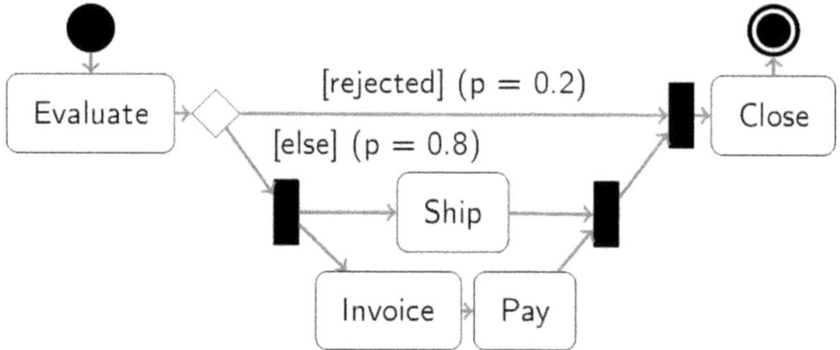

Figure 2. Running example for the inference algorithms

merge nodes, respectively.

3.2. Example model

In order to illustrate the algorithms, we will use the order processing workflow shown in Figure 2:

1. Upon receiving an order, it is evaluated: 20% of all orders are estimated to be rejected, and the remaining 80% are accepted.
2. If the order is accepted, perform these activities in parallel:
   2.1. Create and post a shipping order.
   2.2. Create an invoice and receive payment for it from the client.
3. Close the order.

3.3. Inference of concurrent requests

This first algorithm infers how many requests each task in the workflow must be able to handle per second in order not to become a bottleneck. The algorithm propagates the global performance constraint starting from the initial node, traversing the nodes in ascending order of *depth*. The depth of a node is defined as the length of the longest path from the initial node to it.

For the most part, the number of requests per second (C) is propagated without changes from the initial node. There are two special cases when the propagated value changes:

- Flow edges with probability p lower than 1 propagate pC.
- Join nodes which connect mutually exclusive branches (split off at a decision node) propagate the sum of the values from each branch. Otherwise, the branches split off at a fork node and the minimum value from all branches is propagated. Finding out where the branches split off requires computing the lowest common ancestor of every branch. Using the naïve approach recommended by Bender et al. [12] works best for the sparse graphs commonly used in workflows.

As an example, let us consider the activity "Pay" in Figure 2, when the whole process is required to handle 5 requests per second. This global constraint is propagated mostly unchanged, except for the conditional flow edge from the decision node to the fork node which precedes "Pay". Since its probability is 0.8, "Pay" must handle $0.8 \cdot 5 = 4$ requests per second.

3.4. Inference of time limits

This algorithm infers the time limit for each task from the local and global information available in the workflow, meeting the three following properties:

1. The sum of the time limits in every path from the initial node to any of the final nodes meets the global constraint.
2. Available time is split equally among the nodes, to distribute the load equally across the system. We are also interested in avoiding services that are too coarse-grained or fine-grained.
3. Time limits should be as lenient as possible.

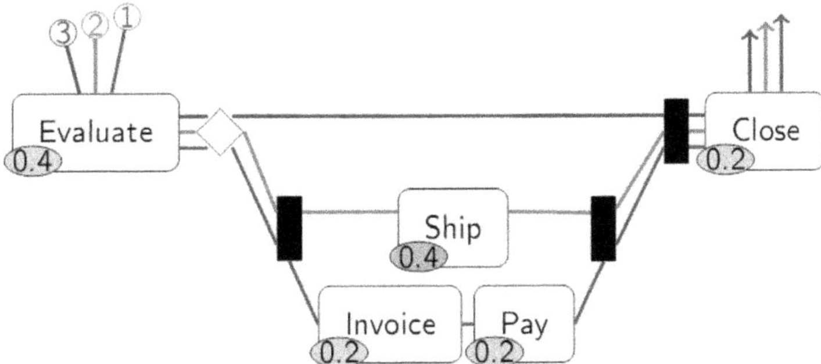

Figure 3. Inference of time limits for the running example

Before describing the algorithm, we need to define this minimal set of terms:

- $F(p)$ is the set of all activities in the path p which lack manual time limits (they are *free*).
- $T_L(G)$ is the time limit for every path in the workflow graph G.
- $T_M(p)$ is the sum of the manual time limits in the path p.
- $S_A(p,G) = T_L(G) - T_M(p)$ is the *slack* left unassigned by the manual times on path p of the workflow graph G.
- $T_E(p,G) = S_A(p,G)/(1 + |F(p)|)$ is an estimate of the magnitude of the time limit each task would obtain with only the information of the path p of the graph G. The lower the value, the more restrictive the path is.

The algorithm follows these steps:

1. Previous automatic constraints are removed.
2. Store all paths from the initial node to a final node in L, and sort them in ascending order of $T_E(p,G)$.
3. For each path p in L, check its slack, $S_A(p,G)$:
    - 4.1. If it is negative, that means that the manual times exceed the global constraint. The algorithm is aborted.
    - 4.2. If it is zero, check if there are any free nodes. In that case, we cannot assign a time limit to them, and an error should be reported. Otherwise, continue.
    - 4.3. If it is greater than zero, distribute it equally among the unrestricted activities in the path.

The correctness of the algorithm relies on the ordering defined on the paths: since paths are traversed from most to least restrictive and previously inferred annotations are not overwritten, the algorithm will ensure the global constraint still holds for the previous paths. When all paths are done, the global constraint holds for all of them.

To illustrate the algorithm, let us assume that all requests to the process in Figure 2 should require 1 time unit or less of processing time, and that "Evaluate" is assumed (according to prior knowledge) to take never more than 0.4 seconds. In this case, the bottom path is the most restrictive, followed by the middle and top paths. The bottom path is visited first and the 0.6 seconds of slack are split equally among the 3 unrestricted nodes. The middle path is visited next, and the remaining 0.4 seconds of slack are assigned to "Ship". Finally, the top path has no unrestricted nodes, so we are done.

4. Propagation of performance constraints

In the previous section, we described how the existing global and local information can be used to augment the activities in the workflow with target values for their performance indicators.

In order to describe each activity in further detail, the user can automatically expand them with a new subgraph including a single initial node, a final node and a single subactivity. The local performance constraints in each activity are now the global constraints for their subgraphs.

Therefore, all the user needs to do is repeatedly design the workflows, annotate them, and expand them. Once the high-level business activities have been broken down into simple actions, the process can be executed in a WFMS, using a workflow language such as WS-BPEL 2.0 [13]. These executable actions can be monitored to ensure that the key performance indicators of each high-level business activity can be reached.

5. Conclusions and future work

Manufacturing firms need to collaborate with others, and to do that they need to formalize their business processes and integrate their information systems. Workflows can model business processes at each level of abstraction. In a service-oriented architecture, the information system is a portfolio of services which can be reused as building blocks for the workflows. However, checking that the desired performance for the business process is obtained from the concrete activities is difficult.

In this work we have presented a top-down approach which ultimately derives the expected performance of the lowest-level activities from the target key performance indicators of the process. Two algorithms fill in the details at a certain level and a transformation lets the modeller proceed.

The algorithms have been successfully implemented as part of the SODM+T toolset [14], which is a set of plug-ins for the Eclipse integrated development environment [15]. The code is freely available under the Eclipse Public License.

However, the algorithms right now only operate on a simplified workflow language, and the time limit algorithm requires all paths in the workflow to be enumerated. We intend to optimize the performance of the algorithms and extend them to popular languages such as BPMN or WS-BPEL 2.0. Later on, the algorithms will allow for finer-grained control of the inference process.

Acknowledgements

This work was co-financed by the Junta de Andalucía. The University of Cádiz is an Academic Associate of the FP6 I*PROMS Network of Excellence on Innovative Production Machines and Systems.

References

[1] K. Johansen, M. Comstock, and M. Winroth, "Coordination in collaborative manufacturing mega-networks: A case study," *Journal of Engineering and Technology Management*, vol. 22, Sep. 2005, pp. 226-244.

[2] C. Piddington, "SME interoperability in the global economy: A discussion paper," *Proceedings of the 16th IFAC World Congress*, P. Zítek, Ed., Czech Republic: 2005.

[3] J. Browne, I. Hunt, and J. Zhang, "The Extended Enterprise (EE)," *Handbook of Life Cycle Engineering: Concepts, models and technologies*, A. Molina Gutiérrez, A. Kusiak, and J.M. Sánchez García, Eds., London, United Kingdom: Kluwer Academic Publishers, 1998, pp. 3-30.

[4] Object Management Group, "Business Process Modeling Notation 2.0 - Beta2," 2010. Available at http://www.omg.org/spec/BPMN/2.0/Beta2/.

[5] OASIS, "Web Service Business Process Execution Language (WS-BPEL) 2.0," Apr. 2007.

[6] R.S. Aguilar-Savén, "Business process modelling: Review and framework," *International Journal of Production Economics*, vol. 90, Jul. 2004, pp. 129-149.

[7] Computer Systems Laboratory of the National Institute of Standards and Technology (NIST), "FIPS 183: Integration Definition for Function Modeling (IDEF0)," Dec. 1993.

[8] T. Erl, *SOA: Principles of Service Design*, Indiana, EEUU: Prentice Hall, 2008.

[9] W3C, "Web Services Glossary," 2004. Available at http://www.w3.org/TR/ws-gloss/.

[10] A. Tharumarajah, A. Wells, and L. Nemes, "A comparison of the bionic, fractal and holonic manufacturing concepts," *International Journal of Computer Integrated Manufacturing*, vol. 9, 1996, pp. 217-226.

[11] Object Management Group, "Unified Modeling Language (UML) 2.2," 2009. Available at http://www.omg.org/spec/UML/2.2/.

[12] M.A. Bender, G. Pemmasani, S. Skiena, and P. Sumazin, "Finding Least Common Ancestors in Directed Acyclic Graphs," *Proceedings of the 12th Annual ACM-SIAM Symposium on Discrete Algorithms (SODA'01)*, 2001, pp. 845-853.

[13] OASIS, "WS-BPEL 2.0 Primer," 2010. Available at http://docs.oasis-open.org/wsbpel/2.0/Primer/wsbpel-v2.0-Primer.html.

[14] A. García-Domínguez, "Homepage of the SODM+T project," 2010. Available at https://neptuno.uca.es/redmine/projects/sodmt.

[15] Eclipse Foundation, "Eclipse.org home," 2010. Available at http://eclipse.org/.

Innovative Production Machines and Systems
D.T. Pham, E.E. Eldukhri and A.J. Soroka (eds)

Interaction- and Event-Based Management of Processes in Service-Oriented Infrastructures

T. Schlegel[a], S. Dusch, K. Vidackovic[b]

[a] SEUS Group, Technical University of Dresden, Nöthnitzer Str. 46, 01062 Dresden, Germany
[b] Fraunhofer IAO, Nobelstr. 12, 70569 Stuttgart

Abstract

Service infrastructures often have different service providers and require management by an organisation that provides a common platform. When business processes are executed using mechanisms like multiple BPEL engines, it is often impossible to monitor and manage the execution of these processes. We present an approach and a system implementation that uses a complex event processing (CEP) engine to integrate different BPEL engines and other applications into a common service platform. As interactions with the processes and services running are crucial in such an infrastructure, we provide also a concept for creating ad-hoc user interactions on a monitoring dashboard, which allows platform managers as well as stakeholders in the processes to interact with the platform and the processes executed – independent of their context of execution.

Keywords: Service Platform Management, Complex Event Processing, Decentralized Processes, Interaction

1. A new complexity in service infrastructures

Service infrastructures today become more and more complex as different services emerge each day, which are again incorporated in processes that use them to achieve their goals building on the services' functionalities.

For example, insurance companies relying on a service-oriented architecture (SOA) will need to monitor their cross-company processes and events that occur in their infrastructure. In Virtual Enterprises or even fully decentralized service platforms, monitoring activities, processes and exceptions becomes a key problem.

In such a scenario, architecture has to support the exchange of information and triggers between different processes as well as a management platform that is capable of monitoring a process and service landscape, also generating interactions with users in different roles (e.g. platform managers and process participants).

2. Process Execution, Events and Interactive Control

2.1. Process modelling and execution

Where automatable, IT-based business processes are concerned, commonly the workflows are modelled and executed using the Web Services Business Process Execution Language (WS-BPEL) [1].

WS-BPEL processes have the advantage of being executable and using web service calls directly in the processes. Unfortunately, monitoring of processes is not easy even on one engine instance running. It currently becomes nearly impossible, if multiple processes are executed on different servers / engines.

Therefore, it is necessary to achieve monitoring with the processes in order to monitor complex service infrastructures.

2.2. Complex Event Processing

Complex event processing (CEP) has been invented already in the 1950s [2] to achieve handling of events and dependencies in a complex environment. In the 1990s research focused more on dependencies and relations between events, first building proprietary software, while later SQL-like queries were offered.

Events are routed to software either using data streams (e.g. Stream [3]) or the CEP engine is embedded into an application and receives events directly from an API (e.g. Esper [4]).

EDA (event-driven architecture) [5] is an architecture pattern that is often applied in software development, if events are produced and consumed in different parts of a software product. One big advantage for dynamic and decentralized systems is the fact that producers of events do not necessarily have to know the consumers of an event produced by them (loose coupling). The event-driven concept is often implemented using publish-subscribe architectures.

The architecture pattern consists of three different parts. An event source produces events, while event channels transmit them to subscribed event sinks. Components in an event-driven architecture may take the role of event sinks and event sources at the same time. This is especially the case, when a CEP engine is involved which processes received events from different sources and reacts on identified event patterns by sending new events to downstream components [6].

WS-Eventing [7] definies the contents (body) of a SOAP message that is used for subscribing and unsubscribing to specific events. These events are then also sent to the recipient within the SOAP body. This makes it possible to register applications on events generated by a BPEL engine.

Therefore, we are using a CEP engine to enable routing and rule-based interpretation of events between different processes and BPEL engines.

With regards to service and process infrastructures, complex event processing engines offer a valuable approach to dispatch information, status changes and actions across multiple processes and platforms [8].

2.3. Interaction with service / process infrastructures

Approaches to integrate human activities explicitly in the process have been led to the specification of BPEL4People [9], which aims at including human actions into a WS-BPEL process.

In conjunction with BPEL4People, WS-Human Task (WS-HT) [10] serves as a basis for including human tasks into the WS-BPEL process. WS-HT is based on WSDL, which is used to describe service interfaces, making it possible to describe also human users (roles, timeout, interactions) as part of the process.

BPEL4People and WS-HT serve as a means for modelling and integrating interactions and generally human tasks into a BPEL process, executed in a process engine. We refer to this kind of interactions as **explicit interactions**.

Explicit interactions comprise all interactions that are forseen in the process, i.e. they are expected to occur and have to be accomplished to execute the process.

Implicit Interactions [11] form a second category of interactions that is not modelled in the process but becomes necessary when unforeseen situations, interactions between processes or interactions outside the process (e.g. management of the service/process platform) occur.

We will use events in the CEP engine to trigger interactions and return results. A dashboard is then used to show events and execute the interactions.

The figure below shows a screenshot of the graphical user interface based on XAML (Extensible Application Markup Language). The dialog asking for acceptance of a task is generated from a specific event received from the CEP Engine.

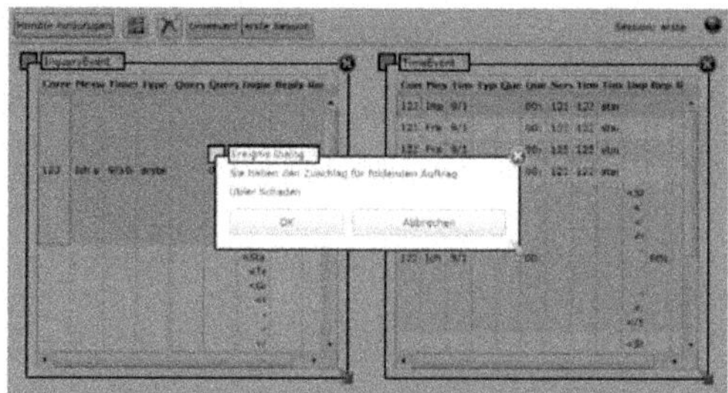

Fig. 1. Dialog interface of the client application

2.4. System description

First, we present the basic model used to describe events and transmit them in the system. Then we describe the system architecture, which uses multiple BPEL engines that transmit events from their processes to the complex event processing system.

The client application, used to monitor, manage and interact with the processes, receives events from and also transmits events to the CEP engine.

3. Event Model

Events are the main concept that helps in distributing processes over multiple engines in a platform but also facilitates integration by enabling the decentralized communication, i.e. clients and process engines do not have to know each other but will nevertheless be interconnected by the CEP engine.

The event schema consists of all the events that can be created by the BPEL engines. It describes the properties of each type of event. To allow for easy use and exchange of the event schema and the events, we have defined them using XML Schema.

The simplest event is a plain notification, which is used for example to notify the CEP engine or interactive client about status changes. It contains a timestamp, which is also used to order events and therefore is essential. The timestamp is also necessary, because not all the available BPEL engines are capable of reliably sending events on a defined point in time and also the time needed for transmission may differ. To enable users understanding a message more easily, a message text is included in the notification event, which can be displayed in the client for explanation purposes or be used for debugging purposes.

Each event has a specific event type. This allows defining semantic queries and filtering also by users to receive and interpret only specific types of events.

3.1. Time Events

Additionally, the event itself can contain a query, which is evaluated by the CEP. Only if this query returns a result, the event is passed to the CEP. This offers developers of business processes the possibility to define events also depending on the current context. A timeout defined for the queries ensures that queries finish in time and do not block or invalidate events.

Each event also contains a correlation, which relates it to the specific process and context it originated from, making it possible to return results to the originator of an event. The XML Schema with the properties described, can be found below.

```
<xs:complexType name="TimeEventType">
 <xs:sequence>
  <xs:element name="Message" type="xs:string" />
  <xs:element name="CorrelationID" type="xs:string" />
  <xs:element name="Timestamp" type="xs:dateTime" />
 </xs:sequence>
 <xs:attribute name="Type" type="xs:string" />
 <xs:attribute name="Query" type="xs:string" />
 <xs:attribute name="QueryTimeout" type="xs:duration" />
</xs:complexType>
```

3.2. Interaction Events

Interactions in most cases produce results. The InquiryResultEvent is used to communicate these results or information, entered by a user, back to the BPEL engine(s).

For the prototype, we have used a simple mechanism to derive interactions from events: the client requires embedded interactions in the events, so it can directly display them for gathering information from the user. This also offers the opportunity of providing an individual interface for each type of event.

The event must also include a property that defines, who is allowed to see and enter information or more generally use the interactions provided with the event – according to the role selection in BPEL4People. Selection of the correct role has been implemented by using XPath: The XPath expression is being evaluated for an XML document, which exists as a profile for each user who can connect to the client application. This XML document defines all the roles of the specific user. Its structure is also defined by an XML Schema.

If results have been requested by a specific stakeholder or part of the system (e.g. a BPEL engine) a URL (Uniform Resource Locator) specifies this intended recipient of the information. At this URL, an application has to be available, which implements a defined interface for receiving the interaction results. The user interaction follows this schema:

```
<xs:complexType name="InquiryEventType">
 <xs:complexContent>
  <xs:extension base="mytns:TimeEventType">
   <xs:sequence>
    <xs:element name="InquiryForm"
type="mytns:InquiryFormType"/>
     <xs:element name="ReplyToUrl" type="xs:string"/>
    <xs:element name="RoleQuery" type="xs:string"/>
   </xs:sequence>
  </xs:extension>
 </xs:complexContent>
</xs:complexType>
```

The user interface is defined in the element InquiryForm, which contains the specification of the user interface and the return values to be sent back as a result. The event can carry user interface descriptions in any kind of specification language, which is identified in an attribute. As long as an interpreter exists or is newly built for it, which generates the user interface from this description, any language can be used. The parser/interpreter is also responsible for reading the return values from the dialog or dashboard.

The return values are defined in a way that they describe also the dialog element and its properties. If no return values are specified, only the information contained is displayed to the user. The definition of the interactions reads as follows:

```
<xs:complexType name="InquiryFormType" mixed="true">
 <xs:sequence>
  <xs:element name="Content">
   <xs:complexType>
    <xs:sequence>
     <xs:any namespace="##any" processContents="skip" />
    </xs:sequence>
   </xs:complexType>
  </xs:element>
  <xs:element name="ReturnControlNames" >
   <xs:complexType>
    <xs:sequence>
     <xs:element name="ReturnControlName" type="xs:string"
minOccurs="0" maxOccurs="unbounded" />
    </xs:sequence>
   </xs:complexType>
  </xs:element>
 </xs:sequence>
 <xs:attribute name="lang" type="xs:string" />
</xs:complexType>
```

Further events have also been defined using XML Schema.

4. Technology and Implementation

The architecture and implementation has to build on existing technologies in order to be applicable to practice.

4.1. Selection of the Framework

First, a framework is needed, which allows for communicating events between BPEL engine, CEP engine and client application.

OpenESB [13] needs a BPEL engine, but also offers a specialized BPEL engine that can be used in the implementation. The BPEL engine can execute processes defined in WS-BPEL 2.0. Other BPEL engines can be connected to OpenESB by implementing its specific interface, which is a disadvantage compared to other solutions that allow integration as a service. A BPEL engine connected to the Enterprise Service Bus (ESB), can produce and consume events directly, if the BPEL engine selected is capable of doing this. If not, this is a problematic restriction.

Using the BPEL engine of the Apache Software Foundation, **Apache ODE**, processes can be defined in WS-BPEL 2.0, too. It offers fast an easy installation and deployment of business processes, but lacks some stability in the development environment and does not offer direct connection to a CEP engine. Being open source, Apache offers the possibility of integrating a CEP engine directly in the code, which would mean high efforts and a proprietary solution.

Esper is a CEP engine, which currently offers the most benefits [14]. It offers a runtime for Java and .NET. Esper can be integrated in any application by transmitting information via a web service call or an ESB to the CEP engine – independent of a specific BPEL engine and also offering connection of processes not executed on BPEL engines. This complies with the goal of offering a decentralized and product-independent solution.

4.2. Graphical User Interface Language

As already described above, the user interface and especially the event-driven interactions on it are defined within the events transmitted and processed via the CEP engine. Therfore, they are already generated when an event is triggered – depending of the type, context and origin of it.

This means that after creation of the event, interactions are already defined. Other solutions like the use of object-oriented / semantic models for describing them [12] are more flexible – especially regarding their interpretation by the client application – but introduce another complexity for software and process developers as well as they can cause unforeseen reactions of the user interface. So we chose to use a interface description language that is directly displayable and executable by a runtime environment.

For this reason, the language has to be a standard language and also allow for embedding it into the payload of an event, i.e. being incorporated into the XML structure of an event.

This requires a representation of the model, which does not need to be compiled before execution in the runtime environment. Like HTML, XAML can also be interpreted in a runtime environment, based on Microsoft .NET and Silverlight.

5. Architecture

The system consists of a **server**, which starts the CEP engine and collects the events generated by BPEL engines or other applications integrated with the system. These events are then forwarded to the CEP engine for further processing. The server also handles registration of new users and elements in the user interface / dashboard.

In the user interface, new queries can be defined for monitoring and managing the whole plattform. These queries are also forwarded to the CEP engine. If a query returns results, the server also deals with transferring them to the user interfaces registered with it to achieve notification in due time.

From the user interface, input of users can be sent back to the BPEL engines. These user inputs are also passed to the server, which establishes a connection to the recipient BPEL engine and transfers the data to it.

Interactive sessions are stored by the user interface management and can be reloaded on login. For this reason, the server stores the user sessions independent of existing connections, as these may change. One session is also capable of supplying data to multiple user interfaces active at a certain point of time.

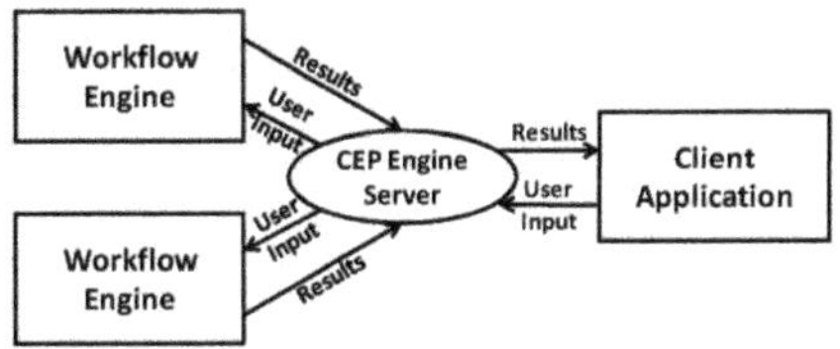

Fig. 2. System architecture

5.1. Server Architecture

The CEP engine server contains interfaces to BPEL engines and other applications, which deliver events to it. This interface consists of only one operation that accepts events.

The CEP engine is the core of the server, collecting events and passing results on to the client application. It also contains a session management component, which stores the sessions of users, including the current connections to the different user interfaces used for input from and output to the user.

Fig. 3. System architecture

5.2. CEP integration

A big part of the application collaborates with the CEP engine, which is integrated into the server. The server offers a class that reacts on results of the CEP engine and passes these results on to the client application registered with it.

5.3. Client Application

The client application provides the user with a user interface for creating new sessions and queries within them. It also displays output / dialogs and accepts inputs from the BPEL engines.

The window management offers elements for displaying queries in a window-like view. These windows can be resized, moved and closed on the dashboard.

Closing a window deletes the query, which causes the CEP engine to stop sending results for this specific query. The query dispatcher accepts results and passes them to the windows registered for the adjacent query, which are then displayed by the user interface.

6. Test Scenario and Evaluation

We have tested the system using a simple fraud detection process from the insurance sector. The BPEL process is shown in the figure below:

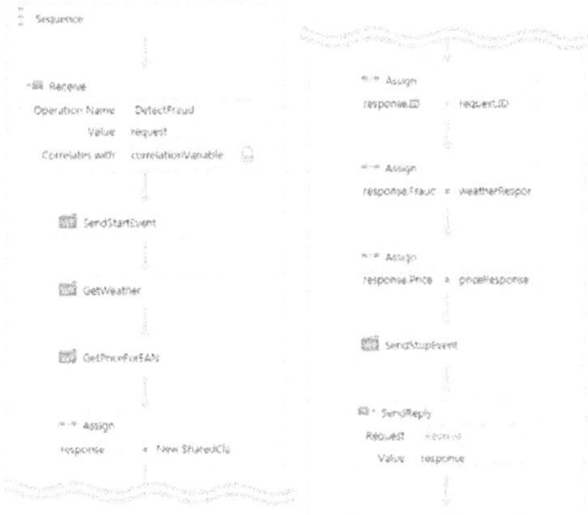

Fig. 4. System architecture

7. Discussion

Executing business processes by using mechanisms like multiple BPEL engines, often make it impossible to monitor and manage the execution of these processes as whole infrastructure. We have therefore presented an approach and a system implementation that uses a complex event processing (CEP) engine to integrate different BPEL engines and other applications into a common service platform. For interactions with the processes and services running in such an infrastructure, we provide also a concept for creating ad-hoc user interactions on a monitoring dashboard, which allows platform managers as well as stakeholders in the processes to interact with the platform and the processes executed – independent of their context of execution.

Being independent of a specific BPEL engine implementation and capable of using multiple BPEL engine instances, the approach benefits especially large and decentralized system structures, as they occur when different organizations and infrastructures have to be integrated on a platform. We also achieved the goal of being able to monitor such a platform and support management task for a platform provider as well as service/process owners in such a system.

Our prototype approach of using embedded XAML has many advantages with regards to simplicity of changes and reduction of efforts for creating a running system. A dynamic user interface creation is not possible when using pure XAML. We are therefore currently evaluating HTML with JavaScript for dynamic changes on the client side.

However, from a research perspective, a model-based approach for the context-based creation of interfaces is a goal for the future. Model-interpretation or transformation could then be integrated either on the server-side (generating target code like HTML or XAML on the server) or on the client-side (sending model fragments interpreted by the client application).

In future, decentralized and multi-engine approaches will gain higher importance due to service infrastructures spanning multiple organizations and integrating different process technologies.

References

[1] Alves, Alexandre; Arkin, Assaf; Askary, Sid; Barreto, Charlton; Bloch, Ben; Curbera, Francisco; Ford, Mark; Goland, Yaron; Guízar, Alejandro; Kartha, Neelakantan; Liu, Canyang K.; Khalaf, Rania; König, Dieter; Marin, Mike; Mehta, Vinkesh; Thatte, Satish; van der Rijn, Danny; Yendluri, Prasad; Yiu, Alex: Web Services Business Process Execution Language. 11.04.2007. Retrieved from http://docs.oasis-open.org/wsbpel/2.0/wsbpel-v2.0.html

[2] Luckham, David: Complex Event Processing | A Short History of Complex Event Processing. Part 1: Beginnings, 2008. Retrieved from http://complexevents.com/wp-content/uploads/2008/02/1-a-short-history-of-cep-part-1.pdf

[3] Widom, Jennifer; Motwani, Rajeev; Arasu, Arvind; Babcock, Brian; Babu, Shivnath; Srivastava, Utkarsh; Wang, Zubin: Stanford Stream Data Manager. 2008. Retrieved from http://infolab.stanford.edu/stream/ on 24.07.2009

[4] Herreman, David: Esper - Complex Event Processing. 2009. Retrieved from http://esper.codehaus.org/ on 24.07.2009

[5] Michlmayr, Anton; Rosenberg, Florian; Leitner, Philipp; Dustdar, Schahram: Advanced event processing and notifications in service runtime environments. In: Proceedings of the second international conference on Distributed Event-Based Systems (DEBS '08), pp.115-125. Rome, 2008

[6] Bruns, Ralf; Dunkel, Jürgen: Event-Driven Architecture | Softwarearchitektur für ereignisgesteuerte Geschäftsprozesse. Springer, Berlin, Heidelberg, 2010.

[7] World Wide Web Consortium: Web Services Eventing (WS-Eventing). 15.3.2006. Retrieved from http://www.w3.org/Submission/2006/SUBM-WS-Eventing-20060315/ on 30.03.2009

[8] Vidačković, Krešimir; Kett, Holger; Renner, Thomas: Event-Driven Service Chain Monitoring for the Internet of Services. In: Proceedings of the eChallenges e-2009 Conference, 2009.

[9] Agrawal, Ashish; Amend, Mike; Das, Manoj; Ford, Mark; Keller, Chris; Kloppmann, Matthias; König, Dieter; Leymann, Frank; Müller, Ralf; Pfau, Gerhard; Plösser, Karsten; Rangaswamy, Ravi; Rickayzen, Alan; Rowley, Michael; Schmidt, Patrick; Trickovic, Ivana; Yiu, Alex; Zeller, Matthias: WS-BPEL Extension for People. 25.06.2007. Retrieved from http://download.boulder.ibm.com/ibmdl/pub/software/dw/specs/ws-bpel4people/BPEL4People_v1.pdf on 27.04.2009

[10] Agrawal, Ashish; Amend, Mike; Das, Manoj; Ford, Mark; Keller, Chris; Kloppmann, Matthias; König, Dieter; Leymann, Frank; Müller, Ralf; Pfau, Gerhard; Plösser, Karsten; Rangaswamy, Ravi; Rickayzen, Alan; Rowley, Michael; Schmidt, Patrick; Trickovic, Ivana; Yiu, Alex; Zeller, Matthias: Web Services Human Task (WS-HumanTask). 06.2007. Retrieved from http://download.boulder.ibm.com/ibmdl/pub/software/dw/specs/ws-bpel4people/WS-HumanTask_v1.pdf on 27.04.2009

[11] Schlegel, Thomas: An Interactive Process Meta Model for Runtime User Interface Generation and Adaptation, MDDAUI 2010 at CHI 2010, 2010

[12] Schlegel, Thomas: Object-oriented interactive processes in decentralized production systems. In: Proceedings HCI International 2009, 2009

[13] OpenESB community: IEPSE - OpenESB: the Open Source ESB for SOA & Integration. 2009. Abruf von http://wiki.open-esb.java.net/Wiki.jsp?page=IEPSE retrieved 09.09.2009

[14] Dusch, Sebastian; Fonden, Christian; Harbich, Florian: Datenstrommanagementsysteme. Student work on the University of Stuttgart. Retrieved from http://w3studi.informatik.uni-stuttgart.de/~fondencn/fs_dsms_dusch_fonden_harbich.pdf on 15.10.2010, Stuttgart, 2008

Innovative Production Machines and Systems
D.T. Pham, E.E. Eldukhri and A.J. Soroka (eds)

Developing and Implementing a Multi Faceted Business Improvement Programme in a Manufacturing Environment

Richard Barton [a] & Andrew Thomas [b]

[a] *Newport Business School, Newport University, Allt-yr-yn Campus, Allt-yr-yn Avenue, Newport, NP20 5DA*

[b] *Faculty of Engineering, Coleg Sir Gar, South Wales, SA15 4DN*

Abstract

Increasing globalization and customer demands have led businesses to produce a growing variety of products and services, with superior features according to customer preferences [1]. In order to meet these demands, manufacturing companies are following new and emerging trends in manufacturing improvement, often without establishing stability, thus benefit, of the previously improved systems, with the result that potential benefits are rarely achieved. However there are several notable cases in which companies have grown against probability or expectation due to the methodology of improvement being more continuous and building on the status quo rather than redesigning it at each change of emphasis. This paper describes the background to such an approach, thus defining how the successful companies have differed from a paradigm specific approach to their own benefit. Subsequently, a diagrammatic representation is proposed to visually demonstrate this approach, such that other companies could identify the most appropriate scope and scale of improvement activities for their particular instance, achieving the best available fit to strategic objectives.

Keywords: Continuous improvement, lean thinking, strategy, mapping, fit

1. Introduction

The competition posed by low labour cost countries, along with the mass customisation requirements of a much more consumer conscious society, is forcing UK industry to become increasingly lean and more responsive to customer requirements [2]. UK manufacturing industry must be able to develop and introduce increasingly sophisticated products more quickly and frequently to the marketplace through reduced product design and development times, and that this must be supported by an ability to rapidly reconfigure their manufacturing systems as well as being able to ensure high product quality and low manufacturing costs. In order to achieve these aims, the development and implementation of new and innovative manufacturing technologies [including manufacturing paradigms] is critical to the success of a company [3].

Over recent years, there has been a steady turnover of improvement paradigms proposed by practitioners and academics alike as new elements of best practice are identified by correlation with financial stability and success. Subsequently, business leaders move toward implementation of tools and techniques in the hope of emulating successful proponents of the tools and achieve reduced costs, improved product functionality and higher levels of customer satisfaction.

Between 2003 and 2007, the authors observed this change process in over 200 companies in the UK, in a variety of sectors, and of varying sizes, with a view to assist the company with operations management or technology implementation projects. The Supporting projects were the Dti funded Manufacturing Advisory Service (MAS) and the Superman Project delivered by Cardiff University's MEC. Of these companies there were obviously a significant variety of improvement practices being adopted, with associated levels of improvement culture. Since this time, the authors have had opportunity to maintain lengthy relationships with these companies and observe how improvements activities have changed, how they have affected product characteristics and delivered financial rewards. This is the first description of the observed trend, demonstrated in a particular case study referred to as Company A for confidentiality purposes.

Without exploring the full history of these paradigms, which can be seen in other, more lengthy, publications such as Burgess [4] and Ramesh & Devadasan [5], this paper considers Business Process Re-engineering (BPR) to be the epitome of system redesign at the convergence of improvement practices.

The following sections explore the concepts of process reengineering, lean thinking and strategic objectives, before considering how they can be best aligned for any particular company. Case studies of good practice are used to exemplify the continuity between practices for sustained benefit, and the various themes entailed are examined briefly individually then brought together visually in a mapping tool to show current and projected positions.

2. Business Process Re-engineering

BPR has often been understood as a collection of good ideas or existing best practices put together for application to business problems; Biazzo [6] states that "BPR clearly seems to be a re-invention and a re-assembly of elements that had already been present, for some time, in management thinking". This supports the suggestion that business lifecycles are aligned (often out of phase) with activities undertaken to overcome problems in the business. E.g. Lean manufacturing is employed to reduce cost as the company is starting to secure market share and ideally would also need to improve process flexibility and new product introduction, each of which could be considered to be at odds with waste reduction activities. This is supported by Smallbone et al [7], who find that most successful firms are not "production-led" but rather were characterized by an ability to make production process changes to complement an active market development strategy.

However, it is difficult to dismiss Lean as a business improvement framework, with much compelling evidence to promote its early adoption in a business improvement process. In addition, it has proven to be a critical step in company growth in the case study, providing a strong framework from which the company can draw on tools and techniques as required.

3. Lean Thinking as a framework for Improvement

As mentioned, academic interest and widespread consultancy activities have pushed Lean Manufacturing to the forefront of manufacturing company improvement initiatives. There are a variety of publications to summarise the virtues of lean thinking; Sanchez [8] describes six lean indicators which are considered drivers for a proactive company:

1. Elimination of non-value-adding activities,
2. Continuous improvement,
3. Multifunctional team,
4. JIT production and delivery,
5. Integration of suppliers, and
6. Flexible information system.

While Richards [9] contends that lean producers set their sights explicitly on perfection:

- Continually declining costs,
- Zero defects,
- Zero inventories, and
- Endless product variation.

This focus is supported by other authors [10] [11] [12] who also characterise a lean production system by an additional level of detail:

- Single piece production,
- Zero defect,
- Customer involvement,
- Flexibility, and
- Autonomous system in operation.

In our test company, Lean manufacturing was introduced to the operational areas around 1999, introducing 5S, standard operations and waste reduction practices. As a standalone exercise it seemed to be well received and fitted well, and naturally, with the scientific nature of the people working in the operational areas. Evidently there is a significant gap between technique utilisation and achievement of the above process characteristics, so Company A has gone on to consider the impact of areas throughout the business, and is currently rolling out a programme if lean training to all operational and service functions.

4. Lean Manufacturing, Service and Strategy

It is proposed by exponents of lean thinking that it is not just an approach to be used in improvement of the manufacturing environment, but that it is applicable to service and strategy definition. Naturally, service and strategy specialists resent this, and suggest that 'tool heads' should not apply lean techniques to a system which they do not suit [13]. Voss [14] goes on to explain that when using lean thinking to determine manufacturing strategy and maintain a competitive edge, failure to achieve continuous improvement in all areas only serves to neutralise that competitive edge. In effect you have to improve to stand still. A more effective strategy is a balance with other activities to react to customer requirement and build on core business capabilities to achieve a true competitive advantage.

In practice, company A, referenced lean thinking in respect of policy deployment, but not considered to be the key tool in determining the strategic approach for future growth and meeting the demands of the various stakeholders. However, it was a critical part in identifying the current state of the business process, and comparing this with the future state requirements, determined by the demands of the afore-mentioned stakeholders.

5. From Current State to Future State

The principal benefit of using lean thinking was to provide a framework for modelling the current state, by means of value stream mapping and projecting this forward to a future state of the business which is capable of meeting customer demand, thus identifying the gaps and projects required to improve capability. The future state map was also represented diagrammatically for ease of communication and consistency.

Again, in practice, Company A reflected the common trend of starting with a traditional textbook approach but then digressing from the theoretical 'learning to see' questions [15], instead choosing to use cause and effect analysis of the gaps identified and categorising them across four major themes.

Although the headline titles of the themes vary between instances in other companies, the emerging trend is to consider aspects of improvement in generic terms of: people, process, product and structure. Each of these umbrella themes then has its own component 'bones' to its cause and effect which reveal the detail and move closer to the root causes of poor performance in each area. Evidently this is useful in identifying definite and tangible projects to improve performance in this theme. Figure 1 shows how the current state mapping was translated to a current-state of performance in the four areas, which are described briefly below.

People; representing the culture of improvement and engagement, this reflects the degree to which silo mentality is hindering cross-functional co-operation and creating pockets of excellence in localised areas. In Company A this is closely related to the 'structure element described below, and a good example of how interactions must be considered.

Process; representing the capability of the production process(es) to meet customer demand, on time, in full, with low levels of scrap. As indicated in Company A's example in figure 1, this shows that there are high levels of individual skill in the business, which can be seen as a response to high levels of variability in raw material and finished product performance.

Product; representing material throughput and leadtime, it has serious impacts on working space and work in progress (WIP) values within Company A.

Structure; Representing the degree to which infrastructure of the business, system complexity, interactions, functional co-operation (and also this transformation process) are managed in the business. In Company A, this is the most significant variable, and not previously characterised to the degree of detail uncovered by this iteration of the current/future state mapping event.

However, the argument being presented in this paper is that the thought process taking the company to this point is not necessarily represented by project delivery. That is to say that a project identified in terms of improving product throughput or people engagement for example, may either fail to achieve the anticipated benefits or even act against a project identified in one of the other strands. One such example could be that of visual management which could be used to improve communication but also greatly increase system complexity and impact areas such as material control or purchasing, thus throughput. The ideal then, would be to plot each theme in context of the others in a style force field analysis in multiple dimensions; scale of 'severity', scale of penetration and also a vector of change from current to future state.

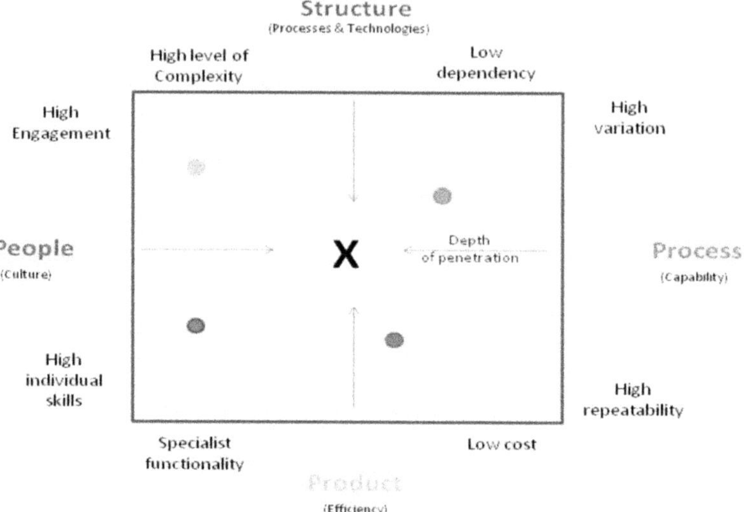

Fig 1. Current state performance in the generic improvement themes

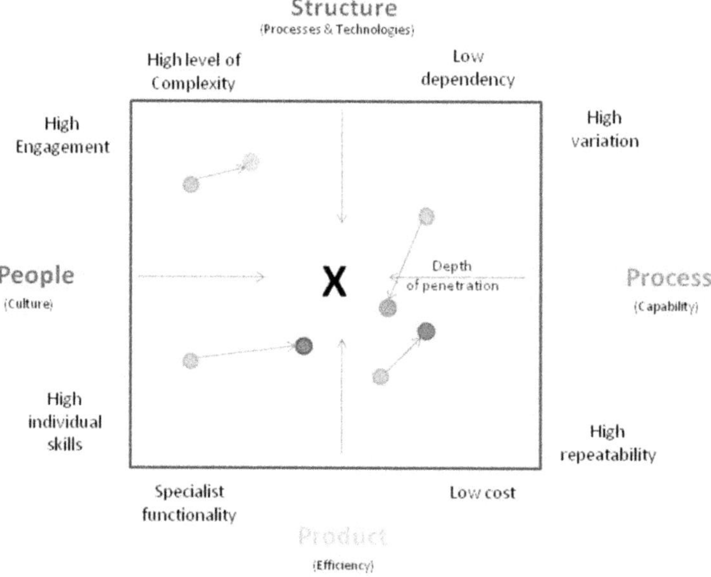

Fig 2. Transformation from current state to future state performance in the generic improvement themes

The transition from current to future state is a clear communication of corporate strategy, and a representation of the strategic objectives determined to create or maintain a competitive advantage, so evidently a critical success factor for the business. Further to this, the various themes must be triangulated to consider the impacts of activities undertake to achieve the projected change, and avoid conflict between intentions.

Currently, goal deployment sessions have been radically improved and altered to accommodate these considerations, and there is a much more focussed effort on achieving 'goodness of fit' of manufacturing characteristics with strategic objectives and business process improvement projects. The key to success is now appreciated to be the interactive facility provided, moving away from top-to-bottom communication or cascade of goals into activity.

A further improvement which has not yet been utilised in company A would be the introduction of a measure or metric to calculate this goodness of fit, which could subsequently be used to promote or relegate activities according to their relative contribution to achievement of the business objectives. This has been highlighted when just such a consideration has been debated at some length when considering the status of a new project idea.

6. Conclusions

Business Improvement techniques have varied in name and focus over recent years, and tools and techniques have shifted accordingly, to identify metrics and parameters around which the characteristics of manufacturing companies can be benchmarked and adjusted to achieve improvement against that theme. Naturally enough, the leaders of these businesses have followed the latest trend in search of continued improvement without considering the sustainability (or suitability) of both current and future improvement practices.

The most consistent improvements, seen in many cases and demonstrated by Company A, have been a result of a multi-faceted approach, implemented over a significant time but building on previous activities instead of replacing them. In the traditional way of thinking, these companies may not be progressing well against the latest benchmarks, but in their own context they continue to progress steadily and return good profits. The missing link, then, is to be able to determine the companies improvement context correctly in the first instance.

In company A, this process is seen to follow the fundamental lean thinking methodology; to improve flow, establish pull and pursue perfection, by means of focussing on themes of people, product, process and technology. The relative current maturities, capabilities and requirements can be plotted respectively in terms of depth of penetration of the theme, position of the company currently (independent) and where the ideal optimum is anticipated to lie (dependent). This paper proposes a collective methodology to diagrammatically plot this in this test bed companies.

7. Future Development

Evidently, many further test bed companies are required to validate this methodology, triangulating historic, current and future improvements longitudinally, against the financial reward achieved.

There may also be potential for growth of scope of consideration by means of movement toward outward (external environment) facing functions, such as those modelled by Porters five forces, [16] or the PESTLE analysis developed from the original PEST criteria [17], but maintaining balance with the current state to prevent continuous re-invention of the system, losing the benefit of previous activities by jumping between discontinuous themes in search of the panacea to production optimisation issues. The generic themes identified, which span the improvement paradigms available, are typified in this paper in terms of: people, product, process and structure. Company A has utilised this process to ensure goodness of fit to the strategic objectives, however this is not currently measured.

Analysis tools or techniques could be used to measure this fit, and could include;

- An auditing technique to determine the goodness of fit between characteristics, projects and objectives,
- A complexity model to define interactions and conflicts between the above elements, with a view to mitigating or minimising compromise,
- A characterisation plotting tool to locate the current and anticipated company profile in terms of maturity against improvement themes such as those outlined above; people, product, process and technology,
- A forcefield style tool to determine the relative impact of the external forces, which may vary according to location of the company in the characterisation plot,

References

[1] Ashkenas,R (2010) "Simply Effective: How to cut through complexity in your organization and get things done" Boston: Harvard Business Press. pp 82, 93-107

[2] Thomas, A.J, And Pham, D.T., (2004) "Making Industry Fit: The Conceptualisation Of A Generic 'Fit' Manufacturing Strategy For Industry", Proceedings 2nd Ieee Int Conf On Industrial Informatics, Indin 2004, Berlin, June 2004, R Schoop, A Colombo, R Bernhardt And G Schreck (Eds), Pp 523-528, Isbn 0-7803-8513-6.

[3] Pham, D.T, Thomas, A.J, (2005) "Fighting Fit Factories", Iee Manufacturing Engineer Magazine, April 2005, Pp 24-29.

[4] Burgess T. F. (1994) "Making the leap to agility: defining and achieving agile manufacturing through business process redesign and business network redesign", International Journal of Operations & Production Management, Vol. 4, No11.

[5] Ramesh, G., Devadasan, S.R., (2005) "Literature Review On The Agile Manufacturing Criteria", Journal Of Manufacturing Technology Management, 18, 2, 182-201.

[6] Biazzo S., (1998) "A Critical Examination Of The Business Process Re-Engineering Phenomenon", International Journal Of Operations & Production Management, Vol 18, No 9/10, Pp1000-1016

[7] Smallbone, D., Leigh, R., North, D., (1995) "The characteristics and strategies of high growth smes", International Journal of Entrepreneurial Behaviour & Research, Vol 1, No 3, pp. 44-62.

[8] Sanchez, M.S., (2001) "Lean Indicators And Manufacturing Strategies", International Journal Of Operations And Production Management, 21, 11, 1433-1451

[9] Richards, C.W. (1996), "Agile Manufacturing: Beyond Lean", Production And Inventory Management Journal, Vol. 37, No. 2, Pp.60-64.

[10] Mccurry, L. And mcivor, R.,(2001)"Agile Manufacturing : 21st Century Strategy For Manufacturing On The Periphery?", Irish Academy of Management Conference, University of Ulster.

[11]Shah, R. , Ward, P.T., (2007) "Defining And Developing Measures Of Lean Production", Journal of Operations Management, Vol 25, No 4, 785-805.

[12] Phillips, E.J., (2003) "The Pros and Cons 0f Lean Manufacturing for the Small to Medium Size Fabrication Shop", The e-newsletter of Lean Manufacturing : Lean Directions.

[13] Seddon, J., (2007) "Watch out for the Toolheads", Seen at superfactory.com;
http://www.superfactory.com/articles/featured/2007/pdf/0706-seddon-tool-heads.pdf

[14] Voss, C.A., (2005) "Alternative paradigms for manufacturing strategy", International Journal of Operations & Production Management, Vol 15, No 4, pp. 5-16.

[15] Rother, M., Shook, J., (2003) "Learning to See", The Lean Enterprise Institute, US

[16] Porter, M. E., (1985), Competitive Strategy: Techniques for Analyzing Industries and Competitors, New York, NY: The Free Press.

[17]Aguilar, F.J. (1967) "Scanning the business environment", New York: Macmillan.

Innovative Production Machines and Systems
D.T. Pham, E.E. Eldukhri and A.J. Soroka (eds)

Using Technology to Improve The Manufacturing Performance of Companies

A.J Thomas [a], R A Barton [b]

a. *Faculty of Engineering, Coleg Sir Gar, South Wales, SA15 4DN*
b. *Orthoclinical Diagnotics GB, Pencoed, Bridgend, South Wales*

Abstract

UK manufacturing industry is continuing to face strong cost-based competition, and is using an increasingly diverse range of technologies and manufacturing methodologies to drive change and improve company performance. Lean manufacturing is just one example of a paradigm established in recent times to reduce cost and increase business performance, but has come to represent a wide portfolio of tools. As technology capabilities increase, so do the complexities of interface with the business improvement paradigms. This increases the risk of misalignment or interference between them, especially when the active stakeholders are not fully aware of the technology and paradigm parameters. This paper is a review of technologies available, their impact on the range of business improvement practices, and the critical interfaces between process, people and strategy. It is the first outcome of a Knowledge Transfer Partnership (KTP), and aims to provide a valuable framework for learning for future projects in this popular area.

1 Introduction

UK manufacturing industry is experiencing significant changes in its strategic and operational systems [1]. The competition posed by low labour cost countries along with the mass customisation requirements of a much more consumer conscious society is forcing UK industry to become increasingly leaner and more responsive to customer requirements [2]. It is clear that in order to survive, UK manufacturing industry must be able to develop and introduce increasingly sophisticated products more quickly and frequently to the marketplace through reduced product design and development times, to rapidly reconfigure their manufacturing systems as well as being able to ensure high product quality and low manufacturing costs. In order to achieve these aims however, the development and implementation of new and innovative manufacturing technologies is critical to the success of a company [3].

As well as meeting the current challenges of UK manufacturing industry, it is crucial to consider the requirement to develop innovative manufacturing technologies to support future societal demands. The UK DTi Foresight panel report into manufacturing 2020 (2000) [4] highlights the importance of developing new and innovative technologies to support the requirement of 'developing a lifetime service around a manufactured product'. It goes onto highlight that it considers internet technologies as being the major enabler that will initiate a paradigm shift in manufacturing. These issues are further supported by the EU's 'FuTMan' report (2000) [5] and the NRC's 'Visionary Manufacturing Challenges' report (1998) [6] in which it highlights that the development of new and innovative technologies are critical to meeting the six grand challenges for manufacturing in 2020. This paper initially outlines the key challenges facing manufacturing industry as well as identifying the typical enabling technology requirements to assist in meeting these future challenges. Section 3 then shows a table of the various technologies available and their contribution to a leaner and higher performing manufacturing system. Finally this is put in context of a KTP project, which highlights how important these considerations are when developing a sustainable manufacturing system using technological tools.

2 The Challenges Facing Manufacturing Industry.

Through analyzing the key documentation surrounding the challenges facing UK, EU and US manufacturing industries [4], [5], [6], [28] it is possible to summarise the major challenges and also identify some of the key manufacturing obajectives that are required to achieve business sustainability. Table 1 highlights these challenges.

Table 1 Key Manufacturing Challenges

Manufacturing Challenges	Key Manufacturing Objectives [4], [5], [6], [28]
Achieve rapid and consistent delivery of new products into the marketplace by developing responsive design and engineering capabilities in companies based around new and sophisticated manufacturing technologies and techniques	Design to manufacture through time compression and the achievement of adaptable and reconfigurable products that meet customer requirements
	Timely, affordable, one-of-a-kind or niche products which can be quickly produced to meet specific customer requirements
	Business Intelligence for agile production plants. The development of "real-time" data systems seamlessly integrating with business processes and adjusting the company's focus on the market needs in an agile manner
	Dynamically Integrated Manufacturing Systems
	Collaborative software and Technology systems
Develop, enhance and extend human competencies , knowledge and technical capabilities within companies in order to create a shift away from 'manufacturing only' capabilities towards the manufacture of high value added products and services to a global market.	Development Technology Centres and R+D Cells Developing centres of excellence and knowledge bases which add high level value to products and services the company produces
	Information Systems Design and Integration
	Manufacturing Systems Modelling and Optimisation
	Improve the management and capitalisation of operational knowledge
	Competency management through a resource based perspective
Develop responsive and precise knowledge management and business intelligence systems in order to facilitate quicker and more accurate decision making within companies	Synthesis and architecture technologies for converting information into desired knowledge
	Methods towards developing worker empowerment through the development of effective educational methods
	Knowledge Management, Knowledge Engineering, Organizational Learning & Competence Management
	Resource Based View of Operations Management
	Expert Knowledge & Advanced Industrial Information Systems
Minimise environmental damage and operate towards energy neutral manufacturing environments including the development of highly responsive yet energy efficient local supply chains and logistics systems.	Waste- free processes and manufacturing systems The development of new and modified processes that must be able to produce zero waste, to consume the minimum amount of energy and to do both economically.
	Energy efficient low consumption manufacturing equipment and Eco- Manufacturing Systems
Achieve rapid enterprise reconfiguration through being able to rapidly reconfigure a company's manufacturing capabilities as well as its supply chain and logistics systems.	Adaptable and reconfigurable manufacturing processes and systems
	Address issues of information sharing in Supply Chains (security, confidentiality...)
	Defining a new balance between centralised and distributed organisations
	Investigate the social and legal aspects of cooperation in Supply Chains
	Rapid Reconfigurable Supply Chain Systems. Developing local and established long supply chains
Develop innovative products, processes and services by driving down product lifecycle times and continuously developing and enhancing the new product development and introduction systems (NPD/I) in companies.	Inventory control methods for the development of responsive and competitive companies.
	Systems reliability and maintenance
	Improve the management and development of highly customized products
	Developing new and accurate Total Cost of Ownership and Acquisition Modelling
	Resource based management
	Integrated Product and Service Management systems and "intelligent product" in the production management field
Develop closer and more meaningful collaborations between universities, colleges and industry including the development of improved collaborative design, research and manufacturing environments.	Collaborative Production Process Design Environments
	Developing technological and Innovation bases for manufacturing processes and supporting systems
	Intelligent Quality Systems
	The Development of new manufacturing management paradigms capable of meeting new manufacturing challenges of the future
Seek to develop new manufacturing management paradigms so as to create a more flexible, responsive and competitive manufacturing organisations that are able to continuously evolve and adapt to change and to develop effective change management and leadership methods to drive such systems.	The Fit Manufacturing approach aims at integrating the traditional concepts of Lean, Agility and Sustainability to develop a company's ability to survive and prosper in a competitive manner
	Development of Agile Virtual Enterprises
	Development of Change Management and resilience Paradigms including improving the communication between the tactical (sales and operation planning) and the operational (execution) level of POM.
	Technologies and management approaches to ensure products and processes achieve repeatability high quality and reliability

Develop and enhance digital networks, systems and technologies in order to enhance the digital economy.	Digitally connected supply chains and virtual enterprises
	Investigate how new digital information technologies could lead to totally new production management methods and practices
	Enhanced learning models and advanced pedagogic systems created through digital product manuals

3 Role of technology in sustainable manufacturing

In order for companies to achieve the competitive advantage described in the reports mentioned above, many choose to focus on improvement of a particular production capability, e.g. by means of new technology implementation to increase customer perceived value, or a best practice adoption to reduce waste and subsequently cost. However, it is proposed that these should be two-fold considerations, i.e that a single technology utilisation has the opportunity to add value through increased product functionality while providing an ideal opportunity to reduce waste from the product realization process. The nature of Lean Manufacturing implementation has varied in trend of both management tendency and tool/technique selection since its earliest utilization and subsequent 'translation' into new sectors and industries. Subsequently, practitioners and consultants alike have created new frameworks and strategies to mix and match aspects of the various systems in an attempt to find the ideal recipe for success [7], [8]. A very top level overview is presented in Table 2, the source data for which is not contained in this paper due to space a scope limitations, but is available in other literature reviews such as [9], [10].

Table 2. High level summary of manufacturing improvement strategies.

Era	Paradigm	Objective	Characteristics
1970s	IE	Effective throughout	High WIP, Stock product
1980s	TPS	Lean manufacturing	JIT, Waste reduction
1990s	BPR	Optimum process selection	Increased agility, frequent change
2000s	Six Sigma	Reduced variability	Increased quality, complex systems
2010s	Sustainability	Durability, responsibility	Corporate strategy, Company fitness

In terms of the company fitness and strategy it can be difficult to differentiate between paradigms' principles, with the result that companies now claim to have developed multi-model or integrated manufacturing systems, with buffered or smoothed interfaces to cope with fluctuation and residual uncertainty. One method developed to try and capture this was Business Process Re-engineering (BPR), designed to utilise new technologies and new operating parameters to improve company performance [11]. This had a clear attraction over a pre-defined process, as it could specify the optimum set of tools and techniques in an arrangement bespoke to the host company. Hence it can be said that there is already a history of technology enabling business process redesign for cost benefit. As shown in Table 2, Lean Manufacturing underpins this approach, with strong roots in value engineering and value analysis as well as the core principles of reducing waste and creating flow. To this end it remains to be a good platform for manufacturing improvement. The next stage then, is to determine what impact that these manufacturing technologies have on the achievement of business sustainabilty.

3. Technology and technological systems

In order to meet the requirements of a economically sustainable manufacturing environment of the future, we must concentrate on the innovative technologies that will provide competitive advantage through products and processes that meet the Quality, Cost, Delivery and Flexibility criteria [12]. From this, it can be shown that a series of generic technology categories that can be identified, these are shown in Table 3.

Table 3. Manufacturing technologies and their contribution to Sustainable Manufacturing.

Technology Category	Technology Examples	Contribution to Business Sustainability	References
Design Technologies	CAD, CAE, FEA, CFD, Virtual Prototyping, Manufacturing systems simulation.	Technologies based on concurrent manufacturing approaches aimed at reducing design to manufacture lead times thus enabling companies to: introduce products into the markets quicker and more cost effectively, provide a wider product portfolio and thus increased product flexibility, improved quality and less product design changes once in marketplace, reduction in material usage, improved analysis of operational efficiency thus reducing waste and emissions. Reduction in prototyping costs and product testing and evaluation time.	[13], [14], [15]

Manufacturing Technologies	RP, Micro-manufacture, D/CNC, Robotics, HSM, E-Manufacture	High performance processes capable of rapid reconfiguration, increased responsiveness through greater processing flexibility to meet customer needs, the production of high quality products in an energy efficient and environmentally compatible manner capable of waste free manufacture and material minimization. Technologies capable of reducing prototyping time and product validation costs. Adaptive intelligent systems capable of automatic fault diagnosis and rectification	[16], [17], [18]
Integrating Technologies	Holonic manufacture, distributed manufacturing	Technologies capable of integrating human and technological resources so that knowledge and skills acquisition times are reduced and process efficiency is improved. Technologies capable of integrating the requirements of the internal manufacturing organization (departments within organization) so that; error reduction is achieved, duplication of work is eliminated and the value stream is improved.	[19], [20]
Business Technologies	Internet, E-commerce, ERP, EDI, Extended enterprises	Technologies aimed at connecting the business with both its supply chain and its demand chain (connectivity to the customer) so that inventory reductions are achieved and the voice of the customer is accurately obtained (QCDF)	[21]

4 Linkages Between Technology, Process, People And Strategy

In the modern manufacturing era, the requirement for mass customization and the increasingly complex product lifecycle are not only necessitating development of technologies such as those described above but also increasing the rate at which companies are looking to implement them. D'Aveni [22] used the term "hyper-competition" to describe the condition of rapidly escalating competition that is now changing manufacturing landscape for many industries. The resultant change in the manufacturing landscape requires companies to reconfigure both their business and manufacturing strategies in order to cope with the new pressures.

However, there is little evidence to suggest that companies are realigning their business strategy with their manufacturing strategy. Brown and Bessant [23] provide evidence that companies remain stuck in past paradigms of strategy - most notably on the over dependence of an élite strategy-making group at the top of the organisation's hierarchy. They go onto state, that as a result of this, a state of strategic dissonance (Brown, 2000) occurs not only between the firm and its chosen markets but also within the firm itself, in the mismatch between strategic intent and operations capabilities They suggest that the reason for this is in the changes of manufacturing/operations processes over time and how these became divorced from the firm's strategy process).

It is therefore clear that in order to meet the new manufacturing challenges, a company must ensure that it aligns its business strategy (i.e. its corporate objectives) to its manufacturing strategy (which includes the manufacturing process elements and the technology used in the manufacturing processes).

Kim and Lee [25] define manufacturing strategy as "supporting corporate objectives by providing manufacturing objectives including costs, quality, dependability and flexibility to offer a competitive advantage and focus on a consistent pattern of decision making within key manufacturing resource categories. The objective of manufacturing strategy is to create `operationally significant performance measures' in which the competitive dimensions comprise cost, quality, dependability and flexibility".

Swink and Way [26] define manufacturing strategy as the decisions and plans affecting resources and policies directly related to the sourcing, production and delivery of tangible products. A basic premise of the more recent literature, therefore, is that elements of manufacturing strategy, such as technology management, are most effective when their performance and function relative to the business-level strategy are measured.

Swamidass and Newell [27] define manufacturing strategy as "the effective use of manufacturing strengths as a competitive weapon for the achievement of business and corporate goals". Manufacturing's strengths are developed and sustained by a "pattern of decisions" These are taken in a set of decision areas which encompass manufacturing strategy and are aimed at achieving manufacturing goals that align with business and corporate goals.

Brown and Bessant [23] extend the connection between manufacturing strategy and business strategy in their work in which they investigate the development of agility within the automotive industry. They state that the ability to become agile and to manufacture mass customised products can only be achieved by combining skills, technologies, know-how, processes and alliances with other players, brought about by strategies being in place. The need for manufacturing strategy may seem axiomatic for some academics. However, devising and implementing manufacturing strategies - in particular as a precursor to pursuing mass customisation - is difficult for some firms.

A popular option is to introduce off-the-shelf technology solutions which will enable the company to try and keep up with those top level companies who are already innovating the successor to these packages. The resultant changes required in training, standardisation, roll-out and full exploitation can usually be handled by existing staff, and the strategy can be 'amended' as required. Another option is to employ expert staff, or hire consultants, able to manage an integration of a tailor-made solution, or modify the existing facility, to achieve better alignment to the customer requirement in an attempt to get close to the preferred strategy. Or a third option is to define an optimum manufacturing strategy, understand the gaps in the current capability, and develop a relationship with technology and knowledge providers to close the gaps and achieve optimum results.

5 Discussion & Conclusions

One opportunity to create such a relationship is by way of a Knowledge Transfer Partnership (KTP), a mechanism whereby an associate is employed by the university, but placed full time in a company. This study has been undertaken as part of just such a project, which will be going on to exploit one or more of the technologies in a process design capacity.

It is seen that Lean manufacturing and business process engineering (often by new technology implementation) are recurring themes for KTP projects, and also ones which require the greatest length of induction time due to the holistic nature of the project and the depth of knowledge required by the associate to successfully deliver a project that will have the impact desired by the company. Hence, this paper presents the findings of this projects initial research with a view to providing a framework for future projects in similar areas, and allow for some time compression of the literature aspect of the induction.

References

[1] Williams B.R., (1995) "Manufacturing for Survival" Addison Wesley Publishing Company

[2] Thomas, A.J, and Pham, D.T., (2004) "Making industry fit: the conceptualisation of a generic 'Fit' manufacturing strategy for industry", Proceedings 2nd IEEE Int Conf on Industrial Informatics, INDIN 2004, Berlin, June 2004, R Schoop, A Colombo, R Bernhardt and G Schreck (eds), pp 523-528, ISBN 0-7803-8513-6.

[3] Pham, D.T, Thomas, A.J, (2005) "Fighting Fit Factories", IEE Manufacturing Engineer Magazine, April 2005, pp 24-29.

[4] "UK Manufacturing – We Can Make It Better", Final Report, Manufacturing 2020 Panel. Foresight Manufacturing 2020 Panel Report.

[5] "The Future of Manufacturing in Europe 2015-2020, The Challenge for Sustainability", EU FP6 Funded Programme, (2000).

[6] "Visionary Manufacturing Challenges for 2020" (1998) Committee on Visionary Manufacturing Challenges Board on Manufacturing and Engineering Design Commission on Engineering and Technical Systems National Research Council, National Academy Press, Washington, D.C.

[7] Spell C.S., (1999) "Where do management fashions come from, and how long do they stay?", Journal of Management History, 5, 6, 334 – 348.

[8] Towill D.R. (2006) "Fadotomy – Anatomy of the Transformation of a Fad into a Manufacturing Paradigm", Journal of Management History, 12, 3, 319-338.

[9] Sanchez, M.S., (2001) "Lean indicators and manufacturing strategies", International Journal of Operations and Production Management, 21, 11, 1433-1451

[10] Ramesh, G., Devadasan, S.R., (2005) "Literature review on the agile manufacturing criteria", Journal of Manufacturing Technology Management, 18, 2, 182-201.

[11] Davenport T.E., Short J.E.,(1990) "The new industrial engineering: information technology and business process redesign", sloan management review, summer,31, 4.

[12] Small, M.H, (1999) "Assessing manufacturing performance: an advanced manufacturing technology portfolio perspective", Industrial Management and Data Systems, 99, 6.

[13] Gonzalez-Benito, J., (2005) "A study of the effect of manufacturing proactivity on business performance", International Journal of Operations & Production Management, 25, 3.

[14] Wainwright, C [1995] "Design: a missing link in manufacturing strategy", World Class Design to Manufacture; 2, 3
[15] Provatidis, C G, Vassiliadis, S G (2004) "On the performance of the geometrical models of fabrics for use in computational mechanical analysis", International Journal of Clothing Science and Technology, 16, 5.
[16] Nourelfath M, Ait-kadi D, Isaac Soro W,. (2003) "Availability modeling and optimization of reconfigurable manufacturing systems" Journal of Quality in Maintenance Engineering, 9, 3, 284-302.
[17] Davis, J., (1999) "Fast Track to Waste-Free Manufacturing: Straight Talk from a Plant Manager (Manufacturing and Production)", ISBN / Part Number 1563272121
[18] Pham, D.T., Dimov, S.S., (2000) "Rapid Manufacturing: The Technologies and Applications of Rapid Prototyping and Rapid Tooling" Springer Verlag.
[19] Brown, D., Scherer, W., eds., (1994) "Intelligent Scheduling Systems", Kluwer Publishing.
[20] Joines, J. A., Barton, R. R., Kang, K., and. Fishwick, P. A, eds. (2000) "IMS mission architecture for distributed Manufacturing simulation" Proceedings of the 2000 Winter Simulation Conference
[21] Ball M., Chen C.Y., Zhao Z.Y. (2004) "Available to Promise, in: Handbook of Quantitative Supply Chain Analysis Modeling in the E- Business Era, Series", International Series in Operations Research and Management Science, Vol.74.
[22] D'Aveni, R., (1994) "Hyper-Competition: Managing the Dynamics of Strategic Manoeuvering", Free Press, New York, NY
[23] Brown, S., Bessant, J., (2003) "The manufacturing strategy-capabilities links in mass customisation and agile manufacturing - an exploratory study", , International Journal of Operations & Production Management, Volume 23 Number 7 pp. 707-730.
[25] Kim, Y., Lee, J., (1993) "Manufacturing strategy and production systems: an integrated framework", Journal of Operations Management, 11, 3-15.
[26] Swink, M., Way, M.H., (1995) "Manufacturing strategy: propositions, current research, renewed directions", International Journal of Operations & Production Management, 15, 7, 4-26.
[27] Swamidass, P.M., Newell, W.T.,(1987) "Manufacturing strategy, environmental uncertainty and performance: a path analytical model", Management Science, 33, 4, 509-24.
[28] I*PROMS Deliverable D7.11/D7.12: –The POM Foresight and Grand Manufacturing Challenges Study, I*PROMS POM Cluster Document Oct 2006.

Innovative Production Machines and Systems
D.T. Pham, E.E. Eldukhri and A.J. Soroka (eds)

An optimal surgical aid system for Hip Resurfacing Arthroplasty

L.C.Hieu[a], C.U. Dussa[b], K. Singhal[c], N.V. Hieu[d], M.I. Okereke[a], N. Zlatov[e]

[a] *School of Engineering, University of Greenwich, Kent ME4 4TB, UK*
[b] *Orthopaedische Kinderklinik, Kinderorthopaedie, 83229 Aschau, Germany*
[c] *Prince of Wales Hospital, Wales CF31 1RQ, UK*
[d] *Faculté des Sciences et Technogie, Université Paris-Est,94010 Créteil cedex ,France*
[e] *School of Engineering, Cardiff University, Wales CF24 3AA, UK*

Abstract

Hip Resurfacing Arthroplasty (HRA) is a treatment option for the patients with the advanced hip disease; it is considered as the most technically difficult techniques of all procedures recommended for osteonecrosis of the hip. Technically, the currently applied HRA surgeries lead to unstable and inconsistent results. Surgeons rely a lot on the manual technique and conventional tools as well as their skills to determine the right drilling angle for locating the implant system. Although the robotic and surgical planning systems are available for HRA, the drilling line is still defined geometrically and intra-operatively, not fully considering about the biomechanics aspects of the implant and bone structure. In this paper, an optimal surgical aid system for HRA is proposed. With the integration of the state of the art biomedical modelling, pre-operative planning and personalised surgical tools, knowledge based and expert system, as well as biomechanics modelling and analysis, the precision, safety and speed of surgery are improved, the complexity of surgery is reduced, and therefore the survival rate of the implant is increased. Especially, the proposed system provides a cheap and practically feasible solution with the integration of expertise from both engineering and medicine for improving the treatment quality of the patients.

Keywords: Hip resurfacing arthroplasty, surgical planning, surgical tools, biomedical modelling, implant

1. Introduction

The concept of Hip Resurfacing Arthroplasty (HRA) has evolved directly from the original mould arthroplasty introduced by Smith Petersen in 1948. In the early 1950s, the conventional THR was pioneered by Sir John Charnley, experimented with hip resurfacing using Teflon. In the 1960s, the metal-on-metal (MoM) HRA was developed in Switzerland and France [3, 4]. In the 1970s, cemented systems using a polyethylene acetabular component and a metal femoral cup were introduced in Italy, Japan, England, Germany, and the United States. In the HRA operations, only the diseased or damaged surfaces of the head of the femur and the acetabulum are removed. The femoral head and hip socket are respectively fitted with a spherical shell and a thin spherical cup. Both spherical cups form a pair of metal bearings. The first generation of HRA (1970s – 1980s) showed the disappointing results and the procedure was largely abandoned by the mid 1980s [1, 6]. The failure was essentially a consequence of the use of inappropriate materials, poor implant design and inadequate instrumentation rather than an inherent problem with the procedure itself [2]. The second generation of HRA (1991 to Present) showed the good results. Figure 1 presents the radiograph of the Birmingham Hip Resurfacing and its spherical shell and cup. Early implant loosening and femoral neck fracture now appear to be rare; only 4 % has the risk of the neck of the femur breaking at some time after the operation.

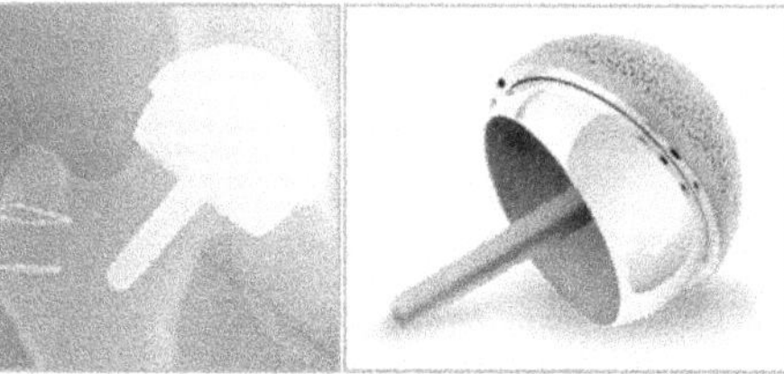

Fig.1. Radiograph of HRA and the spherical shell and cup of a Birmingham Hip Resurfacing system.

Approximately 70,000 primary Total Hip Replacements (THR) are performed in UK each year, of which 15% are revisions, and 15.5% are performed in the younger age group with the mean age of 57.5 years. Rheumatoid arthritis accounts for 6% of the indications for THR and moderate to severe osteoarthrosis for over 75%. The proportion of patients who required device revisions, from HRA to THR, was reported in all but one study and ranged between 0% and 14.3% [1, 5]. The survival rate reports of HRA by different sources of clinical studies [1, 7] are as follows: (i) With the median follow-up of 8 years, 73% of the HRA patients were revised to THR, and 15% of THR were required for the revision; (ii) In the study of 93 patients, the survival rate is 70% at 5 years and 40 % at 8 years. The fracture of the neck is 4 %; and (iii) In the study of 403 patients less then 55 years of age, the survivorship over a 5 years study is 99.7%; and (iv) With a mean follow-up of 9 (range: 1-16) years for the survival of 114 HRA, the 5-year, 10-year and 15-year mean survival were respectively 92%, 47% and 30%.

Currently, HRA is recommended as one option for people with the advanced hip disease who would otherwise receive and are likely to outlive a conventional primary total hip replacement, principally individuals under the age of 65.

In this paper, the challenges and unsolved issues of HRA as well as the proposed treatment method and techniques are presented and discussed. The study is aimed at integration of the state of the art technologies, including Biomedical Modelling, Biomechanics, Intelligent Optimisation and Data Management, and Design and Manufacturing, for enhancing the surgery skills of the surgeons and improving the diagnosis and treatment quality for the patients.

2. Challenges and unresolved issues of HRA

HRA is considered as the most technically difficult techniques of all procedures recommended for osteonecrosis of the hip. The surgical technique requires the skills of well trained hip surgeons. The following are the current challenges for HRA.

2.1. Unresolved issues

Several unresolved issues of HRA have been addressed, including the failure rate, new osteonecrosis and fracture of resurfaced femoral heads, metal ion production secondary to wear debris or corrosion and subsequent hypersensitivity, and appropriate patient selection [1, 8]. The release of metal debris can occur with metal implants. These implants produce corrosion products that are biologically active and may cause chronic inflammatory reactions that can lead to loosening of the implant. It is reported that the concentration of metal debris is higher if the prosthesis is worn or loose, or if the joint is infected. It is not clear what the normal levels of ions in human tissue should be; however, animal studies show that cobalt doses up to 1,000 times normal may be tolerated. Larger doses than that can induce anemia, loss of appetite and weight, and an increase in the number of red blood cells, lesions in mucous membranes, local malignant skin tumors, and death. Additionally, though inconclusive, there is concern that extensive metal ion release may cause changes to the immune function which may lead to lymphomas and leukemia's [8].

2.2. Proper component positioning

There are technical challenges with proper component positioning in HRA. In most of the current HRA surgeries, the pre and intra-operative planning is still based on 2D X-ray images, conventional tools and manual methods, including (i) selection of an implant from standard ones, (ii) measuring the size of the femur head, and (iii) determining the drilling line for preparing the femur head and locating the implant. This leads to the high risk in surgery and unstable treatment quality, which is much dependent on the experiences and skills of the surgeons. The incorrect locating of the implant and the lack of supportive information about the biomechanics aspects between the device and bone structure could be one of the reasons causing the femoral neck fractures and reducing the survivorships of the spherical shell and cup.

2.3. Surgical planning and tools

Although several software and systems are available for surgical planning based on CT/MRI data, to transfer this surgical planning information into the operation rooms for implementation is not easy and straightforward. The computer assisted surgery therefore has been an interesting area of research in the recent decades. Since 2004, many efforts have been concentrated on developing the 3D surgical planning for HRA, including the following two main systems: (i) ORTHOsoft® (Zimmer Ltd., UK) - A navigation system for HRA, and (ii) Acrobot (The Acrobot Company Ltd, UK) - Surgical systems for computer-assisted 3D planning, surgical navigation and surgeon-controlled robotic surgery. The specialised surgical planning software allows the surgeon to visualise patient anatomy, decide on implants and their positioning. The surgical navigation allows the surgeon, during surgery, to guide tools into place and ensure that implants are positioned as planned. A robotic system is able to work with the surgeon to ensure that bone resection is performed optimally in line with the plan to accept implants and that cutting tools are confined to the regions requiring resection.

However, it is not clearly seen the benefits that the current medical image processing (MIP) software, robotic surgery and navigation systems contribute to HRA in term of the cost, diagnosis and treatment quality as well as the technology transfer to hospitals. The computer-assisted surgical systems are expensive and not always available in the most of the hospitals. In addition, the use of these systems requires the special training as well as the know-how and skills of the surgeons. Therefore, the manual surgical method and tools are practically and commonly used for pre- and intra- operative planning and surgeries of HRA in most hospitals, without the support of the computer system, 3D imaging techniques, and personalised surgical tools.

Additionally, although the position of the implant is determined based on MIP, currently, surgeons still base on their experiences and skills as well as the available conventional tools to determines the drilling line and prepare the femoral head for the implant insertion. Moreover, the methods used by the robotic surgery and navigation systems such are done intra-operatively; therefore it is time consuming, and it can not take the advantages of the pre-operative planning results. Especially, the implant angle and the position are determined based on the reference points collected by the surgeons during the operation, with the aid of the MIP system. These lead to the long operation time, and the surgical procedure becomes more complex. Finally, as mentioned above, the investment of the robotic surgery systems is expensive; and it requires the use of both the MRI or CT data of the patient and the robotic system.

It is also noted that, most of the robotic surgery and navigation systems and MIP software for HRA are based only on the geometrical analysis of the femur for surgical planning. There has been no surgical system that is available for optimsing the implant position in term of biomechanics's point of views, and then transferring the analysis and planning information into the actual surgery accurately. If the implant is not optimally located, and the aspects of biomechanics between the device and bone structure are not carefully considered, the survivorships of the implant system is reduced, and it may cause the fracture of the femoral neck. More over, if the personalised surgical aid tools are not available, it would be difficult for the surgeons to obtain the best surgical accuracy, and the treatment quality is still highly dependent on the experience and skills of the surgeons.

2.4. Knowledge based system for HRA

There has been no knowledge based system and tool existed for supporting surgeons in pre-operative planning, advising and supporting in diagnosis and treatment, managing the patient data, and monitoring patient database in HRA. Most of the current surgical planning systems for HRA are focusing on guiding and helping the surgeons about the surgical procedures and obtaining the surgical planning parameters. Especially, the survival of the HRA implant system is up to 15 years [1, 7] or more. Therefore, it would be necessary to construct a knowledge based computer system and tools to support surgeons not only obtaining the good surgery, but also managing the patient data and monitoring patient treatment progress, as well as obtaining the right information from the database for a new treatment.

3. An optimal surgical aid system for HRA

In order to overcome the above mentioned key challenges, an optimal surgical aid system for diagnosis and treatment of HRA (HIPOS) has been developed as shown in Fig. 2, with the integration of 4 surgical aid units, including (i) Pre-operative planning,

(ii) Personalised surgical tools, (iii) Biomechanics modelling and analysis, and (iv) Knowledge based and expert system. These 4 surgical aid units share the same database and work together to support the surgeons all necessary information and surgical tools in order to obtain the best diagnosis and treatment for the patients. In this way, the proposed optimal system for HRA is able to obtain the following objectives: (i) Enhancing the surgical skills of the surgeons and improving the treatment quality for the patients; (ii) Increasing the accuracy and reducing the time of the surgery and (ii) Providing a cheap technical and practical solution as well as improving patient and data management for diagnosis and treatment of hip disease.

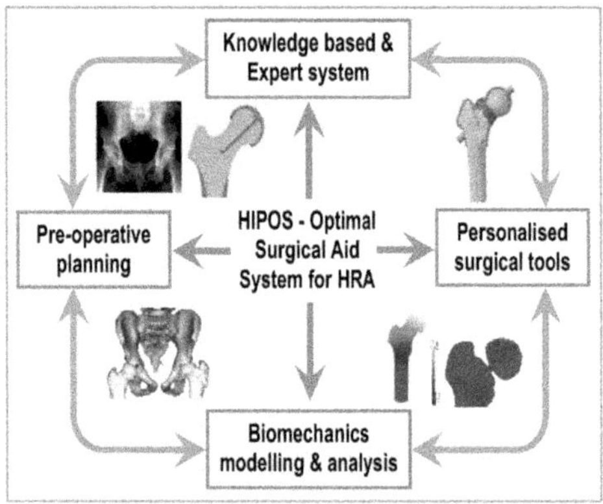

Fig. 2. An optimal system for HRA.

3.1 Enhancing the surgical skills of the surgeons and improving the treatment quality for the patients

With the use of the state of the art MIP and biomedical modelling technologies, complex surgical planning can be implemented preoperatively. In addition, the surgical risks and time can be reduced dramatically, and the skills of the surgeons are improved, since the surgeons can use the biomodels for surgical planning and rehearsals before the real operation is done [9, 10, 12]. Surgical tools such as the drilling guides proved the remarkable contribution of the biomedical modelling in improvement of the treatment quality for the patients; especially, for the complex surgeries.

A preoperative planning unit helps surgeons having better understanding of the anatomy and status of the patients, selecting the right size of an implant, determining correctly the optimal implant orientation, conducting simulations and rehearsal of the operation. Finally, the diagnosis, treatment, and surgical planning are implemented via the MIP tool and a knowledge based and expert system (*Section 3.3*).

Figure 3 presents the results about investigation of the influence of femoral component placement on the stress distribution on the femur neck in HRA [11].

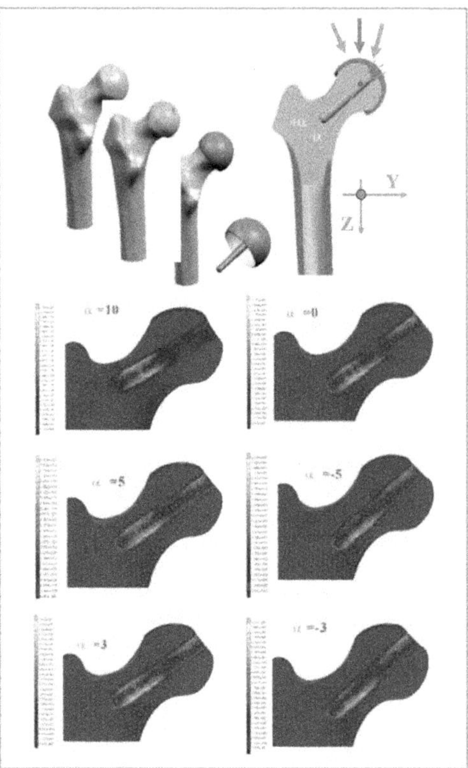

Fig.3. FEA study of the influence of an implant placement on the stress distribution on the femoral neck in HRA. α is the angle between the stem of the spherical cup and the geometrically optimal centre line of the femoral neck.

It is clearly shown that the stress distribution on the femur changes and influences in the fracture tendency of the femornal neck when the angle of the stem of the spherical cup is changed. Moreover, technically, the traditional and currently applied techniques for determining the centre of the femur neck in HRA surgeries rely a lot on the manual technique and conventional tools as well as surgeons' skills to determine the right drilling angle for locating the implant system. This leads to the inconsistent results. Even when the robotic system is used (*Section 2.3*), the drilling line is still defined geometrically and intra-operatively, not considering about the biomechanics aspects of the implant and bone structure. Finally, there is no surgical procedure and system which is currently and practically applied for HRA with the use of advanced biomechanics computation for determining the optimal orientation of the spherical cup.

With the the biomechanics modelling and analysis unit for computation of the optimal orientation of the implants, surgeons could plan and obtain the best solution based on the specialized Finite Element Analysis (FEA) tool which is integrated into the proposed HIPOS system. Biomedical Engineering (BME) experts can also work with surgeons via the collaborative Computer Aided Design and Engineering (CAD/CAE) and MIP environment. The optimal surgical planning and biomechanics analysis information for the orientation of the implants is transferred to the actual surgery via the personalised surgical tool (*Section 3.2*). In this way, the implant age and the treatment quality for the patients are improved.

3.2 Increasing the accuracy and reducing the time of the surgery with the use of personalised surgical tools to obtain the optimal implant positioning

In the proposed HIPOS system, the personalised surgical tools unit provides essential information and supportive devices for the surgeon to enhance the precision, safety and speed of surgery, as well as to decrease the complexity of surgery. It helps surgeons to implement complex surgeries without requiring high skills. First of all, the surgical planning can be preoperatively implemented to obtain the necessary surgical constraints based on MIP of the patient data in the form of CT/MRI images. From the biomechanics modelling and analysis (*Section 3.1*), the personalised surgical tools for guiding the surgeons to obtain the right orientation and position of the implant quickly and stably are designed and manufactured via the interface between MIP, CAD and CAE.

It is important that the surgical tools are also designed to be integrated with the traditional and standard HRA ones to help surgeons preparing the femoral head for implantation optimally. In this way, the accuracy of the operation is improved, and the operation time is remarkably reduced; especially, less skill is required for such a complex operation.

3.3 Providing a cheap technical and practical solution as well as improving patient and data management for diagnosis and treatment of hip disease

With the use the rapid prototyping (RP) and manufacturing as well as advanced technologies in MIP and CAD/CAE, the HRA based treatment can be done cheaply and effectively. Experts in engineering and medicine can collaborate and share the tasks to accomplish the complex surgery.

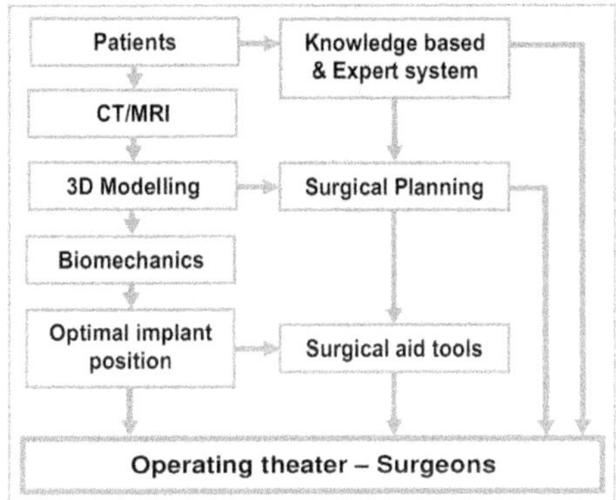

Fig.4. The workflow for practical implementation of the optimal surgical aid system for HRA

Figure 4 presents the key units and workflow for practical implementation of the proposed HIPOS system. The patient data can be sent from the hospitals to BME experts to work on the computational modelling and analysis. The analysis data as well as the clinical constraints generated from the surgical planning process is transferred to design and manufacturing experts to work on development of the personalised surgical tools. Experts in both engineering and medicine are able to share the surgical planning and database and work together via the collaborative MIP and CAD/CEA environment. Figures 5 and 6 present the FEA simulation of HRA and development of the surgical tool which is used to assist a surgeon to define the right position of the drilling line and install the traditional surgical device in order to prepare the femonal head for implantation of the spherical cup.

As mentioned in *Section 2.4*, the knowledge based and expert system is useful for the hospital in general and surgeons in particular in not only obtaining the best treatment results, but also managing the patient data and monitoring patient treatment progress. Especially the application of HRA for hip disease treatment is only acceptable with the development of the second generation of HRA implants since 1991; and in the United States, the Food & Drug Administration (FDA) only approved the HRA based treatment in 2006. In addition, with the support of knowledge based and expert system, the surgeons and patients are able to obtain the right information for optimal surgical planning, diagnosis and treatment.

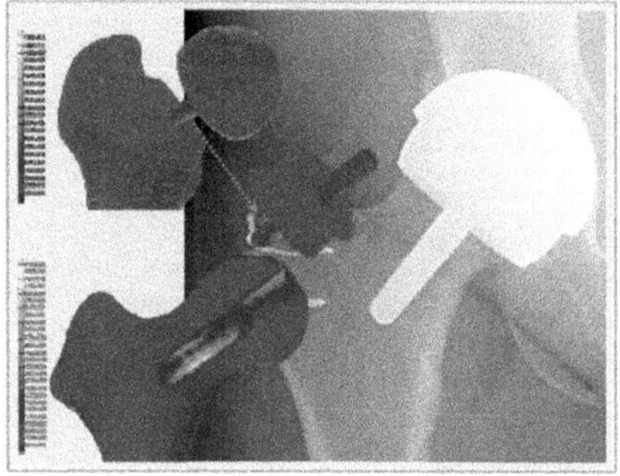

Fig.5. FEA simulation of HRA for determination of the optimal implant positioning

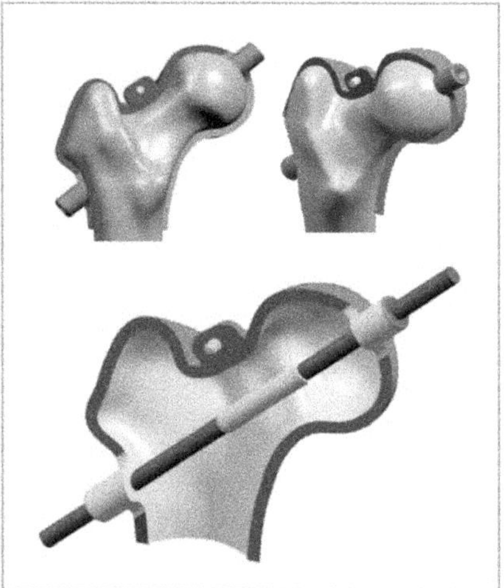

Fig.6. Development of the surgical tool for HRA

4. Conclusion

The traditional and currently applied methods for the HRA based treatment lead to expensive, unstable and inconsistent results which are highly dependent on the manual techniques and conventional surgical tools, as well as surgeons' skills and experieces. Although the robotic and surgical planning systems are available for HRA, the drilling line for locating the implant is still defined geometrically and intra-operatively; the biomechanics aspects of the implant and bone structure are not fully considered. The optimal surgical aid system for HRA was thefore developed to solve the current challenges in HRA; it integrates the state of the art biomedical modelling, pre-operative planning and personalised surgical tools, knowledge based and expert system, and biomechanics modelling and analysis, in order to obtain the high precision, safety and speed of surgery. Especially, the proposed HIPOS system provides a cheap and practically feasible solution with the collaborative involvement of experts from both engineering and medicine for improving the diagnosis and treatment quality for the patients.

References

[1] Guidance on the use of metal on metal hip resurfacing arthroplasty. UK National Institute for Health and Clinical Excellence, 2002, www.nice.org.uk.

[2] Roberts J, Roberts P, and Grigoris P. Metal-on-metal hip resurfacing. SMJ, 2005, 50(1), pp 10-12

[3] Muller ME. Lessons of 30 years of total hip arthroplasty. Clin Orthop., 1992, 274, pp12- 21.

[4] Gerard Y. Hip arthroplasty by matching cups. Clin Orthop., 1978, 134, pp 25-35.

[5] Tennent TD and Goddard NJ. Current attitudes to total hip replacement in the younger patient: results of a national survey, Ann R Coll Surg Engl, 2000, 82: 33-38.

[6] Debbie A, and Sharon Q. Hip resurfacing: Past, present and future. Journal of Orth. Nursing, 2005, 9, 87–94.

[7] Duijsens AW, Keizer S, Vlieland TV, and Nelissen RG. Resurfacing hip prostheses revisited: failure analysis during a 16-year follow-up. Int Orthop., 2005, 29(4), pp 224-228.

[8] Hip Resurfacing. National Medical Policy, 2007, www.healthnet.com.

[9] Hieu LC, Vander Sloten J, Bohez J, Khanh L, Binh PH, Toshev Y, Zlatov N. Medical Rapid Prototyping Applications and Methods, Assembly Automation Journal, 2005, 25(4), pp 284-292.

[10] Hieu LC, Bohez E, Vander Sloten J, Phien HN, Vatcharaporn E, Binh PH, and Oris P. Design for Medical Rapid Prototyping of Cranioplasty Implants, Rapid Prototyping Journal, 2003, 9 (3), 175-186.

[11] Dussa CU, Hieu LC, Singhal K, Sloten JV. Influence of femoral component placement on the load distribution on the neck in surface replacement arthroplasty. In proceedings of XXIII SICOT/SIROT Triennial World Congress, 2005, Istanbul, Turkey.

[12] Van Cleynenbreugel J, Schutyser F, Goffin J, Van Brussel K, and Suetens P. Image-based planning and validation of C1-C2 transarticular screw fixation using personalized drill guides. Comput Aided Surg., 2002, 7(1), pp 41-48.

Innovative Production Machines and Systems
D.T. Pham, E.E. Eldukhri and A.J. Soroka (eds)

Automated simulation of stochastic part variation to identify key performance characteristics of assemblies

M. Mazur, M. Leary, A. Subic

School of Aerospace, Mechanical and Manufacturing Engineering, RMIT University, Melbourne, Australia.

Abstract

The functionality of a manufactured assembly is dependent on the stochastic variation of part parameters. The effects of manufacturing variation are typically assessed through tolerance analysis using virtual product assembly models. However, the definition of a tolerance analysis problem requires a formal definition of the Key Product Characteristics (KPCs) which indicate whether a given manufactured assembly will conform to intended functional requirements. Identifying assembly KPCs can be a challenging problem for the designer as assembly behaviour in the presence of stochastic variation may be difficult to predict. The authors have developed an automated Process Integration and Design Optimisation (PIDO) tool based platform to aid designers in identifying assembly KPC within the native CAD design environment. To demonstrate the capabilities of the developed platform, a test case involving a commercial product assembly is presented.

Keywords: robust design, design for manufacture, tolerance analysis, Key Product Characteristics

1. Motivation

Manufacturing processes impart stochastic variation to part parameters. Such variation can compromise assembly functionality through unforeseen and unintended part interactions which are divergent from the intended design. Such unintended part interactions can be assessed by empirical inspection. However, assemblies may consist of many interacting parts, each of which includes a number of part features, each of which is be subject to a number of stochastic dimensional variations. The resulting combinatorial explosion of potential geometric variation results in a very large number of assembly permutations, which makes manual, empirical inspection impractical.

The effect of manufacturing variation on assembly functionality is typically analysed through tolerance analysis using virtual models (analytical or computational) of the product assembly. Numerous analytical and numerical methods have been proposed for addressing tolerance analysis problems in complex product assemblies [9]. Also, commercial Computer Aided Tolerancing (CAT) tools have been developed that offer tolerance analysis capabilities utilising CAD models of product assemblies [6].

However, a fundamental requirement for the definition of a tolerance analysis problem is the establishment of the functional boundaries of the assembly design. Assembly parameters must be formally defined which indicate whether a given manufactured assembly will conform to the intended functional requirements. The parameters of particular relevance to functionality are referred to as Key Product Characteristics (KPCs), and are traditionally geometric, such as clearances or nominal dimensions.

An assembly response function defines KPCs in terms of the parameters of an assembly. The assembly response function may be explicitly defined by an algebraic expression, or may be implicitly defined within a numeric model, for example a CAD assembly or CAE model. Specification limits applied to KPCs define targets beyond which product functionality is compromised. An assembly which conforms to the specification limits of all KPCs is one which satisfies all functional requirements.

Identifying the assembly KPCs can be a challenging problem for the designer as it is seldom obvious how deviations of the part parameters from the nominal intended values affect the performance of the assembly, and which assembly parameters are indicative of compromised functionality.

Furthermore, the competing objectives of achieving acceptable manufacturing cost and desired product quality, as well as the functional constraints imposed by the product design requirements, may require the designer to identify the relative importance of the identified KPCs.

To aid designers in identifying assembly KPCs, the authors have developed an automated platform to identify unforeseen and unintended part interactions within the native CAD product design environment.

The method utilises Process Integration and Design Optimisation (PIDO) tools and functionality commonly available within commercial CAD software, to simulate the effects of manufacturing variation on the part parameters of an assembly.

2. Proposed method

2.1. Platform architecture

A need has been identified by the authors to aid designers in rapidly identifying KPCs that need to be controlled during subsequent tolerance analysis of the assembly. The KPCs should be assessed from the CAD assembly available at the design stage, within the native CAD product design environment. A PIDO tool based platform was developed which integrates with commercial CAD software to assess a CAD assembly for unforeseen assembly behaviour (e.g. clash or interference) when subject to part parameter variation as expected during manufacturing. The platform methodology consists of the following phases:

1. Definition of CAD models for each part of the product assembly.
2. Specification of expected variation and tolerance types for part parameters of interest.
3. Definition of part relationships that constitute the assembly such as mating conditions and assembly sequence.
4. Definition of clearance measurements between part features within the assembly in regions considered important by the designer.
5. Interfacing the CAD assembly parameters with a PIDO workflow using scripting capabilities (Section 2.2).
6. Initialisation of a feasible set of part parameter values from a random sample of their respective probability distributions using Monte Carlo sampling (Section 2.5.2)
7. Sequentially updating the CAD model part parameters for each value combination.
8. Updating the CAD model assembly and executing a clash/interference analysis.
9. Recording assembly output parameters such as clearance measurements as well as undesired part interferences or contacts instances including information detailing the nature of the interference.
10. Recording of images of the assembly and detail view of any interference scenarios. If a clash condition is identified the penetration distance is also recorded.
11. Storing of the CAD assembly configuration evaluated for reference.
12. Evaluation of the sensitivity of each input part parameter to a particular interference case. Sensitivity analysis identifies which part parameters are most likely to compromise assembly functionality when subject to the expected manufacturing variation.

The capability of the platform is demonstrated on a commercial product assembly in Section 3.

2.2. PIDO tools

Multi-disciplinary CAE design is increasingly common within product design problems involving disparate engineering disciplines. A range of PIDO software tools such as ESTECO modeFrontier have been developed to aid in multi-disciplinary CAE design. PIDO tools integrate standalone CAD and CAE capabilities to enable automated parametric analysis, statistical analysis and multi-objective optimization [3, 5].

Standalone CAD and CAE software interaction with PIDO tools is achieved through commonly embedded scripting capabilities (based on JavaScript, Visual Basic, Python or DOS script) and can be structured using serial or parallel connections or conditional switches. For example, the CAD software CATIA accommodates the ability to execute Visual Basic language scripts. Scripting capabilities of CAE and CAD tools allows for autonomous:

- modification of CAD model parameters
- initialisation of CAE simulations
- recording of the obtained simulation results

Problem formulation within the PIDO tool is defined by a logic-based workflow which describes the intention of the simulation and integrates with the system model to be analysed. The workflow establishes links between external CAE models (through scripting capabilities), the model input and output parameters under analysis, and a design of experiments (DOE) for an automated parametric study of the model under analysis. During a simulation, each of the parameter combinations established as part of the DOE is automatically applied to the external CAE model and any affected model output parameters of interest are read and recorded within a database. At the completion of the simulation, statistical tools embedded in the PIDO tools can be applied to analyse the recorded results.

In this work, the authors have applied the process integration, DOE, data logging and statistical analysis capabilities of PIDO tools to enable the automated statistical analysis of CAD assemblies subject to manufacturing variation. This platform aids the designer in the identification of unforeseen part interaction and allows identification of the associated KPCs.

2.3. CAD tools

The modelling approach in typical CAD software is inherently parametric. Three-dimensional solid geometric models are created from parametric two-dimensional sketches which are subject to operations such as extrusions, sweeps and lofts. If any part feature needs to be altered, the relevant relationships or dimensions can be parametrically addressed and updated.

CAD assemblies are defined by multiple CAD parts whose interaction is constrained to restrict the degrees of freedom between parts. If any part parameters are modified, the assembly constraints are re-evaluated to rebuild the assembly model.

Commercial CAD tools also feature the ability to carry out scripted instructions (i.e. macros) without user input. The CAD software CATIA for instance is able to execute macro instructions programmed in Visual Basic script; SolidWorks can execute Python script instructions. The parametric design and scripting capabilities provide a means of implementing autonomous analysis of CAD model parts and assemblies.

CAD tools also typically feature assembly clash detection capabilities which report the location and magnitude of unintended part contact or interference. This interference detection capability can be utilised in the proposed platform to assembly regions where unforeseen part feature interference may violate functional requirements.

2.4. Potential limitations

Although the platform enables insight into unforeseen and undesirable part interactions, it may suffer from limitations associated with applying parametric CAD for simulating the effect of manufacturing variation:

- Some forms of geometric variation defined in GD&T standards are not readily accommodated using parametric CAD models [9].
- Parametric CAD models may have inherent limitations in representing intermittent part contact.

Commercial Computer Aided Tolerancing (CAT) tools have been developed that can overcome these issues, however, this approach requires the development of a specialised CAT model of the parts and assembly [6]. The proposed platform is applied directly to the CAD models developed for concept design and development, and can be directly integrated with the design process without the need for additional modelling or separate CAT software.

2.5. Manufacturing variation

The fundamental nature of manufacturing processes results in stochastic process outputs. This variation can occur within geometric parameters such as dimensions of a part feature. Tolerances are specified on part features to describe the level of manufacturing variation that can be tolerated. Acceptable tolerances are referred to as the upper specification limit (USL) and lower specification limit (LSL). The quality of a manufacturing process can be quantified by how consistently and accurately it produces the desired process outputs. Manufacturing process quality is assessed using process capability indices (PCIs).

2.5.1. Process capability indices

Manufacturing processes which exhibit a normal distribution can be analysed to determine the mean, μ, and standard deviation, σ of the process output. Various PCIs have been developed to quantify these statistical attributes in terms of the required target [8]. These indices compare the specification limits to the 6σ limits of the manufacturing process distribution, i.e. 99.73% of the predicted population, where a higher process index indicates a more accurate process.

$$C_p = \frac{USL - LSL}{6\sigma} \quad (1)$$

$$C_{pk} = \min\left\{\frac{USL-\mu}{3\sigma}, \frac{\mu - LSL}{3\sigma}\right\} = \min\{C_{pl}, C_{pu}\} \quad (2)$$

$$C_{pm} = \frac{USL - LSL}{6\sqrt{\sigma^2 + (\mu - T)^2}} \text{ with, } T = \frac{1}{2}(USL + LSL) \quad (3)$$

Where

C_p Compares process distribution with the specification limits assuming centred process mean.

C_{pk} Accommodates non-centred distributions, but does not provide explicit data on the location of the mean within the specification limits.

C_{pm} Measures the ability of a process to achieve any nominal value, T, and the specification limits (for example, Figure 1) [1].

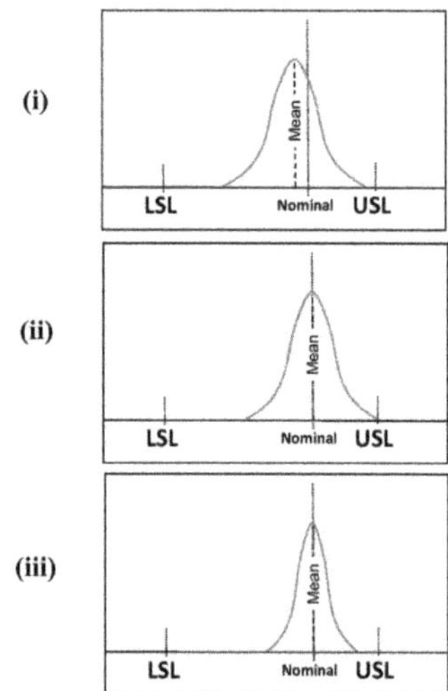

Fig.1. Example process output distributions.
C_{pm} increasing from upper to lower image:
(i) large standard deviation, mean not equal to nominal,
(ii) large standard deviation, mean equal to nominal,
(iii) small standard deviation, mean equal to nominal.

PCI data will be applied to quantify the expected tolerance distribution for each parameter.

2.5.2. *Uncertainty propagation*

To identify the effects of part parameter variation on the assembly in a virtual model environment, a method of propagating the expected part parameter distributions through the assembly response function is required. A variety of Uncertainty Propagation (UP) methods have been presented in the literature [7].

Monte Carlo simulation was applied in this work as it is well documented and readily implemented within the proposed platform. The Monte Carlo method consists of the generation of a feasible set of part parameter values from a random sample of their respective probability distributions, and the evaluation of the resulting assembly response. As the number of evaluations increases, the resulting sample distribution of assembly responses approaches that of the population. Studies have identified that approximately 1000 samples are required to provide good accuracy in an assembly tolerance analysis problem [4]. Monte Carlo simulations of 3000 evaluations were applied in this work to ensure reliable outcomes.

3. **Application**

The proposed platform was applied to a commercial assembly to aid in task of identifying unintended part interactions and associated assembly KPCs.

4.1. *Process data*

To estimate expected dimensional variations, the existing moulding process used for the manufacture of the test case assembly components was analysed. The manufacturer specifies assembly component dimensional tolerances on injection moulded components according to standard recommended specification limits [2].

The capability of the manufacturer's injection moulding process was determined by inspection of 1600 production part samples from a commercial component of comparable geometry and material to the product under analysis (Figure 2). The production parts are manufactured using multiple moulding cavities, each having its own variation characteristics. Process capability indices were determined within each cavity and across multiple cavities (Figure 3). The measured PCIs were used as input for the test case analysis.

Fig.2. Production mould cavity used to establish PCIs.

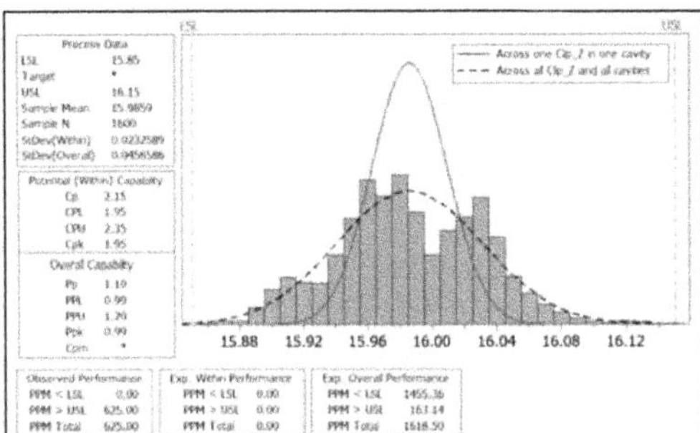

Fig.3. Mould cavity PCI data.

4.2. *PIDO integration*

The parametric CAD assembly model of a commercial actuator assembly developed by the designers (Figure 4) was interfaced with a PIDO workflow (Figure 5) in accordance with the proposed platform (Section 2.1). A feasible set of part parameter values was initialised from a random sample of their respective probability distributions and subjected to a Monte Carlo simulation (3000 samples were used) consisting of the stages detailed in Section 3.

4.3. *Results*

A number of undesirable part clash and contact instances were identified as part of the simulation. The interference data recorded for each identified scenario detailed the nature of part interactions (whether a contact or clash). An image was recorded for each interaction showing the relevant region of the assembly. The frequency of interference between part pairs was determined (Figure 6). A Student t-test was conducted to quantify the sensitivity of the number of interference scenarios occurring to part parameter variation (Figure 7).

The simulation results identified assembly regions prone to unwanted part interference (Figure 8). The results were used by the designers to specify assembly KPCs which avoid the identified interference scenarios. The specified KPCs were used in subsequent tolerance analysis of the assembly.

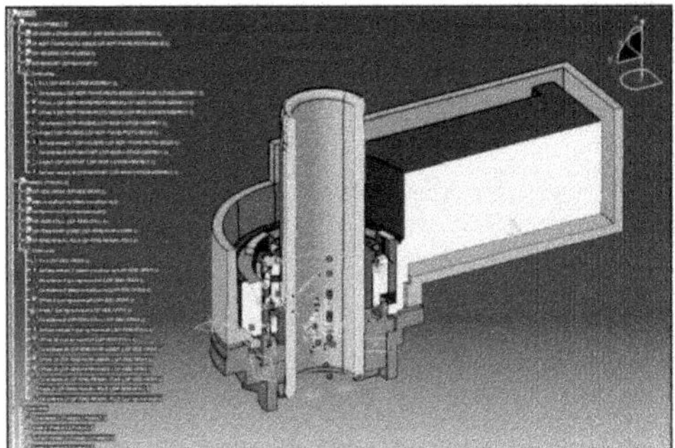

Fig.4. Parametric CAD assembly model.

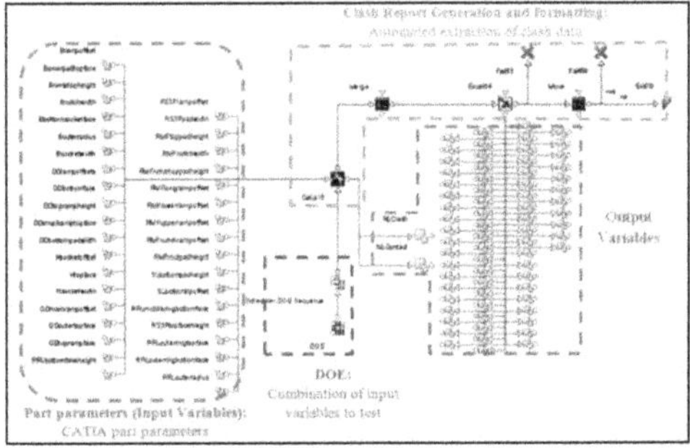

Fig.5. PIDO workflow model.

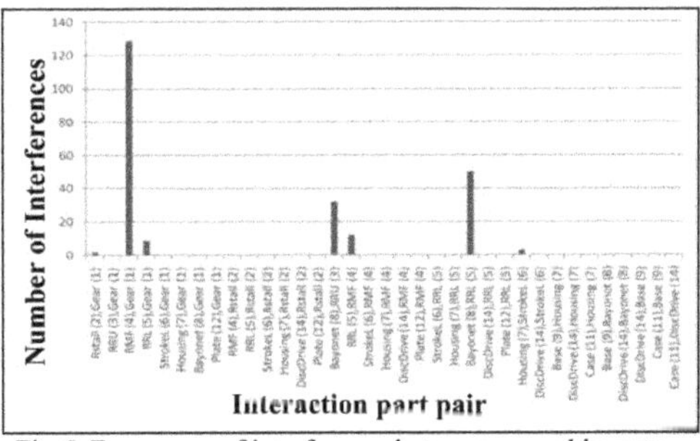

Fig.6. Frequency of interference between assembly parts.

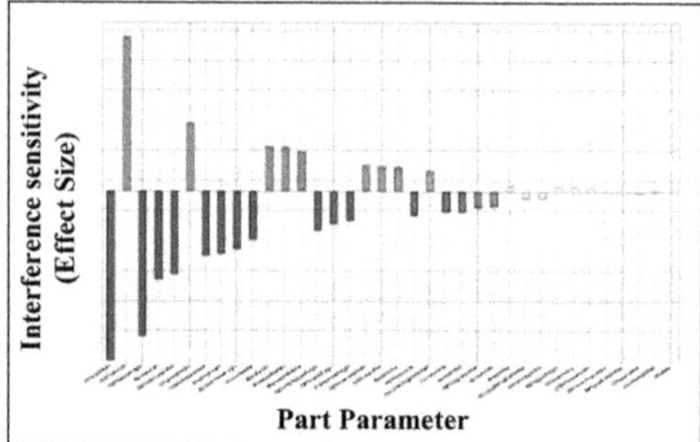

Fig.7. Interference sensitivity to part parameter variation. A positive effect size indicates that an increase in the parameter value results in interference, a negative effect size indicates a negative relationship. Sensitivity is indicated by magnitude.

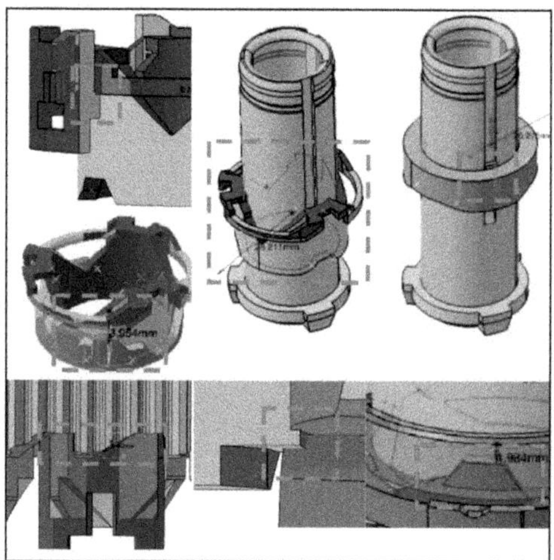
Fig.8. Assembly regions identified as prone to unintended part interference. KPCs were defined to avoid the identified interference scenarios.

5. Conclusions

Definition of the assembly KPCs required for tolerance analysis can be a challenging problem for the designer as assembly behaviour may be difficult to predict. The authors have developed an automated PIDO tool based platform to aid in the identification of KPCs. The platform allows for:

- Rapid identification of unforeseen part interaction due to stochastic variation.
- Identification of the sensitivity of the identified KPCs to assembly function.
- Direct integration with the native CAD models developed for concept design without the need for additional modelling or separate CAT software.
- Integration with all mainstream CAD software.

To demonstrate the capabilities of the platform, a test case involving a commercial product assembly was presented. The platform identified a number of undesirable part interactions that were otherwise overlooked and quantified their associated effect size. Furthermore, as the platform is compatible with the native CAD models used to develop the prototype assembly, the application is straightforward and no additional modelling was required.

Acknowledgments

This original research was proudly supported by SMR Automotive Australia Pty Limited, and the Commonwealth of Australia, through the Cooperative Research Centre for Advanced Automotive Technology.

References

[1] Chan LK, Cheng SW, Spiring FA. A New Measure of Process Capability: C_{pm}. Journal of Quality Technology 1988; 20(3):162-75
[2] Deutsches Institutfür Normung (DIN). DIN 16 901 - Plastics Mouldings - Tolerances and acceptance conditions for linear dimensions. Berlin: DIN; 1982.
[3] Flager F, Welle B, Bansal P, Soremekun, G, Haymaker J. Multidisciplinary process integration and design optimization of a classroom building. Electronic Journal of Information Technology in Construction 2009;14: 595-612
[4] Gao J, Chase KW, Magleby SP. Comparison of assembly tolerance analysis by direct linearization and modified Monte Carlo simulation methods. Proc. of ASME Design Eng. Tech.Conf1995:353–60.
[5] Hiriyannaiah S, Mocko G. Information Management Capabilities of MDO Frameworks. ASME Conf. Proc. 2008:635
[6] Makelainen E, Heilala J. Assembly Process Level Tolerance Analysis for Electromechanical Products, IEEE International Symposium on Assembly and Task Planning 2001: 405-410.
[7] Nigam SD, Turner JU. Review of statistical approaches to tolerance analysis. Computer Aided Design 1995; 27(1):6–15.
[8] Pearn WL, Kotz S. Encyclopedia And Handbook of Process Capability Indices. New Jersey: World Scientific Publishing: 2006.
[9] Shah JJ, Ameta G, Shen Z, Davidson JK. Navigating the Tolerance Analysis Maze. Computer Aided Design and Applications 2007;4(5):705-718.

Innovative Production Machines and Systems
D.T. Pham, E.E. Eldukhri and A.J. Soroka (eds)

Intuitive Interaction with Multifunctional Smartphone Interfaces

A. Britton, R. Setchi

School of Engineering, Cardiff University, Cardiff, CF24 3AA, UK

Abstract

This paper investigates intuitive interaction in the context of touchscreen smartphones. The experimental studies conducted confirm that intuitive interaction requires some unconscious use of existing experiential knowledge; it only occurs between a first time user and a complex artefact if proficiency in its key functions is gained with minimal cognitive processing power. Two experimental studies were conducted. The first one compared participant interactions between touchscreen and button phones to evaluate the significance of the touchscreen. It was found that participants who had never owned a touchscreen handset were still inclined to utilise experiential knowledge of existing handsets for all of the tasks performed. The second study used the Immersion® CyberGlove® to record participants' interactions with a prop in place of a touchscreen phone when carrying out various tasks. The aim was to determine which image schemata were prevalent in the use of a touchscreen handset. Six image schemata were identified, and it was concluded that these image schemata could improve the efficiency of inclusive design if taken into account. This research suggests that individual image schemata can be related to the continuum of knowledge sources in order to create design solutions that take advantage of the user's basic sensorimotor experience.

Keywords: Design for intuitive use, intuitive interaction, mobile phone, touchscreen interaction, image schemata, intuitive interface

1. Introduction

Recently, mobile phones have evolved into much more than communication devices, with touchscreen smart phones becoming more affordable and more powerful. People are using them to perform tasks, which previously would have needed laptop computers. With the ever-growing number of features incorporated into these devices, it is becoming increasingly necessary for designers to consider intuitive interaction when designing their handsets.

'Intuition' is defined as "the power of knowing or understanding something immediately without reasoning or being taught" [1]. In the context of mobile phones this definition can therefore be interpreted as the ability to attain knowledge about using an interface without intervention of the use of reason. An intuitive decision would therefore consist of the identification of various patterns relating to the dynamics of the present situation. These unconscious decisions could be affected by a person's age, previous technological experiences, and cultural differences. The fact that they could vary from person to person so greatly is certain to be a complication involved in intuitive design of products. With this in mind mobile phone designers must engineer every aspect of their products from the size and shape of a device to the positioning of items on the menu to meet the criteria necessary to make them intuitive.

Many of the modern phones currently on the market have touchscreen interfaces, all of which having very similar designs. In 2008 23% of all mobile phone sales were touchscreen smart phones, compared to only 13% in 2007. The figure rose above 45% in the final quarter of 2009 [2]. There could be any number of reasons for the growing popularity of these devices, including fashion trends, marketing techniques or recent lower costs; however, it is likely that the touchscreen allows for new ways in which users can interact with handsets intuitively. The increasing market share that touchscreen smart phones are experiencing makes them an area of interest with regard to intuitive design.

As intuition is not a physical or tangible variable, experimental procedures and technologies from many fields, such as psychology, science and ergonomics may have to be combined to implement a suitable study. Such a study would be continuous as each individual would have had different experiences, and hence a different level of intuition. This is not to suggest that it is impossible to define a set of parameters which would be intuitive to all, but perhaps that this level of intuition would be too basic to be implemented solely into an interface without some level of learning required for a completely effortless interaction. Patterns of intuition may be exposed between generations and cultures from which the participants came, possibly resulting with a number of salient directions of intuition.

This paper continues previous research on investigating the impact intuition, as a psychological phenomenon, has on the way consumers interact with mobile phone interfaces [3]. The aim is to gain more understanding of what aspects of these devices are intuitively designed and what influences people's behaviour whilst interacting with them. Comparisons will be made between touchscreen interfaces and more conventional button layouts.

The paper is organised as follows. Section 2 outlines current research in design for intuitive use. Section 3 introduces the experimental procedures employed and the main hypotheses tested. Section 4 contains results and a discussion of the findings. Finally, Section 5 concludes the paper.

2. Literature review

2.1. Intuitive interaction

In a recent comprehensive review of research investigating intuitive interaction, Blackler et al. [4] define intuition as a type of cognitive processing that utilises knowledge gained through prior experience. They describe it as a process that is often fast and non-conscious, or at least not recallable or verbalisable. An application is considered intuitively useable if the users' unconscious application of prior knowledge leads to effective interaction [5]. Products that people use intuitively are those with features they have encountered before [4].

The methodological tools developed in the area include the continuum of prior knowledge [6] and the continuum of intuitive interaction [7].

The continuum of prior knowledge [6] shows that prior knowledge acquired before interaction with the new product comes from a variety of sources: knowledge, embodied interaction, culture, expertise, and use of tools. The higher up the continuum, the smaller the potential number of users possessing this knowledge; lower level knowledge is used more frequently. These are the levels that are more likely to be applied unconsciously and therefore intuitively.

The continuum of intuitive interaction [6,7] outlines three principles of creating an intuitive interface:

1. Use familiar features from the same domain,
2. Transfer familiar things from other domains,
3. Increase redundancy and internal consistency of function, appearance and location within the interface.

Similarly, Hurtienne [5] devises a number of principles for designing intuitive interfaces. These include: suitability for the task, compatibility, consistency, gestalt laws (how the mind groups similar elements to an entity, based upon shape, colour, size and brightness of the elements), feedback, self descriptiveness, and affordances. Some of these principles are related to the use of image schemata.

2.2. Cognitive ergonomic: image schemata

Johnson defines an image schema as a mental pattern that recurrently provides a structured understanding of various experiences [8]. He describes it as "a dynamic pattern which functions somewhat like the abstract structure of an image and thereby connects up a vast range of different experiences that manifest this same recurring structure." His list of schemata consists of twenty seven individual schemata, including centre-periphery, containment, cycle, path, force, link, part-whole, scale, and verticality schemata.

Recently, Hurtienne and Israel [9] suggested the application of image schemata and their metaphorical extensions in the design of intuitively usable interfaces. On creating an interface, its perceived structure should try to fit in as closely as possible to the mind's image schemata. It would appear that the most relevant schemata for an interface are the cycle schema, path and end of path schema and verticality schema, whilst the ergonomics of the interface (e.g. buttons, etc.) will apply to link schema, force schema and scale schema (especially with touch sensitive control).

Further studies indicate that configurations and

patterns can be applied to image schemata [10] thus benefiting from the ability of the mind to group elements when perceiving objects. Five principles often applied in product design are: proximity, similarity, closure, symmetry and common fate. Buttons, for example, grouped together are perceived as having similar functions.

These theories indicate that to teach a mind how to use something, it is the information presented to it that will determine its success. As discussed in [2], the success of this information resulting in the 'correct' cognitive response lies in the similarity of the image schema formed whilst receiving information and the image schema used to implement interaction. Image schemata are flexible structures for the mental organisation of experiences and comprehension.

3. Experimental procedures and hypotheses

Two tests are described in this section: a smartphone comparison test and an image schemata test.

3.1. Smartphone comparison test

3.1.1. Experimental procedure

The aim of this experiment was to gain understanding of what aspects of touchscreen smartphones make them more intuitive or more difficult to use. Common tasks were purposefully chosen in order to gain more consistent results and draw reliable conclusions.

Thirty participants were timed completing various tasks using both a touchscreen smartphone and a conventional button layout smartphone. The phones used were the Apple *iPhone* (touchscreen) and the *Nokia N95* (button). The participants were asked to complete the following tasks:

T11. Contacts task: from the main menu, enter the contacts menu and phone a predetermined contact.

T12. Text message task: from the main menu, enter the messages menu, create a new text message addressed to the predetermined contact and write the message 'hello, how are you?'

T13. Calendar task: starting at the main menu screen, find out what is happening on a specific date via the calendar application.

The phones were unlocked before the tasks took place. The predetermined contact and the dates to look for in the calendar remained constant for all tasks on both phones. The participants were timed completing each task and also observed in order to note any of the key problems they faced. None of the participants observed each other partaking in the experiment so that first time users of the phones remained unaware of their functions.

3.1.2. Hypotheses

The following hypotheses were tested in this experiment:

H11. Participants will enter text messages more quickly using the *iPhone* than a button phone due to the presence of a full QWERTY keypad.

H12. Participants who own or have owned the *iPhone* will perform the tasks more quickly than participants who own or have owned a *Nokia* due to the assumption that the *iPhone*'s touchscreen allows for a more intuitive interface.

H13. It is expected to be a significant gap between experienced users of the *iPhone* and first time users.

The *iPhone* operating system remains a relatively new interface. If this hypothesis is proven to be accurate it will verify previous findings relating experience to intuitive use. This should also mean the longest time recorded for each task is likely to be for the *iPhone*. It is likely most of the users will have had much more relevant experience with phones with interfaces similar to the relatively conventional button layout of *Nokia*.

3.2 Image Schemata Test

3.2.1 Experimental Procedure

The experiment was designed to analyse which of the image schemata shown in Table 1, are utilised in the cognitive process of performing certain tasks on a touchscreen smartphone. A prop of similar weight and size to a touchscreen smartphone was given to participants, and they were asked to perform three tasks whilst wearing the Immersion® CyberGlove®, which recorded the vector movements of their thumbs whilst performing the tasks. The glove was connected to the computer and powered via a transformation box using a 25 pin parallel port. The data collected from the glove were recorded using Autodesk® MotionBuilder®. An algorithm was written using Matlab™ to provide 3D vector graphs for each recording.

The test was carried out using the right hand glove as it is assumed the majority of participants would use their right hand when operating mobile phones and devices. (Only right-handed participants were eligible for the study.) This should increase the likelihood of the unconscious use of image schemata. The subjects

Table 1. Image schemata (as categorised in [11])

Group	Image schemas
Basic	object, substance
Space	center-periphery, contact, front-back, left-right, near-far, path, rotation, scale, up-down
Containment	container, content, full-empty, in-out, surface
Multiplicity	collection, count-mass, linkage, matching, merging, part-whole, splitting
Process	cycle, iteration
Force	attraction, balance, blockage, compulsion, counterforce, diversion, enablement, momentum, resistance, restraint removal
Attribute	big-small, dark-bright, heavy-light, straight, strong-weak, warm-cold

were instructed to operate the device solely using their thumbs whilst the rest of the hand supported the prop. Ten participants chosen in the age range of 18-25 performed the following tasks:

T21. Scroll through the contacts menu

T22. Unlock the mobile phone

T23. Enter the main menu

For each task, subjects were asked to position their thumbs at the bottom centre of the device before the recording began. The participants were told to assume the phone was already unlocked for tasks 1 and 3.

Three of the participants currently owned a touchscreen phone. None of the remaining participants have owned one in the past.

3.2.2 Hypothesis

The aim was to explore if image schemata are formed when performing certain tasks on touchscreen devices. Even in the absence of any buttons or features, people will automatically move their thumbs according to how they would expect to operate existing devices. Participants who own or have owned touchscreen mobile phones are expected to have moulded image schemata in their minds, which predetermine how they react when given a task to perform. The following hypotheses were tested:

H21: The most common movements performed will emulate the features of the smartphones with highest current market share.

H22: In attempting to access the main menu of a phone there will be a clear movement to the bottom centre of the mobile phone where the menu or home button occurs in the majority of existing touchscreen mobile phones. An image schema identifiable with this would be the near-far schema.

H23: In attempting to unlock the mobile phone, a large proportion of participants will perform a swiping action across the screen. The left-right or up-down schema will be easily identifiable if this is the case.

H24: When scrolling through a list, it is expected that participants will swipe up the screen in the positive Y direction. The cognitive processes used are expected to involve the momentum schema when swiping in order to produce a motion and the up-down schema identifiable with lists.

4. Results and discussion

4.1. Smartphone comparison test

Table 2 shows the statistics recorded for each task performed on each phone. Table 3 shows the average time taken to complete each task by experienced users who own or have previously owned either phone.

The most noticeable observation is the difference in the ranges for each task for the *iPhone* and the *Nokia*. For the contacts test, the range was 63 seconds larger for the *iPhone* than the *Nokia* and for the messages test the range was 205 seconds larger for the *iPhone* than the *Nokia*. The range on the calendar test was the same for both phones.

Table 2. Experimental data from smartphone comparison experiment

	iPhone			*N95*		
	T11	T12	T13	T11	T12	T13
Average time taken (s)	27.0	70.2	9.8	10.4	41.5	21.9
Minimum time taken (s)	5	13	4	6	7	7
Maximum time taken (s)	84	284	33	21	54	36
Range (s)	78	252	29	15	47	29

Table 3. Experimental data for experienced users taken part in the smartphone comparison experiment

	iPhone			*N95*		
	T11	T12	T13	T11	T12	T13
Average time taken (s)	12	26	7	7	36	19

Fig. 1 Text message entry

The exceptionally large range for the messages test on the *iPhone* is down to one individual, as the second longest time taken to perform this test was 122 sec quicker. This result amplified a problem which many of the participants had when performing this task. Figure 1 shows the text message screen of the *iPhone*. The majority entered the contact in the 'To:' bar at the top of the screen. However, many of the participants tried to enter the message in the large space highlighted in the figure. Yet it is the bar below this space where the text should be entered. It is conceivable that participants have associated the full-empty image schema with spaces like this when utilising other computer interfaces. Furthermore, a circle in Figure 1 highlights the picture of a camera, which is pressed in order to attach a picture to the message. The placement of this camera image further deterred people from trying to enter text in the bar provided, as many of the participants assumed the text input bar was a feature associated with the camera. It can be concluded that the messaging interface on the *iPhone* is counter-intuitive to new users. It was hypothesized that participants would enter text messages more quickly on the *iPhone*. This hypothesis was made under the assumption that a QWERTY keypad would allow for more efficient typing than a phone keypad. The difference in interfaces was not expected to impact on the results so acutely.

Participants who have owned an *iPhone* were expected to complete the tasks more quickly than those who have owned a *Nokia* due to its touchscreen interface. This was the case for the calendar and messages tests but not the contacts test. This is believed to be down to the ability to skip straight through the contacts list to the alphabetical letter with which the name begins by tapping that letter in on the keypad when searching for a contact. The *iPhone*'s lack of keypad slows people down in this respect.

The hypothesis also stated that it is likely there will be a large gap in times between experienced *iPhone* users and inexperienced *iPhone* users, and the longest times recorded for each task would have been for the *iPhone*. For the first two tests the hypotheses were true for both the ranges and the longest recorded times. The calendar test however showed very similar average time recordings for both phones.

The calendar function on the *Nokia* required the users to enter an applications menu first, whilst the *iPhone* presents the calendar at the top of its primary menu screen. The quickest *iPhone* result was therefore much quicker than the fastest *Nokia* recording for this test, which can be accredited to experienced *iPhone* users. It could be concluded however that a task involving more steps on the *Nokia*, yet still yielding similar overall results to the *iPhone* demonstrates the relationship between experience and intuitive use.

This test showed that while the *iPhone* had a very efficient interface for users with experience, some of the features proved to be counter-intuitive initially, whereas the *Nokia* in general was interacted with very intuitively and quickly. It was perhaps not the aim of Apple to create an interface which participants learned quickly and intuitively, but rather to reinvent the way people interacted with these devices. It is designs like this that may in turn leave people associating entirely new image schemata with various tasks in the same way companies like *Nokia* did at the turn of the century. However, the creation of interfaces with this intention is alienating the older user. It is possibly the opportunity of suffering competitors in the marketplace to increase their competitiveness again through inclusive design to regain market share, rather than remaining focused on younger generations and competing in this already saturated market.

4.2 Image Schemata Test

The vector graphs for thumb movement were produced in Matlab™ for each subject and each task performed (see Fig. 2). The x, y and z axes represent direction of thumb movement, with the x axis being the horizontal axis of the screen, the y axis the vertical, and

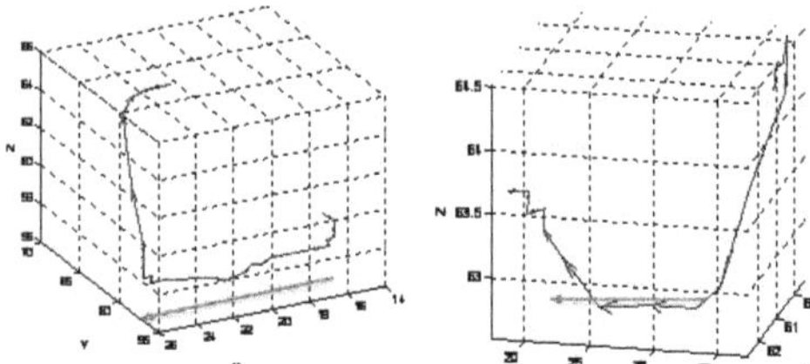

Fig. 2. Examples of vector graphs produced.

the z axis showing thumb movement towards and away from the screen. The units for each graph are in millimetres.

Figure 2 shows two of the vector graphs produced; the left showing a participant trying to unlock the phone by sliding their thumb across the horizontal axis of the screen and the right shows a participant trying to scroll through the contacts menu in the horizontal axis of the screen.

It was hypothesised that participants conducting the first task, scrolling through the contacts list, would perform an upwards swiping motion on the screen. Six of the ten participants performed a swiping motion, four of which were in the positive y direction as hypothesised. The remaining four participants pressed down in one place rather than swiping it. One participant showed a very steep approach angle in the negative y direction possibly implicating a 'striking' action in order to create a downwards momentum. The hypothesis was not fully supported although all of the individuals who own a touchscreen phone performed the expected positive y axis motion. It was shown that it is common for individuals who do not own a touchscreen mobile to still utilise schemata associated with their button phones despite the knowledge that the device they are interacting with has a touchscreen.

In the second task, unlocking the mobile phone, six of the ten participants performed the swiping action as hypothesised, four in the x axis and two in the y axis. All of the participants who owned a touchscreen phone performed a swiping action, one in the vertical y axis and two in the horizontal x axis. One participant clearly demonstrated an expectation of unlocking a touchscreen phone by pressing two on-screen buttons consecutively, imitating a method used with many existing button phones. This was linked to the path schema. This particular participant has indicated the application of existing experience in interaction very clearly.

The six participants who performed a swiping action are expected to have created image schemata through experiential use of existing touchscreen phones. The 4 participants who performed a horizontal swipe across the screen may have used image schemata associated with the Apple *iPhone*. This would coincide with the hypothesis that participants will emulate actions required to unlock the market leading phones as Apple currently holds the leading market share for touchscreen handsets.

The Google Android operating system currently holds the fifth largest share of the smartphone operating system market. The technique to unlock it is to slide a central bar on the screen upward. The two individuals who slid their thumbs across the screen in the y direction may have had some experience with this operating system or similar.

The three participants who pressed down in a single place on the screen may or may not have drawn upon previous experience when performing the test. However, the Samsung s8000 is an example of an interface which requires this motion. The Samsung interface does not currently hold a large share of the worldwide touchscreen operating system market, making it unlikely all three participants have had experience with this particular phone. It may be that one or more of these three participants simply pressed down in one place as a conscious guess, implying the touchscreen was too unfamiliar to draw upon any existing unconscious cognitive thought processes.

Having analysed the results for the third task, entering the main menu, it is apparent that two participants appeared to show thumb movement away from the screen rather than towards it, with no negative z direction movement. The main assumed reason for this was that the participants had began the movement before the recording had officially taken place, meaning relevant frames needed for analysis were missing.

Out of the eight participants with coherent results, five pressed directly down on the screen from the primary position and three pressed down after movement to the right. None of the participants showed any notable movement in the y direction meaning all of them expected the menu to be at the bottom of the screen. This corresponds with the hypothesis that the near-far schema will be associated with the menu button in most user's cognitive thought processes due to its importance. The button is never assumed to be far away from the natural thumb position. Three of the participants moved to the right before pressing down.

The three dominant smartphone manufacturers of 2009 were *Nokia*, RIM Blackberry, and Apple, Mfyall (2009). All of the touchscreen handsets manufactured by these companies at the time positioned their main menu buttons at the bottom centre of the screen. The evidence that participants have emulated this in the experiment further implicates the use of existing experience in intuitive interaction. These findings also correspond to the hypothesis that participants' actions would imitate the existing interfaces currently holding highest market share. A popular main menu button position on conventional button mobile phones was under the bottom right of the screen. This is the positioning of the main menu button on the *Nokia N95*.

Table 4 shows the total amount of uses of each image schema clearly identified in the test. The assumed uses of the up-down schema in the contacts task for people who indicated to simply press down in one place were not included. The most frequent were the near-far and up-down schema, followed closely by the left-right and momentum schema. The recognition of individual image schemata relating to specific tasks may in this context agree with claims that intuitive decision making is a heuristic process which utilises clear identifiable 'rule of thumb' concepts gained through trial and error.

Table 4.
Total identified uses of image schema

Tests	Left-right	Up-down	Momentum	Counter-force	Near-far	Path
Scrolling	-	6	6	1	-	-
Unlock	4	2	-	-	-	1
Main menu	3	-	-	-	8	-
Total	7	8	6	1	8	1

Listed below are four schemata identified as specifically relating to three particular touchscreen phone functions:

Scrolling – up-down, momentum
Unlocking – left-right
Main menu – near-far

It is suggested that designers can produce studies like the CyberGlove® experiment in order to identify schemata, in relation to each function performed by individuals using a touchscreen device. When implementing a function into the interface, designers could use these studies to guarantee that they implement a method which utilises the schemata identified in the study. For example, when adding a scroll function to the interface, the use of the up-down and momentum schemata could be included as a minimum requirement. Identification and implementation of schemata in the design process provides a clear cut quantifiable way of improving intuitiveness of individual functions. Inclusive design is one of the main areas of design where a process like this would be applicable. The use of the CyberGlove® to record these motions rather than simply taking a video recording for observation demonstrates a method of graphing the results and analysing them in more detail, making the observations less ambiguous.

Image schemata can be used in design solutions to help identify each cognitive process drawn upon by the user in completing a task. An interface can then be designed which provides a solution geared to trigger individual image schemata and tap into the user's most sensorimotor unconscious knowledge.

It would conceivably be beneficial to incorporate the analysis of image schemata into inclusive design of mobile phones. If each individual image schema involved in performing each function on an interface is identified, the opportunity to maximise the number of processes that only draw upon sensorimotor sources of knowledge arises. The use of these techniques in engineering design could potentially widen the target customer base of multifunctional smartphones, directing more attention to less technologically experienced users who would previously have been alienated by them.

Although it is possible to maximise the use of sensorimotor knowledge this way, many of the complex technological functions featured in modern day mobile phones will certainly require some use of image schemata only obtained from prior experience to be used intuitively. It is evident that many mobile phone interfaces have attempted to visually represent their functions in a way that could make them obvious to users such as a picture of an address book representing the contacts list or an envelope representing the text messages menu. It is highly improbable however that a user with no experience with mobile phones would draw upon existing image schema and intuitively know what the functions were.

It can be concluded through literature review that the holistic definitions of intuition are not directly relevant to intuitive interaction with technological products. Age and cognitive ability have a large influence on intuitive use, leading to the development of inclusive design. It is evident that experience must be taken into account when defining intuitive use. The theory regarding the continuum of knowledge sources demonstrates that user interfaces, which tap into sensorimotor knowledge, are more intuitive to use. The use of image schemata in the design process can aid designers in tapping into the user's sensorimotor knowledge, and therefore could prove very helpful in the field of inclusive design.

5. Conclusions

The research reported in this paper confirmed that intuitive interaction only occurs between a first time user and a complex artifact if proficiency in its key functions is gained with minimal cognitive processing power. An intuitive interaction requires some unconscious use of existing experiential knowledge.

The experiments demonstrated that users 'mould' image schemata in their cognitive thought processes when interacting with new devices in order to better understand them next time around. The study identified new image schemata associated with the use of a touchscreen device.

The experiments confirmed that experience was a key factor in intuitive interaction; the most intuitive uses in the studies performed were those of individuals with experience. It was shown that a touchscreen interface is more difficult to grasp among new users than a conventional keypad interface, although for experienced users the touchscreen was more efficient.

The results strongly suggest that it is experience with market leading mobile phones that influence people's interactions with new interfaces.

Identifying image schemata which draw from sensorimotor knowledge on the continuum of knowledge sources is a quantifiable way of designing for intuitive use. Identifying image schemata used to interact with previous successful handsets can also greatly improve the intuitiveness of a design. These methods would surely aid engineers in the process of inclusive design.

The use of image schemata in the design process is not a widely covered topic of study. However, the possibility that this integration of cognitive theory and engineering design could further progress the concept of inclusive design should be assessed in full if new ideas are needed to improve this field of research.

Further work to be carried out may involve the identification of image schemata for a wider variety of tasks performed on touchscreen phones.

If a particular available handset was identified to embody a large proportion of these image schemata into its interface, it could be possible to compare its level of intuitive interaction with other existing handsets which do not embody as many image schemata into their functions. A comparison like this could validate the theory that image schemata can indeed be used in the design process in terms of inclusive design.

The use of the Immersion® CyberGlove® and a prop in place of a touchscreen handset was an effective way of graphing interactions so that they could be analysed and displayed in the form of a technical report. Further work could include developing techniques to identify image schemata in a measurable way in order to efficiently improve the design process and further understand the relevance of cognitive psychology in the field of design for intuitive use.

References

[1] Oxford English Dictionary, Oxford University Press, 2005.

[2] Mfyall, M. Smartphone sales up 27%. Available at: http://metrics.admob.com/page/3/ [Accessed 14th Sept 2010]

[3] Marsh A. and Setchi R. Design for intuitive use: a study of mobile phones. 4th Virtual I*PROMS Conference, I*PROMS' 08, 1-14 July 2008, Cardiff, UK.

[4] Blackler A. L., Popovic V. and Mahar, D. P. Investigating users' intuitive interaction with complex artifacts, Applied Ergonomics, vol. 41, pp. 72-92, 2010.

[5] Mohs C., Hurtienne J., Israel J. H., Naumann A., Kindsmüller M. C., Meyer H. A. and Pohlmeyer A. IUUI - Intuitive Use of User Interfaces. In: T. Bosenick M. Hassenzahl M. Müller-Prove and M. Peissner (Eds.), Usability Professionals 2006, Stuttgart, German Chapter of Usability Professionals' Association, 2006, pp. 130-133.

[6] Blackler A. L. and Hurtienne J. Towards a unified view of intuitive interaction: definitions, models and tools across the world, MMI-Interaktiv, 2007, vol. 13, pp. 37-55.

[7] Blackler A. L., Popovic V. and Mahar, D. P. Towards a design methodology for applying intuitive interaction, In: Friedman, K., Love T., Côrte-Real E. and Rust C. (Eds.) Proceedings WonderGround: 2006 Design Research Society Int. Conference, Lisbon.

[8] Johnson M. The body in the mind: The bodily basis of meaning, imagination, and reason. Chicago: University of Chicago Press, 1987.

[9] Hurtienne J. and Israel J. H. Image schemas and their metaphorical extensions: intuitive patterns for tangible interaction. In: 1st International Conference on Tangible and Embedded Interaction (TEI'07), Baton Rouge, ACM Press, 2007, pp. 127-134.

[10] Hurtienne J., Weber K. and Blessing L. Prior experience and intuitive use: image schemas in user centred design, In: Designing Inclusive Futures, Springer, London, 2008, pp. 107-116.

[11] Langdon P. and Clarkson J. The effects of prior experience on the use of consumer products. Universal Access in the Information Society, 2007, vol. 6(2), pp. 179-191.

Innovative Production Machines and Systems
D.T. Pham, E.E. Eldukhri and A.J. Soroka (eds)

Integrated process-machine approach for designing productive and eco-efficient machine tools

J. Zulaika[a], F.J. Campa[b], L.N. Lopez de Lacalle[b]

[a] *Fatronik-Tecnalia, Paseo Mikeletegi, 7 - Parque Tecnologico E- 20009 San Sebastian*
[b] *University of the Basque Country, Department of Mechanical Engineering, Alameda Urquijo s/n E-48013 Bilbao*

Abstract

This paper presents an integrated process-machine approach based on stability lobe diagrams for designing large-volume milling machines bearing in mind their productivity, reliability and accuracy as well as their eco-efficiency. In fact, eco-efficiency of machining processes is an issue of increasing concern among both machine tool builders and manufacturers. Thus, this work introduces a global modelling of the dynamics of milling machines and of milling processes with a goal of supporting engineers in the design of machine tools that result in optimal machining productivity at minimised environmental impacts and costs. This is a result of reducing the machines' material content, and consequently resulting in lowered energy consumption. This approach has been implemented in the design of an actual milling machine, upon which machining tests have been conducted that have shown an increase of 100% in productivity while consuming 15% less energy due to a weight reduction of over 20% in structural components. Thus highly productive machining processes and eco-efficient milling machines have been integrated in one unique system.

Keywords: Machine Tools, Process-Machine interaction, Eco-Efficiency.

1. Introduction

The *eco-efficiency* of a machine tool can be defined as the ratio between the extent to which its functional objectives have been met combined with the environmental impact associated with its total lifecycle. In quantitative terms, machine eco-efficiency can be expressed as the ratio between the *productivity* of the machine and the *environmental impact* associated with its lifecycle.

With the aim of measuring the *environmental impact* associated with the lifecycle of a machine tool, the authors carried out a *Life Cycle Assessment* (LCA) on a milling machine made by "Nicolas-Correa" Company, considering a life span of 10 years for the machine. When conducting the LCA, the authors followed ISO 14040 rules and selected the "Climate change" impact category within the Ecoindicator 99 methodology [1]. The main conclusion reached from the LCA was that much of the environmental impact of a machine tool – around 95% - was associated with its use phase. Furthermore, within the use phase, 95% of its impact was produced by the energy that the machine consumed. Fig. 1 shows these data numerically.

On the other hand, the *productivity* associated with a machine tool or with a machining process is usually measured in terms of *Material Removal Rate* (MRR), which is defined as the volume of material removed per unit of time. The MRR is calculated as the product between the cutting area and the feed velocity of the tool. In the great majority of cases, this productivity is limited by the appearance of self-induced vibrations, commonly referred to as *chatter*. Among the types of self-excited vibrations, attention should be paid to *regenerative chatter* [2]. This self-excited vibration is caused by the regeneration of the metal thickness in systems in which the blade cuts a previously machined surface either totally or partially [3].

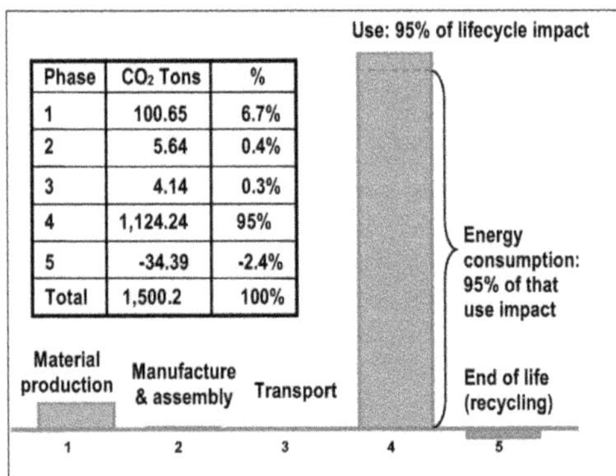

Fig. 1. LCA Study on a milling machine [1].

These two concepts of *eco-efficiency* and *productivity* have been traditionally viewed as being mutually exclusive goals. In fact, current standard practice for machine tool builders is to produce high-performance machines by means of high dynamic feed-drives [4], which forces design engineers to utilize heavy and stiff mechanical structures capable of bearing the higher inertial forces. The increased weight of structural elements in turn creates a requirement for more powerful actuators, leading to higher inertial forces. This is a vicious circle in which the increase of productivity in machining processes has been achieved at the expense of high material and energy consuming machines, increasing both the environmental impacts and costs associated with machining processes.

With the aim of facing this non-sustainable situation of the manufacturing sector, this work presents an integrated 'machine and process' approach for designing machine tools that are both highly productive and eco-efficient by separating the productivity of machining processes from the material intensity of machine tools.

2. Theoretical background for the design approach

The conceptual cornerstone for reducing the material intensity of machine tools lies in defining the exact dynamic and mechanical performances that are required for the machine to assure specific productivity goals, and then in achieving them with the least possible weight of mobile structural elements.

At this point, it is worthwhile to reflect on the functionality of the structural elements of a machine tool [5]:

1. To hold the components and peripherals involved in the machine and to assure their kinematic requirements during machine motions and processes.
2. To withstand the forces produced by the machining process and by the machine motions in a robust manner for assuring the required motion accuracy.

Concerning these two functions, it is feasible to fulfil the former requirement with structural components of much lower weight. An example of this is the case of industrial robots, which are much lighter than machine tools. In quantitative terms, according to a study of the authors, it is possible to assure the kinematic and dynamic performances of machine components with approximately 25% of the weight of the mechanical structure of an average milling tool [6]. This means that the functionality of the remaining 75% of the mechanical structure is only to assure robust performance of the machine during machine positioning and production processes. Therefore, for a specific machining process, if the specifics of the minimum requirements for stiffness and robustness are identified and quantified, and they are achieved using a minimum of structural weight, the result is a wide range of possibilities for reducing the weight of a machine in a significant manner without affecting its productivity.

Following this conceptual approach, the next subchapters will analyse on the one hand the mechanical design requirements associated with specific values of process productivity and accuracy, and on the other hand, will analyse synergetic and complementary approaches for reaching those requirements while using a minimum of material in the machine structure.

3. Design requirements for achieving specific productivity levels

3.1. Requirements of Static Stiffness for Accurate Machine Tools

Table 1 shows the average process forces associated with milling operations when cutting AISI 1045 Steel [7]. These data have been measured experimentally by the authors on various machines in different workshops, and can be used as basis for defining threshold values for static stiffness of a machine tool at the *Tool Centre Point* (TCP) that will assure an accurate and reliable performance of the machine. These results have been combined with accuracy requirements that machine end-users have provided for the different milling operations, which has allowed to develop operational specifications associated with different machining operations.

Table 1
Average forces and acceptable deformations for different milling operations.

Process	Average max. force on TCP	Acceptable deformation
Roughing	Conventional tools: 1.500 N 100-125 mm. diam. tools: 3.000 N	Average: 100 µm
Semi-finishing	Conventional tools: 1.000 N	Average: 50 µm
Finishing	Conventional tools: 200 N	Average: 10 µm

Combining the actual forces with the precision requirements, and taking into account that the main component of machining forces is in the tangential direction of the tool, a static stiffness of 1500N/100µm (i.e. 20 N/µm) at the TCP in the tangential direction can be considered as a reference threshold value for general-purpose milling machines. This means that if a machine has a higher static stiffness, from the static point of view there is margin for reducing the weight of the machine without affecting the accuracy and reliability of machining processes.

3.2 Requirements of Dynamic Stiffness for stable machining processes

In addition to the requirements associated with the static stiffness of a machine tool, it is worthwhile to highlight that, as mentioned above, in the great majority of machining processes the productivity of machining processes is not limited by static deformations but by the presence of self-induced vibrations referred to as *chatter*. Among these self-induced vibrations, special attention should be paid to *regenerative chatter*. The appearance of chatter is linked to the following parameters [8]:
i) The dynamic stiffness of the machine-process system
ii) The geometry of the machining tool
iii) The radial immersion of the tool into the material to be removed
iv) The material to be removed.

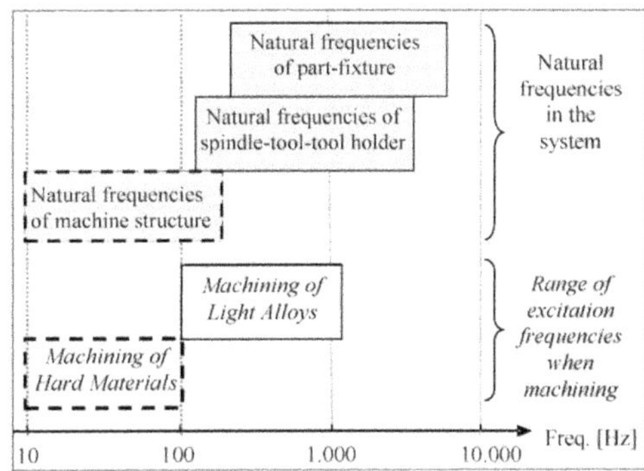

Fig. 2. Ranges for natural frequencies of the system and for excitation frequencies [9].

From this perspective, if the part to be machined is of a hard material such as steel, and its dimensions are below the dimensions of the machine, the dynamic performance of the system will be dominated by the natural frequencies of the machine structure, as shown in Fig. 2 [9].

As can be seen in Fig. 2, the natural frequencies of the machine-process system at low frequencies (below 200 Hz.) are related to structural natural frequencies of the machines, whereas the natural frequencies above 200 Hz. are related to natural frequencies of machine components such as spindle, tool, tool-holder etc. In parallel, the excitation frequencies when machining hard materials are in a range below 100 Hz. Therefore, the stability of machining processes for hard materials will be affected mainly by the modal parameters of the *machine structure*.

Within this view, in the next subchapter a stability model adapted to the machining of parts of hard materials that are smaller than the machine dimensions will be developed, which will serve as basis for translating process stability requirements into design requirements in terms of dynamic stiffness at the TCP.

3.2.1 Stability Model for the Machining Process

As starting point, a stability model has been developed for milling processes that is based on a mechanistic modelling of the milling forces [8], a mono-frequency approach of the stability solution [10], the introduction of the dynamics of the machine tool in modal coordinates as in [11], and the influence of the feed direction as in [12].

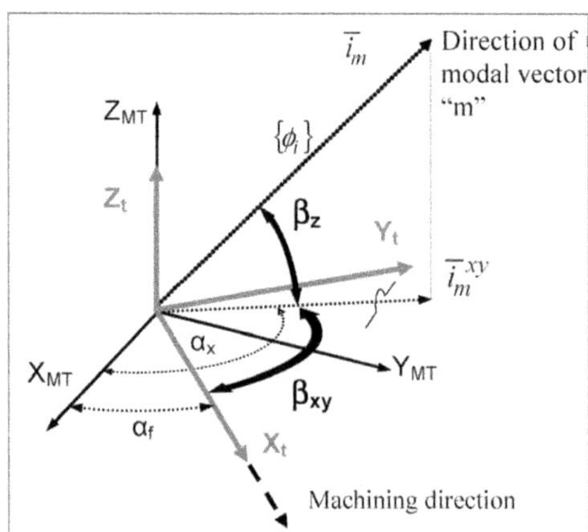

Fig. 3. Orientation of a machine modal vector in machine tool axes *(MT indexes)* and tool axes *(t indexes)*.

In addition, the modal vector at the tool tip is referenced to the machine tool axes X_{MT} Y_{MT} Z_{MT} and to the local axes of the tool X_t Y_t Z_t, coinciding Z_{MT} and Z_t axes and defining X_{MT} and Y_{MT} the working plane of the machine and $+X_T$ the feed direction of the tool, as shown in Fig. 3. $\bar{i}_m$ is the versor that defines the direction of a mode and $\{\Theta_i\}$ the modal vector per mass unity, the projection of the versor over the $X_{MT}Y_{MT}$ plane is $\bar{i}_m^{xy}$, whose angle with respect to the feed direction is β_{xy}, whereas the angle between the modal vector and the $X_{MT}Y_{MT}$ plane is β_z. Hence, the direction of a given machine mode is related to the feed direction ($+X_T$) through the two angles β_{xy} and β_z.

By means of a mechanistic model for the machining forces and a dynamic model of the machine, the dynamic machining forces are projected over the *m*th mode direction i_m, expressing thus those process forces in modal coordinates. In addition, as these modal displacements δ modulate the dynamic chip thickness, the former modal relate to modal displacements by means of the expression in Eq. 1:

$$g = K_t \cdot a_p \cdot A(\phi)\ \Delta\delta/2, \qquad (1)$$

where g represents the milling forces in modal coordinates, K_t the dynamic cutting forces, a_p the axial depth of cut, $A(\phi)$ the matrix of directional factors and $\Delta\delta$ the dynamic displacement of the tool in modal coordinates. Next a mono-frequency solution is applied to solve the stability problem, so that if chatter appears at a frequency ω_c, the dynamic displacement of an *i*th mode in modal coordinates in a tooth passing period τ can be obtained from the modal parameters of a *i*th mode, i.e. natural frequency ω_i, modal vector ϕ_i, damping coefficient ξ_i and effective stiffness $k_{efi}=\omega_i^2/\phi_i^2$, together with the dynamic cutting forces. Taking the average Fourier term of the matrix of directional factors α, equation 2 is obtained, where $H(\omega)$ is a diagonal matrix of transfer functions h_i in modal coordinates and Z is the number of flutes of the tool.

$$g = \frac{Z}{4\pi} \cdot K_t \cdot a_p \cdot (1 - e^{-i\omega_c\tau}) \cdot \alpha \cdot H(\omega) \cdot g \qquad (2)$$

3.2.2 Solution for the Stability Model

The solution of the nth order *eigenvalue* problem in modal coordinates (see Eq. 2) provides the relation between the chatter frequency, the depth of cut and the spindle speed in order to assemble the stability lobes and chatter frequency diagrams, as in [13].

As the equations are expressed in modal coordinates, they are decoupled among them, so that each mode can be analysed in an individual manner. Thus, for the specific case of *one concrete mode*, the critical depth of cut a_{pcrit} is obtained as shown in Equation 3:

$$\begin{aligned} \sigma < 0 : a_{p_{crit}} &= \frac{4\pi \cdot \sin\kappa}{Z K_t} \cdot \frac{1}{\sigma} \cdot \frac{k_{ef}\,\xi(1+\xi)}{1} \cong V \cdot \frac{1}{\sigma} \cdot k_{ef} \cdot \xi \\ \sigma > 0 : a_{p_{crit}} &= \frac{4\pi \cdot \sin\kappa}{Z K_t} \cdot \frac{1}{\sigma} \cdot \frac{k_{ef}\,\xi(1-\xi)}{1} \cong V \cdot \frac{1}{\sigma} \cdot k_{ef} \cdot \xi \\ V &= \frac{4\pi \cdot \sin\kappa}{Z \cdot K_t} \end{aligned} \qquad (3)$$

In Eq. 3, κ is the lead angle of the cutting edge and σ is the *directional factor*, a dimensionless factor that depends on the orientation between the mode and the feed direction as well as on the normalized radial and axial cutting coefficients. If $|\sigma|<1$ this coefficient increases the value of the modal effective stiffness K_{ef}, and in turn if $|\sigma|>1$ it decreases that stiffness.

Within this view, this integral machine+process approach proposes to *achieve the critical depth of cut* a_{pcrit} shown in Eq. 3 *with a minimum possible of involved material content in the machine*. The proposed approach is that, if a specific machine mode does not limit the productivity of the machine, its modal stiffness can be reduced without affecting to the productivity and stability of the machining process.

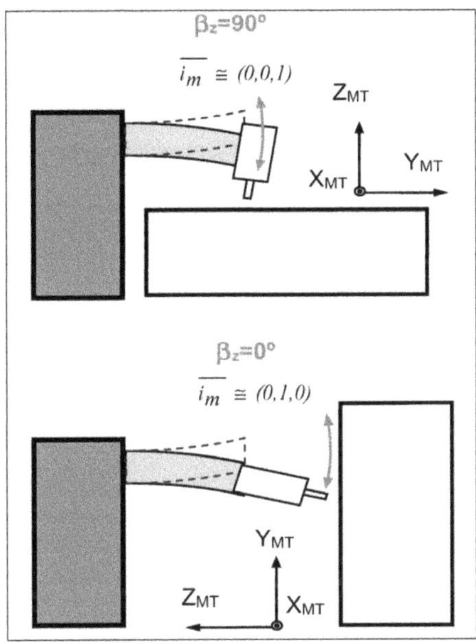

Fig. 4. Change in the orientation of working plane for modifying directional factor σ.

Further, if one specific mode limits the productivity of the machine in one specific milling direction, the productivity of the machine can be maintained while reducing the mass of the machine by means of the following two complementary approaches:
i) By modifying the directional factor σ for reducing its absolute value, e.g. by changing the machine table orientation, as shown in Fig. 4.
ii) By reducing in a deliberated manner the modal stiffness associated to that mode –with a subsequent approximately proportional reduction in the mass of involved machine components - and by increasing in parallel the modal damping in the same proportion by means of active damping devices.

This second approach becomes a very powerful and effective means for improving the eco-efficiency of machining processes, since it states that it is feasible to maintain the productivity of a machine while decreasing by a set percentage the stiffness and mass associated with the mode that limits that productivity, provided that the modal damping is increased by another set percentage for maintaining the product of stiffness and damping, as shown in Eq. 3.

4. Application of the approach for redesigning an actual milling machine

This approach has been implemented in an actual industrial case by redesigning in a comprehensive manner a large horizontal milling machine (30% weight reduction applied). In its light-weighted ram, which has been made of sandwiches of steel and aluminium foam, an Active Damping Device-ADD has been embedded in the X direction, as shown in Fig. 5.

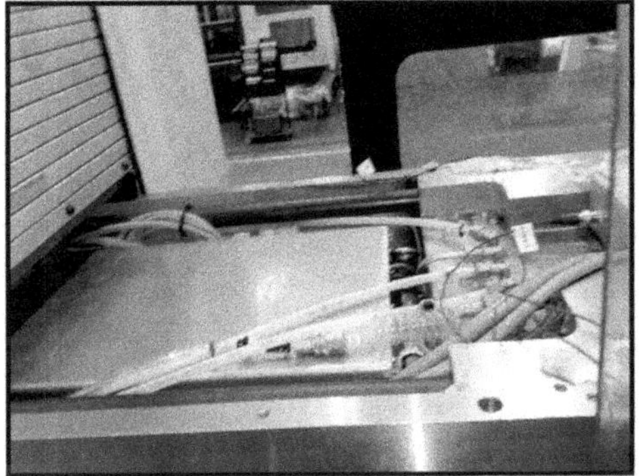

Fig. 5. ADD embedded in a light-weighted machine ram.

This integral re-conception of the milling machine has allowed increasing its critical depth of cut in the X+ direction from the initial 2.2 mm to 4.5 mm. i.e. a *100% increase in productivity*, while at the same time the *energy consumption has been reduced by over 20%* due to the reduced weights of machine components. A similar increase in productivity has been observed when machining in the Y direction (90°), as the critical depth of cut has passed from the initial 2.3 mm to 4.6 mm. Fig. 6 shows these results through an experimental stability lobe diagram in form of a polar plot.

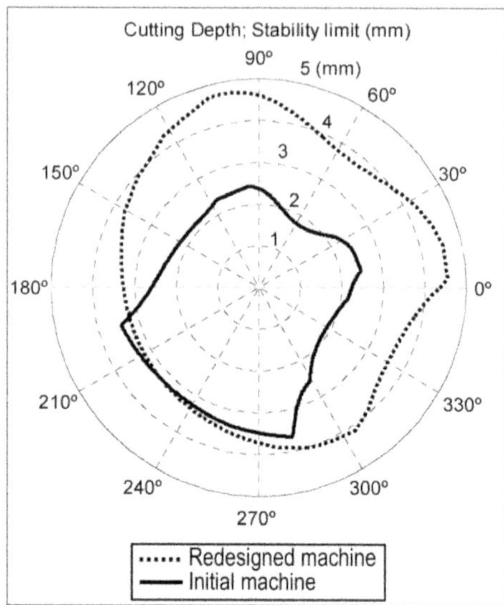

Fig. 6. Productivity of the redesigned machine in different machining directions.

4. Conclusions

This paper has introduced a design methodology aimed at conceiving milling machines in which high productivity is combined with reduced environmental impact – an aspect that is related to a large extent with the consumption of material resources in the construction phase and energy consumption in the phase of use. The approach has been based on breaking the link between a stable machine-part interaction during the machining process and the need to obtain high levels of mechanical rigidity by means of structural components of great weight.

The construction of a Prototype of eco-efficient and productive milling machine, has enabled reductions of 35% to be obtained in the energy consumed in non-load movements in relation to the milling machine referred to above. Parallel to this, increases in the *Material Removal Rate* of 100% have been obtained in some machining directions. On the other hand, the integration of an active vibration control system has enabled reductions in energy consumption to be obtained both in the milling head itself and in the feed servo-drives in terms of the energy consumption of the Prototype, without vibration damping.

Finally, this approach that optimises the ratio between the material content of a machine tool and its productivity level has a beneficial effect both on machine suppliers and their users, because both manufacturing and machining costs are reduced. This mutual benefit may thus enable new business models to be established in the manufacturing sector among machine suppliers and manufacturers by sharing benefits, risks and costs associated with manufacturing processes.

References

[1] Dietmair A. et al., Lifecycle Impact Reduction and Energy Savings through Lightweight Eco-Design of M-Ts. 17th CIRP Int. Conf. on LCE, Anhui, China (2010).

[2] Henninger C., Eberhard P. An investigation of pose-dependent regenerative chatter. Proceedings of the IFToMM 12th World Congress in Mechanism and Machine Science, Besançon, France (2007).

[3] Bravo U., Procedimiento para la predicción de la estabilidad dinámica en el mecanizado a alta velocidad de paredes delgadas (in Spanish). Doctoral Thesis in the University of the Basque Country UPV/EHU. 2007.

[4] Dequidt A. et al., Mechanical pre-design of high performance motion servomechanisms. Mechanism and Machine Theory 35 (2000) pp. 1047-1063.

[5] López de Lacalle, L.N.; Lamikiz, A. (Ed.) Machine Tools for High Performance Machining, 2009, pp. 1-44.

[6] Zulaika J. et al., Eco-efficient and highly productive production machines by means of a holistic Eco-Design approach. Proceedings of the E/E 3rd International Conference on Eco-Efficiency. Egmond aan Zee (2010).

[7] Zulaika J. Campa, F.J. Optimización de los parámetros de diseño de una máquina-herramienta en base a criterios de productividad y ecoeficiencia (in Spanish), XVII Congress on Machine Tools and Manufacturing Technologies, San Sebastián, Spain (2008).

[8] Zatarain, M., Muñoa, J., Peigné, G., Insperger, T., Analysis of the Influence of . Mill Helix Angle on Chatter Stability, Annals of CIRP, Vol. 55, No. 1, pp 365-368 (2006).

[9] Zulaika J. et al. Highly productive machining processes and eco-efficient machine tools, Proceedings of the 4th Int. Conf. of HSM ICHMS2010. Guangzhou (2010).

[10] Altintas, Y., Budak, E., Analytical Prediction of Stability Lobes in Milling, Annals of the CIRP, 44/ 1: 357-362 (1995).

[11] Zatarain, M. etal., Analysis of directional factors in milling: Importance of multifrequency calculation and of inclusion of the effect of the helix angle, 6th Int. Conf. on High Speed Machining, San Sebastian, Spain (2007).

[12] Muñoa, J., et al. Optimization of hard material roughing by means of stability model, 8th CIRP Int. Workshop Model. of machining operations, Chemnitz (2005).

[13] Altintas, Y., Budak, E. Analytical Prediction of Stability Lobes in Milling, Annals of CIRP, 44/ 1:357-362 (1995).

Innovative Production Machines and Systems
D.T. Pham, E.E. Eldukhri and A.J. Soroka (eds)

Training engineers with the aid of 3-D models for robust manufacturing

Bruce Field[a,b], Colin Burvill[b] and Tayebeh Alirezaee[b]

[a] *Department of Mechanical and Aerospace Engineering, Monash University, Australia*
[b] *Department of Mechanical Engineering, the University of Melbourne, Australia*

Abstract

During a preliminary investigation of the learning capabilities of injection machine operators, it was found that many possessed limited spatial skills. Previous studies with naïve engineering designers had found that those with low visual skills had difficulties in reading and manipulating data from engineering drawings, and since the advanced training material for machine operators principally comprised technical cross sections, some learning difficulties were anticipated. Some success was gained by using a 3-D wooden model in the operators' training, so the approach was adapted to train engineering designers for the work reported in this paper. The spatial skill of forty-two trainee designers was measured, and a model of an injection-molding machine was made. Half the trainees were instructed on aspects of injection molding design using the model, and half by using conventional 2-D teaching materials. It was found through a post-test that trainees with high visual skills learned equally well from both approaches, but those with low visual skill learned much more effectively when their training involved the use of the 3-D model. In view of the difficulty in developing advanced spatial skills, it would appear that more effective engineering design education across a wider range of abilities might be achieved by a greater use of 3-D teaching tools.

Keywords: Artifacts, visual skill, learning preferences, injection molding design

1. Introduction

From a survey of the Australian plastics industry in 2004, the authors found there was a need to enhance the training of injection mould die designers and injection molding machinery operators. A questionnaire was prepared and distributed to the majority of significant plastic companies in the state of Victoria, Australia (53 companies in total) [1]. Eighteen companies responded to the questionnaire. The average number of employees in the respondent companies was 53, with a median of 50. Larger Victorian companies did not respond, so the largest number of employees in the respondent companies was 170. The responding companies were active in numerous industry sectors, with automotive and building industries identified as significant clients. Other industry sectors included materials handling, electronics, horticulture and marine. Companies were asked to provide a quantitative measure of importance to a series of proposed strategies that may improve the overall performance of the Australian polymer industry sector. There were eight strategies with the highest perceived importance, heavily weighted towards a series of training strategies. An overall weighted score for each of the 8 strategies was calculated by summing the product of each normalized percentage importance and an importance weighting in the range 8 (most important) to 1 (least important). Improved training of die designers had, by far, the largest score (7.28) (figure 1). A secondary, but more immediately accessible goal was the training of machine operators (a score of 3.5). This latter issue was explored in a study using timber model die components and flawed moldings for training machine operators, and the outcomes were reported in 2005 [2]: in particular we identified an inverse connection between the trainees' visual skill and their training outcomes.

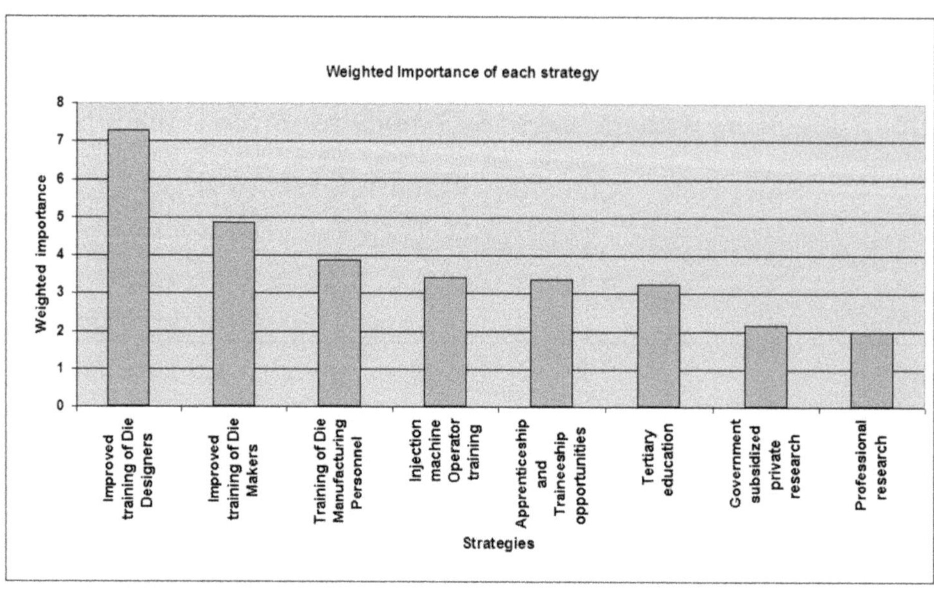

Fig. 1. Outcome from an industry survey that identified the need for training die designers.

The present paper reports on the results of an extended study with trainee engineering designers, using more sophisticated models of an injection molding machine and a die.

2. Training Die Designers

A typical training program for die designers will follow conventional engineering professional education protocols, but there are particular aspects that demand special attributes. Plastic artifacts and die components can have complex 3-D forms, and the dies are, after all, 'negative' representations of the molded components. Consequently, there is a demand for learners to exercise advanced visual capabilities.

Training materials for most engineering programs, and for molding usually contain two dimensional (2-D) representations (in text books, for example), and often include cross sections of those complex three dimensional (3-D) components commonly encountered in die design. Figure 2 is an example of this type of image.

During a 2005 study of trainee molding machine operators, it was reported that many of those operators had difficulty in comprehending the three-dimensionality of the artifacts and injection machine components when those parts were represented by cross sections of the type shown in Figure 2. It appeared that their learning was handicapped by their visual skills. While it appeared more likely that, as a group, trainee-engineering designers would have a higher level of visual skill, there was some evidence that this may not have been the case, or at least, that many of the trainee designers may not possess high visualization skills [4, 5, 6].

3. Spatial Skill in Die Design

The skills associated with the successful solution of holistic spatial problems (e.g. 3-D machines) with incomplete and contradictory data where the form of the solution is not obvious are part of the subset of human skills called the 'right brain skills' [7]. The use of systematic logic (such as mathematical tools) within a narrowly bounded domain draws predominantly on 'left brain skills'.

It is apparent that most of the systematic learning that occurs when the engineering sciences are taught in the traditional manner engages left brain skills, and that students with a highly developed left brain intellect will excel in these courses. It seems as though excellence in design requires a balanced application of skill from both the left and right hemispheres.

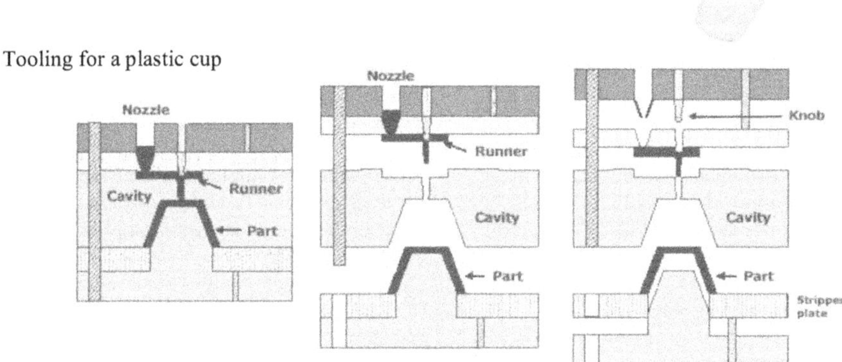

Fig. 2. Typical image from die and moulding training material [3]

During 2004, the authors found that visual capability of trainee engineering designers was correlated with their performance in their design examination [8], so this right-brain skill was included in a study to uncover further related skills. The undergraduate students selected as subjects for the study were drawn from the introductory engineering design course at Monash University (Australia) during the first half of 2006. Most of the 280 design students had completed a 'common' first year in 2005, where they had passed in studies of mathematics, computer programming (both compulsory studies), dynamics, and other fundamental engineering sciences. There were no design studies in that preparatory year.

Grades for four first and second year courses in mathematics and computing were obtained, along with those of the second year design course.

The design grade combined students' separate scores gained from a formal examination comprising an open-ended design and drawing task (40%), individual 2-hour class assignments covering a dozen component skills in design methodology (20%), an individual drafting component within group project work (7%), and the remainder from group scores in two projects (33%).

The students enrolled in design course were tested on two occasions during breaks in the normal lectures. On the first occasion, during the eighth week of the 13-week course, a ten-question version of the Mental Cutting Test (MCT) of spatial visualization skill [9,10], was undertaken. On the second occasion, during the fifth week of the following semester, a management questionnaire [11] with the imbedded question: "When solving a problem (a) I use all of the data available, and get concerned when data seems to be inadequate, or (b) I feel comfortable making assumptions from my own viewpoint" was completed by most of the students.

Results for studies in four mathematics and computing courses with enrolments from 100 to 238 students at two levels were extracted from university records and correlated in pairs. The correlation coefficients for those pairs ranged from 0.59 for level 1 mathematics and computing (238 sets) to 0.67 for a level 2 mathematics course and level 1 computing (151 sets) (all were significant, $p<0.01$). This confirmed that there was a reasonably common basis for performance in the science-based studies, and showed the spread that existed between courses with similar structures. The correlations between the science-based courses were taken to indicate that an underlying academic skill accounted for about 0.6 of a student's grade, and the remaining proportion reflected individual differences in motivation, teaching compatibility, timetabling influences, access to tutors etc.

The correlation for the 97 pairs of results from Design and the grade in second level mathematics was 0.32 ($p<0.01$). This correlation with mathematics was noticeably lower than correlations between the mathematical sciences, and supported anecdotal evidence that many of the academically strong students did not perform as well in design.

The correlation coefficient for the 179 students who completed the 10-question MCT for their visualization score, and their design outcome was 0.32. This is significant ($p<0.01$), and showed either that visualization skill is very useful in design education, or that there was some other common component that underpinned visual and design skills.

The correlation coefficient between response (b) (i.e. comfort in making assumptions) to the management question and the design grade was 0.28 for the 72 students who undertook the test. This is significant ($p<0.01$), so it may be concluded that the better performing design students are more comfortable in making assumptions during problem solving.

A multiple regression of design grade with the 50 complete sets of student results from (a) mathematics, (b) the MCT and (c) the question on comfort with assumptions yielded the regression coefficient of 0.62 and an equation:

$$\textit{Design grade} = 26 + 0.42 \times \textit{Mathematics} + 0.14 \times \textit{MCT score} + 0.056 \times \textit{Assumption score}, \quad (1)$$

with all components measured as percentages. The ANOVA is significant ($p<0.005$) (Figure 3).

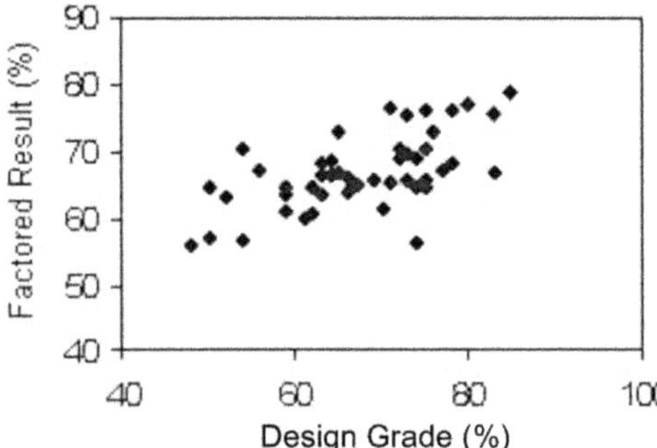

Fig. 3. Scatter plot for 50 students' design grade and composite result from mathematics, visualization and management tests

The study set out to identify possible causes for the spread of design performance in an undergraduate engineering program. It was noted that the average correlation coefficient between mathematics-based courses was about 0.6, providing a baseline target for design course correlation. The design course in the study correlated at only 0.32 with the averaged mathematics courses, showing that there were factors in design performance that were not present in mathematics assessment tasks.

The multiple regression correlation was 0.62 when grades for the mathematics courses, the MCT and the students' willingness to make assumptions were factored together. It was concluded that while a student's academic capability, as demonstrated with left-hemispheric mathematics techniques, was a major correlated factor in their undergraduate design performance at Monash, there were significant design grade advantages if they were also strongly visual, and were comfortable when making approximations during problem-solving.

This finding suggested that some engineering design students would probably find it especially difficult to become proficient in die design where advanced visual skill would appear to be advantageous. Consequently, we set out to apply the findings from our study with injection machine operators, where the use of 3-D artifacts proved to be useful teaching aids, to a more extensive study with potential die designers.

4. Experimental Protocol

The authors hypothesized that trainee designers with weak visual skills would find it relatively difficult to cope with spatial concepts when engaged with conventional 2-D learning materials, such as illustrations and sections like those in Figure 2. However, we expected that those weak visualizers would be more comfortable with actual artifacts that could be physically manipulated.

To test this hypothesis, a simplified model of an injection-molding machine and die set was designed and built (Figure 4), and a set of conventional 2-D learning material was developed to cover the same concepts (Figure 5).

Forty-two volunteer students from the University of Melbourne were recruited and tested with a 15-question version of the MCT to determine their visual capabilities. Their scores ranged from zero to 15 out of 15, with an average of 8. Test subjects of similar visual capability were paired (forming 21 pairs) and scored on a pre-test of their residual knowledge on injection molding: all showed a similar, but very limited basic knowledge. The third author then trained one subject from every pair for a period of 45-60 minutes, using only the model die set. The other subject from every pair was trained for a similar period by the same author, using only the 2-D materials.

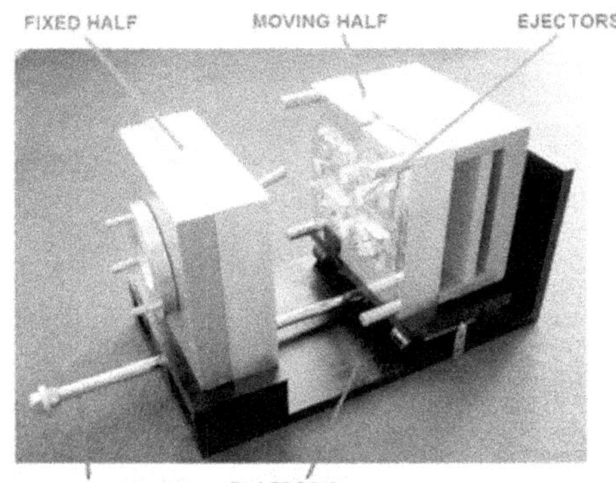

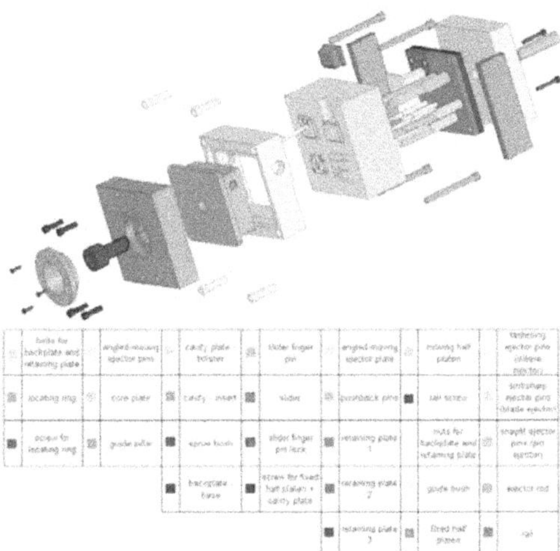

Fig. 4 Model injection molding machine and die set, and features included in the model

Two weeks after they were trained, all subjects undertook a post-test on aspects of injection molding that comprised ten questions to be attempted in ten minutes.

A sample post-test question is illustrated in Figure 6. Most questions contained 2-D images, including cross sections, rendered pictorial drawings from a CAD model and photographs of a die set and molded artifacts, so were expected to favor those subjects who had been trained with 2-D materials.

The outcomes of the post-test are plotted in Figure 7. The linear graphs in Figure 7 show a clear trend. When subjects had high visual capabilities (the subjects toward the right hand end of Figure 7), their post-test scores in injection molding terminology and characteristics were not significantly different whether they had trained with 2-D or 3-D materials. For the subjects with decreasing visual skills, it appears as though those who were trained with the physical artifact score increasingly better than those who were trained solely with the 2-D printed material. When the post-test scores of the ten subjects with strongest visual skills trained on the 3-D artifact are averaged, and compared with the similar average of the strong visualizers trained with the 2-D material there is no significant difference ($p > 0.2$). The post-test scores of the ten subjects with the weakest visual skills trained on the 2-D artifact were some 20% better on average than their counterparts trained with 2-D materials. This difference is very significant ($t = 1.9$, $p < 0.05$).

Ejection method:

The purpose of ejection system is to remove the solidified product from the mould and provide the empty space in cavity for next cycle.

It consists of ejection plate assembly (ejector plate and retaining plate), and ejector rods (surface).

The ejector rod in injection machine actuates the ejection system.

There are different kinds of ejector pins available. The choice of each design selection depends on the product configurations.

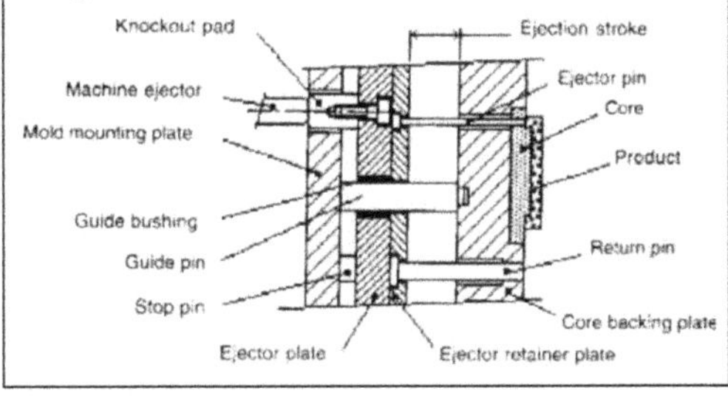

Fig. 5. Sample of 2-D training material

5. Results and Discussion

The principal finding was that trainee die designers with weaker than average visual skills as measured by the MCT benefited from being trained with a model of a die set, whereas there was little, if any disadvantage suffered by strongly visual trainee designers when they were forced to use a model die set. When trained with the model die set, the subjects with low visual skill may have learned less than those with high visual skill, although in the experiment this trend was not significant.

The results are consistent with the parallel experiment where it was found that a trainee designer's grade from their formal study in an engineering design course was significantly correlated with their visual skill as measured by the MCT: this correlation was about the same as that between their university grade in mathematics and their design grade.

While it is apparent that naive engineering designers with low visualization scores would benefit from specific training to develop their spatial abilities, research has shown that the rate of change in visual skill for mature people is slow, even when undertaking a focused course of one to four hours per week for up to thirteen weeks [5, 12]. It would appear, however, that a more rapid development of design skill in those weak visualizers may be forthcoming if the educational programs in engineering design could by-pass the limitations of 2-D teaching materials by using, where appropriate, physical artifacts. Indeed, it is also possible that the concentrated manipulation of 3-D artifacts is one path by which visualization skill is increased [5].

The findings add support to the development of, or retention of undergraduate courses where engineering students handle and become engaged with engineering components and machinery. These courses include machine shop work, laboratory tasks involving the dismantling of machinery, and design-and-build competitions [13, 14, 15].

It is as yet not known whether engagement with 3-D artifacts can be partly replaced with more portable and economic, but realistic representations of those artifacts, such as through current solid modeling software (with, or without full rendering) or various forms of virtual reality. This question provides a basis for useful future research studies.

6. Conclusion

Trainee die designers with low spatial skill achieved greater learning outcomes when their training program was built around the use of a physical model compared with a similarly skilled sub-group who were trained using conventional 2-D printed material.

A control group comprising two separate sub-groups of trainee designers with high spatial skill but trained with the same artifacts or 2-D printed material respectively, achieved learning outcomes that were not significantly different.

It was concluded that training with the use of 3-D artifacts improved the outcomes of designers with low visual skill and did not disadvantage the designers with high visual skill.

The finding has implications for the educational programs for other types of engineering designers, especially since it has been shown elsewhere that trainee designers with low visual skills tend to achieve inferior performance in university design assessments compared to design students of a similar academic capability.

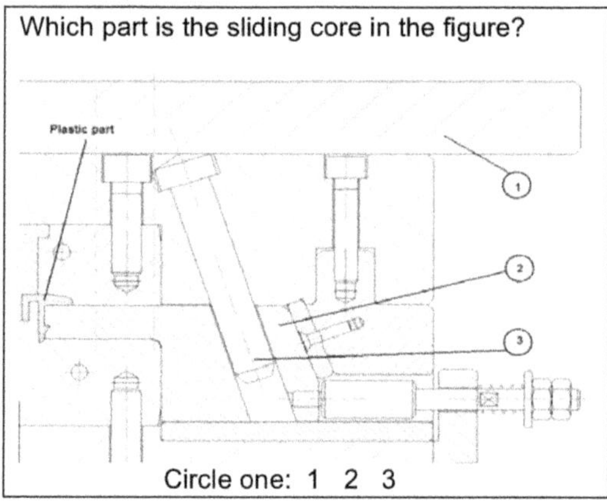

Fig. 6. Sample question from the post-test on injection molding

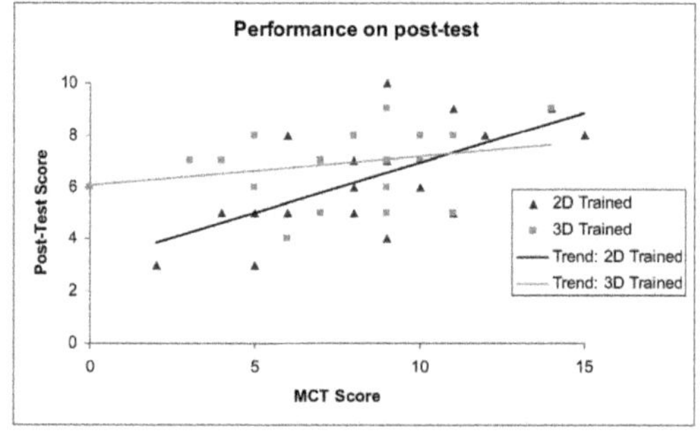

Fig. 7. Results from the post-test on injection molding for the two groups of subjects.

References

[1] Alirezaee, T, Burvill, C., Field, B. and Carton, M. Optimising the mould design and production cycle. In *Proc DETC'04: ASME Design Engineering Technical Conferences and Computers and Information in Engineering Conference*, Sept 28 - Oct 2, 2004, Salt Lake City, USA

[2] Burvill, C.A., Alirezaee, T. and Field, B.W. Identifying and enhancing conceptual design capabilities. In *International Conference on Engineering Design, ICED'05*, Melbourne, Aug 2003, Paper 530.61

[3] SCUDC website: http://www.scudc.scu.edu/cmdoc/dg_doc/develop/design/runner/34000007.htm

[4] Field, B.W., Burvill, C.A. and Weir, J.G. The impact of spatial abilities on the comprehension of design drawings. *International Conference on Engineering Design ICED'05*, Melbourne, Aug 2005, Paper 531.61

[5] Huet, G, Culley S J, McMahon, C A and Fortin, C Making sense of engineering design review activities *AIEDAM* Vol 21 No 3, 2007, pp 243-266

[6] Sorby, S.A. and Baartmans, B.J. The development and assessment of a course for enhancing the 3-D spatial visualization skills of first year engineering students. *Journal of Engineering Education*, 89(3), 2000, pp 301-307

[7] Kokotovich, V Problem analysis and thinking tools: an empirical study of non-hierarchical mind mapping Design Studies Vol 24 No 1,2008, pp 49-69

[8] Field, B.W., Burvill, C.A. and Weir, J.G. 2005. The impact of spatial abilities on the comprehension of design drawings. *International Conference on Engineering Design ICED'05*, Melbourne, Aug 2005, Paper 531.61

[9] CEEB. *Special aptitude test in spatial relations*. College Entrance Examination Board, 1939, USA

[10] Churches, A.E., Magin, D.J. and Barratt, A.J. Reliability and stability of two tests of spatial abilities. *Proc Sixth ICECGDG*, Tokyo, 1994, pp801-805

[11] Robbins, S., Bergman, R., Stagg, I. and Coulter, M. *Management*. 3rd Edition, 1995, Sydney, Prentice Hall

[12] Field, B. W. Visualization, intuition and mathematics metrics as predictors of undergraduate engineering design performance. *ASME Journal of Mechanical Design,* 129(7), July 2007, pp 735-743

[13] SAE International. http://students.sae.org/competitions/formulaseries/ accessed November 2010

[14] Churches, A. and Magin D. Student design and build projects revisited. *International Conference on Engineering Design ICED'05*, Melbourne, Aug 2005, Paper 128.61

]15] Samuel, A.E. *Make and test projects in engineering design: creativity, engagement and learning*. Springer, 2006

Innovative Production Machines and Systems
D.T. Pham, E.E. Eldukhri and A.J. Soroka (eds)

Analysis of Process Parameters in Roll Drawing

M. Durante, A. Formisano, A. Langella, F. Memola Capece Minutolo

Department of Materials and Production Engineering,
University of Naples Federico II, Piazzale Tecchio 80, Naples 80125, Italy

Abstract

Roll drawing is a technological process that allows to obtain a notable reduction of the drawing force in comparison with the traditional drawing process with converging dies. Consequently, no thermal or chemical treatment of the wire is required and no lubricant is employed, with significant environmental advantages. The drawing force and the geometrical dimensions of the wire are critical parameters for optimal process design. They have been evaluated by analytical models but, owing to the complexity of the process, a three-dimensional FE model has been developed to obtain a more precise method able to control the process.

Keywords: Roll drawing, Analytical models, FEM analysis

1. Introduction

In wire drawing, the negative effects due to friction phenomena influence fundamental technological aspects of the process: wear of the dies, increase in drawing force, heat generation and excessive metal flow close to the contact surface with the die. To cope with these effects, studies on lubricants and lubrication conditions in metal forming processes have been conducted [1-6] and innovative systems have been studied in order to increase the efficiency of lubricants [7].

In classical wire drawing, the use of lubricants is essential to reduce friction, but nowadays the request for decreasing the environmental burden, as the utilization of chlorine-free lubricant, determines the search for semidry or dry processes. One of these could be the roll drawing process in which the surface of the die is given by the external cylindrical surfaces of a number of rolls. Bayoumi noted a force reduction of about 40% compared with the one required for drawing with converging dies, employing a set of symmetrically arranged flat idle rolls [8].

The study of these new process types could be conducted by examining the various contributions to the global drawing force and the resulting geometry of the wire after drawing. This last point is a typical aspect studied in rolling processes; many research works have been carried out both experimentally and numerically on shape rolling for design improvement. FEM analysis is very effective in calculating the surface profile and the cross sectional area of the workpiece; accordingly, several researchers [9,10,11] used FEM models for sequence design and optimization of rolling processes.

However, to simulate these processes a 3D analysis is necessary, requiring considerable computing time for numerical simulations and making analytical modelling highly desirable. In the case of rolling, in several studies [12,13,14] analytical formulae were developed for calculating the spread of the workpiece. Starting from the Shinokura-Takai's theory [15], Min et al. [16] calculated the spread in a three-rolls rolling process for round-hexagonal and hexagonal-hexagonal passes. In order to calculate the drawing force, a number of analytical models have been proposed to correlate the different process parameters. In particular, the two formulae of Avitzur [17] (lower and upper bound) and the Siebel's formula [18] calculate the drawing force as the sum of three terms (internal-power, shear-loss and friction-loss) and determine the contribution of each term to the global force.

In this work, the study of a roll drawing process carried out without lubricant, using three rolls at 120° to each other, is presented. The aim is the evaluation of the global drawing force and the geometry of the output wire as a function of the distance between rolls and the initial diameter of the wire. For their simple formulation, the Siebel and Shinokura-Takai's analytical models were selected, whereas the software code used for FEM simulation is LS DYNA, particularly suited to the analysis of metal forming processes.

2. Case study

The three rolls used for the roll drawing case study were cylinders with diameter of 120 mm and height of 15 mm, realized in cemented carbide. The features of the groove on the side of the rolls are shown in Fig. 1, where d is the distance from the centre of the rolls depending on the section desired. Two cases were analyzed:

- wire with diameter D = 7 mm and distance d = 3.1 mm (reduction 7-6.2);
- wire with diameter D = 5.5 mm and distance d = 2.4 mm (reduction 5.5-4.8).

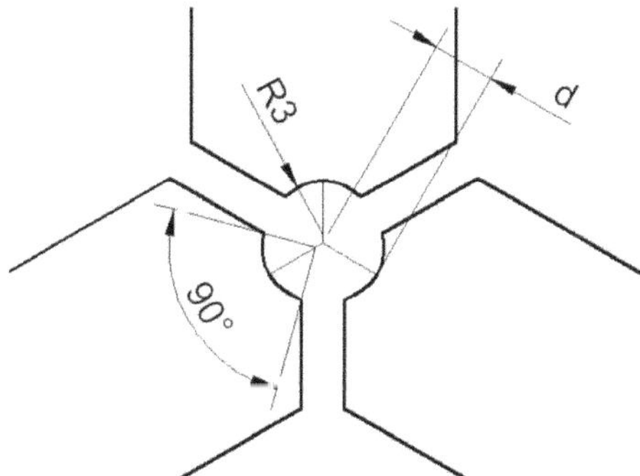

Fig. 1. Schematization of the three rolls used for the roll drawing process.

The wire material is AISI 201 steel for the 7 mm diameter and AISI 301 steel for the 5.5 diameter, mechanically characterized by tensile tests carried out before and after the roll drawing process, since the selected models need a yield strength value equal to the average of yield strength before and after drawing. The work material presented an elastic-plastic behaviour with limited work hardening and the average yield strength for the two wires was equal to 485 and 350 MPa for diameter 7 mm and 5.5 mm, respectively. For both cases, a Young modulus of 205 GPa and an ultimate elongation of 40% were measured, whereas the average ultimate tensile strength was respectively equal to 600 and 430 MPa.

A length of ten meters of wire was tested for each case study. The wires were mechanically surface treated before the process to eliminate the oxide layer from their surface. The drawing force was measured by a load cell. The experimental tests were carried out at the FAR SUD plant, Casoria (Naples), Italy, that produces electric welded wire mesh. Fig. 2 shows the sections obtained for the two cases analyzed. It is possible to observe that the reduction 7-6.2 presents an incorrect filling with work material coming out of the rolls.

3. Analytical models

According to Siebel's theory, the force necessary for the wire drawing process can be derived from the total deformation work made up by the following components: internal-power, shear-loss and friction-loss. The analytical formulation that allows to describe the drawing force is the following:

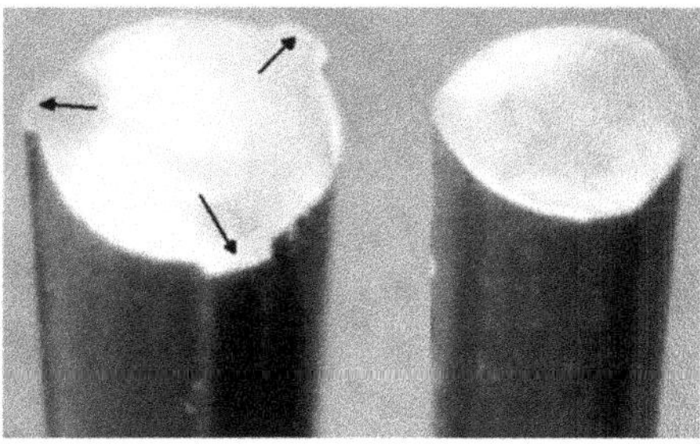

Fig. 2. Products obtained by reduction 7-6.2 (left) and 5.5-4.8 (right). The arrows indicate the work material coming out of the rolls for reduction 7-6.2.

$$F = \sigma_m S_u \left[\frac{2}{3}\alpha + \left(1 + \frac{2\mu}{\sin 2\alpha}\right)\varphi_{max}\right] \quad (1)$$

where F is the drawing force, σ_m is the yield strength equal to the average of the values before and after roll drawing, S_u is the area of the circle inscribed in the opening described by the positioning of the rolls, φ_{max} = ln S_i/S_u with S_i representing the area of the initial section of the wire, α is the contact arc between roll and wire and μ is the friction coefficient of the wire-rolls contact. The study of the roll drawing process using the analytical models is based on the hypothesis that the influence of friction is very low compared with the traditional wire drawing process [19]. This hypothesis is based on the different kinds of friction present in the process: the contact between rolls and wire develops a rolling friction component notably lower than the sliding friction verified in traditional wire drawing.

The Siebel's model modified for the roll drawing process allows to neglect the friction contribution to the drawing force that, therefore, can be calculated with:

$$F = \sigma_m S_u \left(\varphi_{max} + \frac{2}{3}\alpha\right) \quad (2)$$

The values of the parameters in Eq. 2 for the two cases analyzed are reported in Tab. 1.

With reference to the spread of the wire due to the process, the Shinokura-Takai's model modified for a three-roll rolling process [16] was analyzed. The spread formula utilised in this study can be written as:

$$\Delta b = \alpha \frac{F_O}{F_H} L_c \quad (3)$$

where Δb is the spread, α is a calibration coefficient, F_O and F_H are the areas respectively dotted and in grey in Fig. 3, and L_c is the length of the contact arc between rolls and wire.

This model was analyzed to evaluate the calibration coefficient α specific for the roll drawing process.

4. FEM model

The process under examination presented three planes of symmetry because of the geometry and the loads. Thus, only one sixth of the entire process was simulated through LS DYNA (Fig. 4).

From Fig. 4, it is possible to note that two different element types were used for the discretization of the parts interested in the process: solids for the wire and shells for the roll. In the latter case, only the groove was considered, since it is the only zone in contact with the wire. It was not necessary to use an adaptative remeshing, since the deformations were not very considerable.

A contact typology for forming processes, named contact forming surface to surface, was specified by assigning a static friction coefficient of 0.3 and a dynamic one of 0.15 (typical of metal-metal contact), and the law of motion was assigned

Table 1
Parameters used for the evaluation of the drawing force by Siebel's model

TEST	σ_m (MPa)	S_u (mm^2)	φ_{max}	α (rad)
7-6.2	485	30.19	0.24	0.058
5.5-4.8	348.5	18.09	0.27	0.054

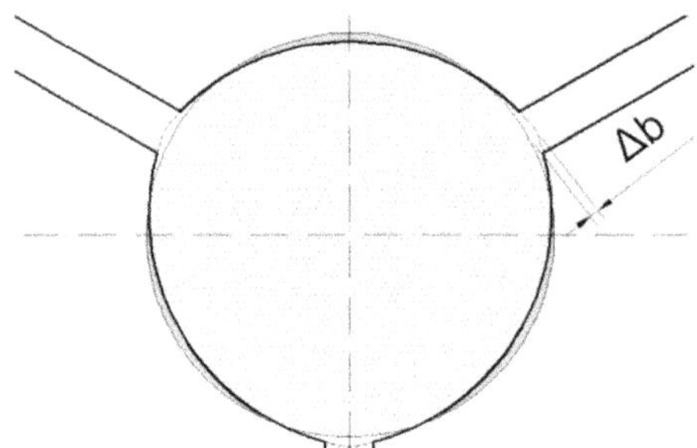

Fig.3. Features for the Shinokura-Takai's model. Dotted area: F_O. Grey area: F_H. Δb: wire spread.

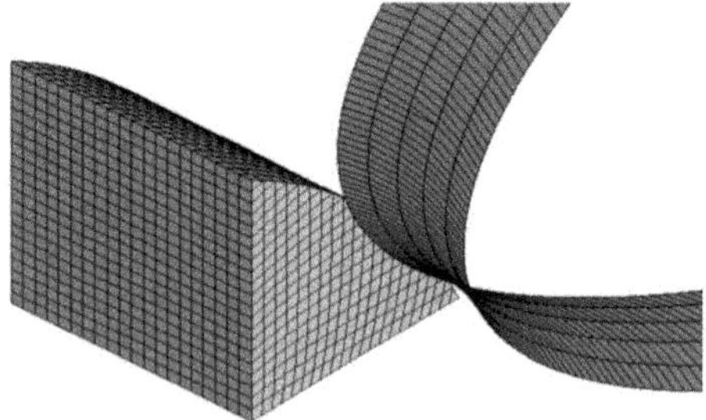

Fig.4. Discretization of the roll and the wire for FEM simulation of the roll drawing process.

to the front section of the wire, whereas the control of the drawing force was conducted on the nodes of an internal section.

The material models were: rigid for the rolls, as their deformations during processing is totally negligible, and elastic-plastic for the wire.

The rolls can idly turn around their axes as the actual rolls of the equipment employed were fixedly mounted on ball bearings. Moreover, appropriate constraints were assigned to the section nodes interested in the symmetries.

An explicit approach was used and the calculation time was about 50 minutes for each simulation.

5. Results

Both the drawing force and the sections at the end of the process were evaluated for the cases under examination. As regards the drawing force, the average values out of the ten meters of wire tested for each of the two reductions are reported.

5.1. Forces

The drawing force values, evaluated by the Siebel's model (Eq. 2), the FEM analysis and the experimental procedures, are reported in Tab. 2.

Siebel's model yielded a drawing force value little different from the experimental one for the reduction 5.5-4.8, whereas a quite lower value was obtained for the other reduction case. This result is justifiable since in the second case an incorrect filling of the rolls was verified and the model couldn't but underestimate the roll drawing force.

The FEM analysis predicted optimally the drawing force for both cases, with deviations within ±7%.

Table 2
Siebel's model, FEM analysis and experimental roll drawing force values (in Newton)

TEST	SIEBEL	FEM	EXPERIMENTAL
7-6.2	4124	5304	4974
5.5-4.8	1942	1728	1805

5.1. Geometries

As regards the geometry obtained by the roll drawing process, Fig. 5 shows the models of the simulated wires at the end of the process, whereas Fig. 6 shows the comparison between experimental and FEM sections for the two cases analyzed.

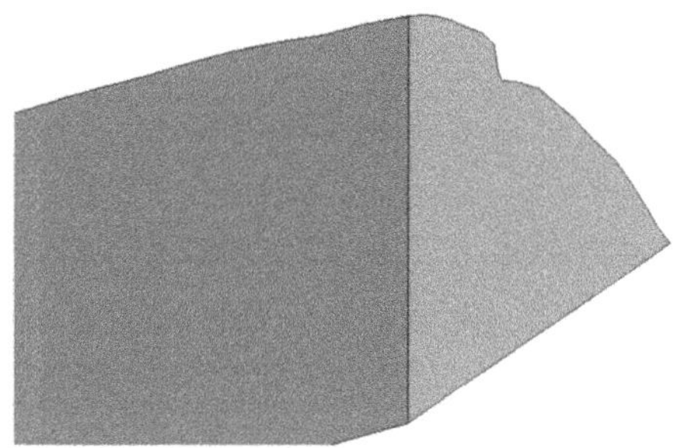

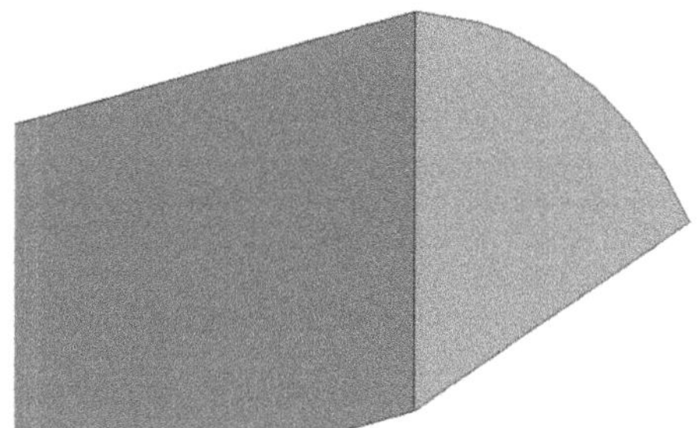

Fig.5. FEM simulation of reduction 7-6.2 (up) and 5.5-4.8 (down).

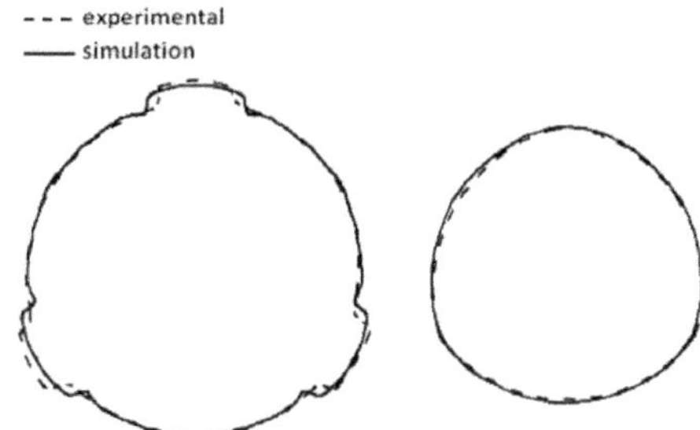

Fig.6. Comparison between experimental and simulated sections for reduction 7-6.2 (left) and 5.5-4.8 (right) cases.

A very good prediction by the FEM analysis of the geometries realized was highlighted. This result is particularly important in order to predict, by simulations, the correct filling of the rolls and the correct choice of the distance at which they should be set. For such predictions, the evaluation of calibration coefficient α was carried out only for reduction 5.5-4.8. By the support of a CAD software, the Shinokura-Takai's parameters of Eq. 3 were evaluated as: Δb = 0.1002 mm, F_O = 19.6174 mm^2, F_H = 4.1409 mm^2 and L_c = 6.33 mm.

It was possible, then, to evaluate the calibration coefficient α for the case of roll drawing: $\alpha = 0.075$.

6. Conclusions

In this work, a comparison in terms of drawing force and wire geometry between experimental, analytical and FEM simulation values, obtained for a roll drawing process of steel wires, was carried out.

The analytical evaluation of the drawing force was realized by the application of the Siebel's theory, modified for the case of roll drawing. The Shinokura-Takai's model was analyzed in order to adapt it to the roll drawing process. LS DYNA software code was used for FEM analysis aiming at drawing force and wire geometry definition. In order to evaluate the effectiveness of the above mentioned models, two test cases were performed on steel wires.

The modification of the Siebel's formula for the case of roll drawing provided good results unless the product presents large distortions. Otherwise it can represent a lower bound, that is a minimum value of the drawing force, for which the equipment could be dimensioned in order to make possible the correct implementation of the technological process object of the study. As regards the spread, the Shinokura-Takai's model was specified for the process under examination, by evaluating the calibration coefficient α.

With reference to the results obtained by the FEM simulation, the model provided more reliable results compared with those of the modified Siebel's model. Furthermore, it allowed to predict also the geometrical quality of the process, principally in terms of shape of the drawn product. In this manner, it allowed to optimize the choice of the process parameters in order to obtain a shape of the drawn product very close to the designed one.

In conclusion, the analyzed models were able to provide good indications on the products and the drawing force values useful for the dimensioning, during the planning stage, of the roll drawing equipment. These results can be considered a good starting point for a deepened study of such complex, and yet so streamlined, process as the roll drawing process.

References

[1] Wistreich JG. Lubrication in wire drawing. Proceedings of the Institute of Mechanical Engineers Conference on Wear, Paper 4 (1957) 505.

[2] Rowe GW. Lubrication in drawing of steel wire. Wire industry (1978) 229.

[3] Sutcliffe MPF and Johnson KL. Lubrication in cold strip rolling in the "mixed" regime. Proceedings of the Institution of Mechanical Engineers 204 (1990) 249-261.

[4] Munther P and Lenard JG. Tribology during hot, flat rolling of steels. Annals of the CIRP 44 (1995) 213-216.

[5] Lenard JG. The effect of temperature on the coefficient of friction in flat rolling". Annals of the CIRP 40 (1991) 223–226.

[6] Azushima A and Kudo H. Direct observation of contact behaviour to interpret the pressure dependence of the coefficient of friction in sheet metal forming. Annals of the CIRP 44 (1995) 209-212.

[7] Weinhold H, Kieselstein S, Kurze B, Zschorn G and Heubnerr A. New pressure lubrication system for wire drawing. Wire 5 (1996) 46-50.

[8] Bayoumi LS. Round-to-square section drawing through flat idle rolls. Mechanical sciences 41 (1999) 1323-1338.

[9] Shin HW, Kim DW and Kim NS. A study on the rolling of I-section beams. International Journal of Machine and Tool Manufacture 34 (1994) 147-160.

[10] Yuan SY, Zhang LW, Liao SL, Jiang GD, Yu YS and Qi M. Simulation of deformation and temperature in multi-pass continuous rolling by three-dimensional FEM. Journal of Materials Processing Technology 209 (2009) 2760-2766.

[11] Komori K and Koumura K. Simulation of deformation and temperature in multi-pass H-shape rolling. Journal of Materials Processing Technology 105 (2000) 24-31.

[12] Saito Y, Takahasi Y and Kato K. Calculation of spread, elongation, effective roll radius, roll force and torque when rolling in the square-diamond, square-oval and round-oval passes. Journal of the Iron and Steel Institute of Japan 2 (1978) 66-75.

[13] Shinokura T, Takai K and Meguro H. A new method for predicting speed in bar rolling. Advanced Technology of Plasticity 2 (1984) 1299-1304.

[14] Lee Y, Choi S and Kim YH. Mathematical model and experimental validation of surface profile of a workpiece in round-oval-round pass sequence. Journal of Materials Processing Technology 108 (2000) 87-96.
[15] Shinokura T and Takai K. A new method for calculating spread in rod rolling. Journal of Applied Metalworking 2 (1982) 94-99.
[16] Min JH, Kwon HC, Lee Y, Woo JS and Im YT. Analytical model for prediction of deformed shape in three-roll rolling process. Journal of Materials Processing Technology 140 (2003) 471-477.
[17] Avitzur B. Handbook of metal forming processes. Ed. John Wiley and Sons, 1983.
[18] Siebel E. Le façonnage des metaux par deformation plastique. Ed. Beranger, 1936.
[19] Lange K. Handbook of metal forming. McGraw-Hill Book Company, 1985.

Innovative Production Machines and Systems
D.T. Pham, E.E. Eldukhri and A.J. Soroka (eds)

Laser cutting of CFRP sheet by pulsed Nd:YAG source: Influence of pulse energy, pulse duration and overlapping on kerf geometry and HAZ

N. Pagano[a], C. Leone[a], V. Lopresto[a], I. De Iorio[b]

[a]*Department of Materials and Production Engineering, University of Naples "Federico II", Naples, Italy*
[b]*Department of Aerospace Engineering, University of Naples "Federico II", Naples, Italy*

Abstract

Carbon Fibre Reinforced Plastic (CFRP) materials are used for structural applications in different fields. Even though near-net manufacturing of composite materials is possible, cutting and drilling remain unavoidable operations with conventional cutting methods. Laser cutting represents a promising alternative: it does not involve any mechanical cutting force and tool wear. Moreover, thanks to the small spot, laser beam allows to realize very small and complex shape too. However, laser cutting is based on thermal interaction between laser beam and the composite material. It results in the presence of fibres pull out and thermal degradation of both fibres and matrix.

The aim of this work is to investigate the influence of the process parameters: pulse energy, pulse duration and overlapping percentage, on the kerf geometry and HAZ extension, in cutting operation of 1 mm CFRP plates using a lamp pumped Nd:YAG working in the regime of short pulses. Experimental results showed that a pulsed Nd:YAG laser, with a nominal power of 150 W, can be successfully used to cut 1 mm CFRP plates up to 11 mm/s. However, an accurate selection of the process parameters is necessary in order to obtain high quality cutting and narrow HAZ.

Keywords: Composites, laser cutting, kerf geometry, HAZ.

1. Introduction

Carbon Fibre Reinforced Plastic (CFRP) materials are generally used for structural applications in different fields. Even though near-net manufacturing of composite materials is possible, cutting and drilling will remain unavoidable operations. Cutting and drilling of CFRP composites with conventional cutting methods produce tool wear and damages such as delamination and fibre pull out [1-3]. Moreover, machining is not suitable when a very small complex shape is required. Laser cutting represents a promising alternative: it does not involve any mechanical cutting force and tool wear, thanks to the small spot laser beam that allows to realise complex shape as well.

However, laser cutting is based on the thermal interaction between the laser beam and the composites that produces a heat affected zone (HAZ). Thermal damage is a direct consequence of the large difference in thermal properties of the carbon fibre and the epoxy matrix [4-5]. In fact, during the beam-material interaction, the time that elapses before reaching the vaporisation condition for carbon fibres is larger than that for the resin. During this period, thanks to the good conductivity of the fibres, a great amount of heat is absorbed by the matrix, thus producing matrix burning [4-5]. As a result, the HAZ is characterized by the presence of fibres debonded from the matrix and by thermal degradation of both fibres and matrix [4-5]. Obviously, the HAZ dimension strictly depends on the adopted laser source and the working parameters [4-10].

The use of pulsed Nd:YAG laser in cutting operation of CFRP was investigated in [11]. It was shown that the high beam intensity and the better focusing behaviour of the pulsed Nd:YAG lasers, compared to the CO_2 ones, lead to a smaller thermal load and consequently a narrow kerf and a small HAZ.

Since the HAZ and the kerf width are related to the energy release-time dependence, a strategy to decrease them could involve the use of high pulse energy released in a very short time, as in the case of short or ultra-short laser pulsed source. The high peak power released in short time rapidly heats the material and directly leads to evaporation and/or plasma [11-12], therefore reducing the heat absorbed by the matrix and then the HAZ extension.

The aim of this work is to investigate the influence of pulse duration, pulse energy and spot overlapping of a 100 W pulsed Nd:YAG laser in cutting of CFRP laminates. For this reason, cutting tests were carried out on a 1 mm thick CFRP plate obtained by autoclave cure. During the tests, different pulse durations and pulse energies were adopted while maintaining the mean power at the constant value of 95 W. For each process condition, the maximum cutting speed was determined. Then, starting from this value, the cutting speed was decreased until a spot overlap value of about 98 % was achieved.

The kerf sections of all the samples were examined through microscopic analysis by measuring the top and bottom kerf width, the taper angle and the HAZ extension. At the end, the kerf geometry and the HAZ were related to the corresponding process condition.

2. Equipment, materials and experimental procedures

In the experimental tests, an industrial multimode lamp pulsed Nd:YAG (Starcut 150 from Rofin), working at the wavelength of 1064 nm, with a maximum nominal power of 150 W, was used.

The laser source was mounted on a CNC machine (StarCut System Universal from Rofin) which allows the generation of the geometric patterns and the setting of the following process parameters: pump lamp voltage (V), pulse frequency (F), pulse duration (D), beam speed. Mean power (Pm), pulse energy (E) and pulse power (Pp) depend on the lamp voltage, the pulse frequency and the duration. The spot overlap, on the contrary, depends on the cutting speed (S), the pulse repetition frequency and the kerf width (W) by mean of the equation (1).

$$\text{Spot overlap}\,(\%) = 100 \cdot \left(1 - \frac{S}{F \cdot W}\right) \qquad (1)$$

All the tests were performed using a cutting head (PRECITEC AK YK52) with a focal length of 80 mm and a nozzle of 0.8 mm in diameter. In these conditions, the focused beam diameter is about 170 μm. During the tests, nitrogen at 10 bar of pressure and 30 l/min of flow was used as assist gas.

With the aim to select the pulse energy and the duration to use in the study, the laser source was first characterised. In figures 1 and 2,–mean power and pulse energy are reported as a function of the pulse frequency for the different pulse durations used in this work at the maximum lamp voltage (700 V). From figure 1, it is possible to observe that the mean power depends on both the pulse frequency and the pulse duration.

Since the HAZ and the kerf geometry are affected by the energy release-time dependence, a strategy to decrease the HAZ and the kerf width could involve the use of a very short interaction time. For this reason, the duration was selected within the range of 0.05 ÷ 0.20 ms, while the lamp voltage was selected at the maximum value (700 V) to ensure the highest mean power.

Then, for each duration value, the two values of pulse frequency corresponding to a mean power of about 95 W (95 W is the maximum value obtained for a duration of 0.05 ms) were taken into account. They were located before and after the maximum peak respectively. The selected conditions are reported in figures 1 and 2 as closed dots.

At least eight conditions of pulse energy were considered. The adopted process parameters are reported in table I, where the grey rows indicate the low pulse energy conditions.

Table I
Parameters setting used for the laser source.

Durations (ms)	**Frequency (Hz)**	**Pulse energy (J)**
0,05	1790	0,054*
	2000	0,049*
0,10	597	0,164
	1600	0,059*
0,15	300	0,317
	1500	0,063*
0,20	185	0,514
	1350	0,066*

* High frequency and low pulse energy.

It is important to notice that the selected pulse energies are in the interval from 0.05 to 0.51 J. This choice assures that the influence of the pulse energy on the HAZ and the kerf geometry is highlighted. The different overlapping percentage values were obtained by fixing the pulse frequency and changing the cutting speed. The maximum cutting speed was first determined for each process condition, then, starting from this value, the cutting speed was decreased till a spot overlap value of about 98% was achieved.

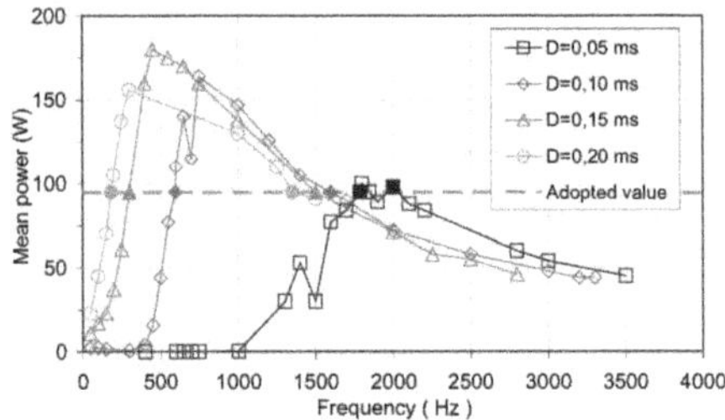

Fig. 1: Beam mean power as a function of the pulse frequency for different durations and lamp voltage of 700V.

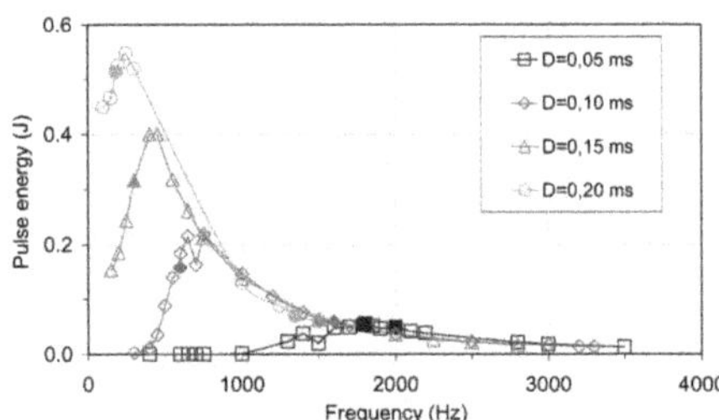

Fig. 2: Pulse energy as a function of the pulse frequency for different durations and lamp voltage of 700V.

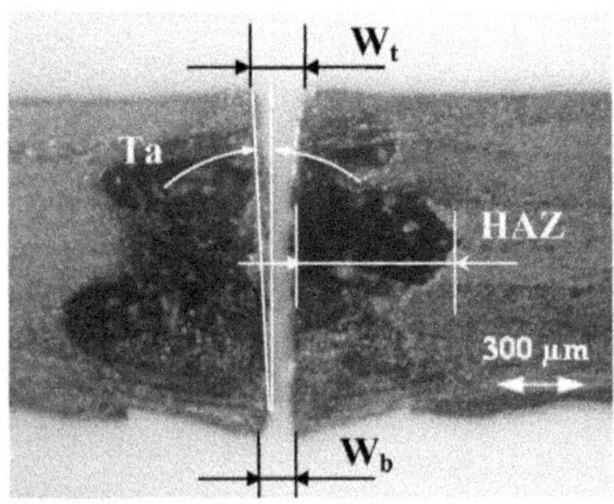

Fig. 3: Examples of kerf geometry. Wt: top kerf width; Wb: bottom kerf width; Ta: taper angle; HAZ: extension of the Heat Affected Zone.

The investigated material was a CFRP plate, 1 mm in thickness, obtained by autoclave cure. The employed reinforcing material was T400 carbon fibre. The lay-up was: 2 external plies of fabric 193 gr/mm^2 each one, with 50% fibre in the warp and weft directions, and eight internal unidirectional plies of 145 g/mm^2 in $(45_2/\text{-}45_2)_s$ sequence. The matrix was HMF 934 epoxy resin.

In order to verify the cut quality, the top and the bottom kerf width, the taper angle and the HAZ extension were measured by microscopic analysis as proposed by Tagliaferri et al. [5]. At the end, the kerf geometry and the HAZ were related to the process conditions. In figure 3, the section of a kerf is reported together with the analysed parameters. These are: the top kerf width (Wt); the bottom kerf width (Wb), the taper angle (Ta) and the extension of the Heat Affected Zone (HAZ). At last, no less than fifteen cuts and profiles were analyzed for each test condition.

3. Results and discussion

3.1 Cutting regions maps

In a previous work [12], the same laser apparatus was employed to cut the same composite using, as working gas, atmospheric air at 10 bar. During the cutting tests, the mean power was varied in the range of 60÷100W, the pulse duration in the range of 0.25÷1 ms and the pulse energy in the range of 0.3÷0.6 J. As a result, it was found that the maximum cutting speed linearly depends on the mean power, while the minimum spot overlapping depends on pulse energy. Furthermore, it was found that both the kerf width and the HAZ increase with the increase of the spot overlapping or the decrease of the cutting speed.

The main results in terms of process windows obtained in [12] are compared to the current results in figure 4 as maximum cutting speed versus mean power. In the figure, the solid dots represent the previously obtained data; the open dots are the results obtained here. As it possible to observe, in the present case, despite the use of a mean power of about 95 W, the maximum cutting speed varies within a large value range (from 4 to 10 mm/s). The latter observation clearly shows that the mean power is a useful parameter for the maximum cutting speed prevision only when the pulse energy values are comparable, like in [12], where the pulse energy was fixed in the range 0.35÷0.65 J. When large variations in pulse energy occur (about 1:10 in the present case), it is not possible to predict the maximum cutting speed by using the mean power.

However, the pulse energy and the minimum value of the spot overlap where cut is possible are strictly dependent one each other. This clearly appears if the spot overlap is reported as a function of the pulse energy, as in figure 5. In that figure, the dashed line represents the boundary between the "cut regime" and the "no-cut regime". As it is possible to see in the same figure, in this case as well, a linear relationship between the minimum spot overlapping and the pulse energy is observable, with the minimum spot overlap that decreases with the increase of the pulse energy. Furthermore, it will be useful to notice that for low pulse energy, cutting is possible only for spot overlapping values higher than 95%.

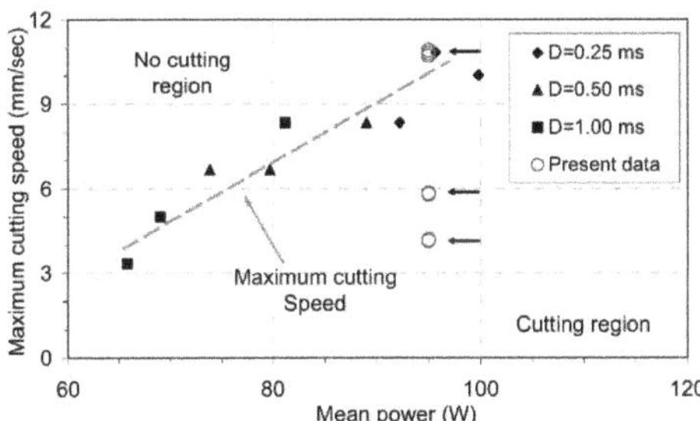

Fig. 4: Maximum cutting speed as a function of Mean power. The dashed line represents the maximum cutting speed obtained in [12].

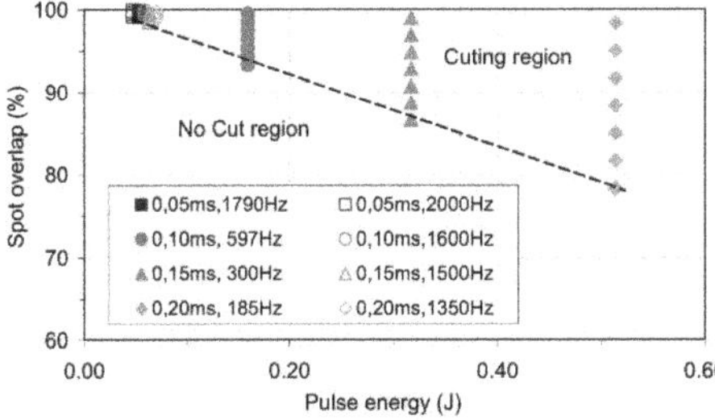

Fig. 5: Influence of the pulse energy and spot overlap on the cutting regions map. The dashed line represents the minimum spot overlap to cut.

3.2. Kerf Geometry

To comprehensively analyze the impact of the process parameters on the cut quality, the kerf profile was examined for all the testing conditions. The results are discussed in the following section as a function of the spot overlap percentage.

In figure 6, the mean values of the top and the bottom kerf width are plotted against the spot overlap. The vertical bars denote the standard deviations, while the red dashed line separates the top from the bottom kerf width. Looking at the figure, it is possible to observe that, in both cases, the data distribution and the presence of data scattering do not allow any particular observation about the influence of the different values of pulse energy and pulse duration. The only observation is that visible reductions of the bottom kerf width occur for the lowest pulse energy (the arrow in figure 6).

Furthermore, the top kerf width values fluctuate in the range of 160÷260 μm. The bottom ones, as regards the tests obtained at the highest pulse energy values, vary in the range of 68÷119 μm, while for the tests obtained at lower pulse energy they vary in the range of 35÷75 μm. However, a slight reduction of the width, compared to the data obtained in [12], was obtained when the spot overlapping overcomes the 85%.

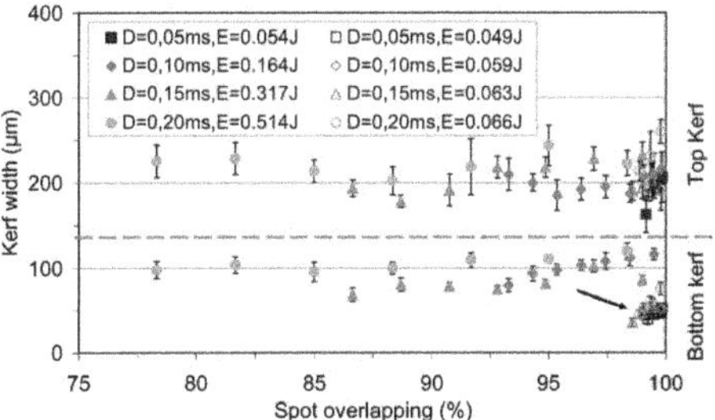

Fig. 6: Top and bottom kerf width as a function of spot overlap for the different process parameters.

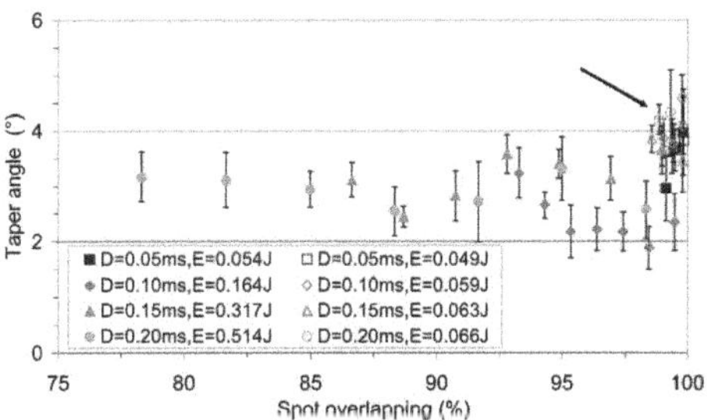

Fig. 7: Taper angle as a function of spot overlap for the different process parameters.

On the basis of the top and the bottom kerf width, the taper angle was calculated and then plotted against the spot overlap in figure 7. From the figure, it is possible to notice that the taper angle is very small, with values in the range of 1 ÷ 4.5 degrees. As expected, the maximum value is obtained in correspondence of low pulse energy (the arrow in the figure).

3.3 Defects and HAZ extension

From the optical microscopy analysis, the presence of typical composites thermal damages such as matrix recession, matrix thermal degradation, fibre bridging and fibre pull out were observed, as reported in figure 8. However, these defects are very limited and the common aspect is similar to the one reported in figure 9.

Despite the limited HAZ external extension, large damage could occur in the centre of the thickness, as visible in figure 3. By observing the section, the HAZ appears as matrix burning with or without the presence of inconsistent fibres and a width larger than the external visible damage.

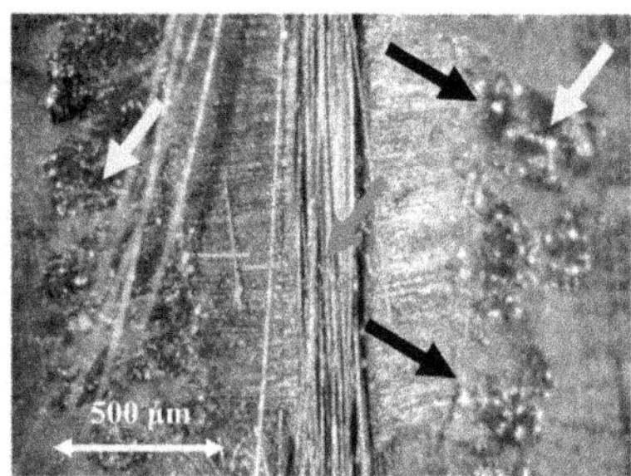

Fig.8: The top side of a "bad quality" cut, with evident matrix burning regression. Yellow arrows: matrix burning; Black arrows: matrix recession; Red arrow: fibre pull out.

Fig. 9: Top side of a "good quality" cut, with limited matrix regression and absence of fibre bridging.

In order to assess the HAZ behaviour as a function of the process parameters, the inner extension of the HAZ was measured and reported against the spot overlapping for the different process conditions, figure 10. From the figure, it is possible to observe that all the points converge to two different curves, the first corresponding to the low pulse energy conditions, the second corresponding to the high pulse energy conditions. In both cases, the curves could be represented by an exponential low (this is in agreement with what observed in [12]).

So, it is clear that the HAZ depends on the energy released and how this energy is released. In order to obtain a single relation that could describe the HAZ behaviour as a function of the process parameters, the data of figure 10 were plotted together with the data obtained in [12] as a function of the following parameters:

$$K = \left(\frac{Pm}{S}\right) \cdot (F \cdot D) \tag{3}$$

where, in the equation, the term (Pm/S) represents the energy released per length unit, while the term (F·D) represents the on time, that is the effective interaction time in 1 second.

From figure 11, it is possible to observe that, despite the different process conditions, included the assistant gas, all the data converge to a single logarithmic curve. Therefore, a unique formulation for the prevision of the HAZ was found.

Hence, since a low HAZ extension is the primary requirement, it is fundamental to work with low spot overlap or low K parameters.

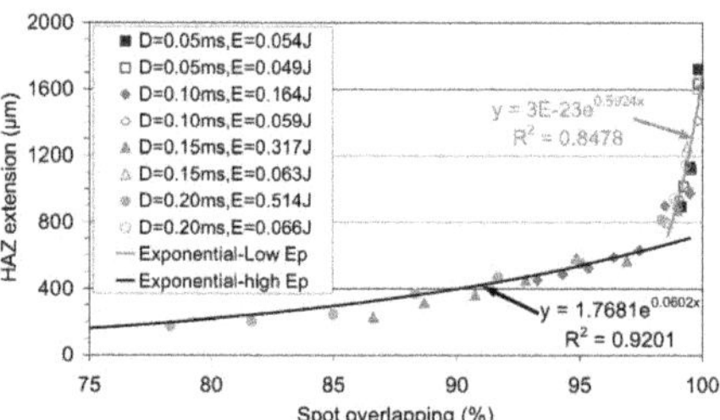

Fig. 10: HAZ extension as a function of spot overlap for the different process parameters.

Considering the previous observations and those made about the cutting regions maps and the kerf geometry, it could be concluded that, in order to obtain a high productivity and reduce the taper angle and the HAZ extension, high speed, high pulse energy and a pulse duration as lower as possible (the latter depends on the behaviour of the laser source characteristic curves) are needed. In this conditions, it is possible to obtain the maximum cutting speed and to minimise both the spot overlap and the k parameters.

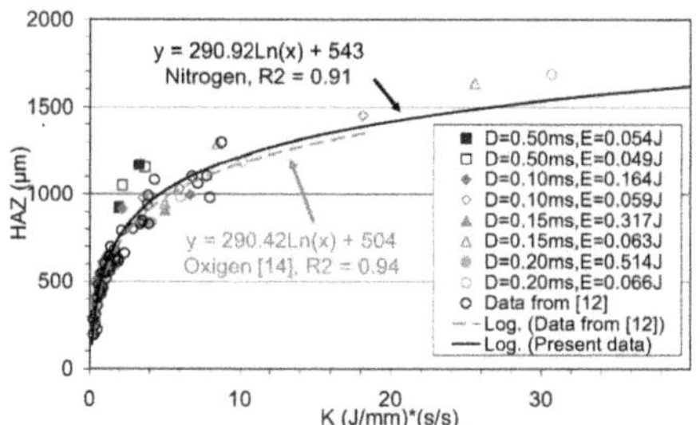

Fig. 11: HAZ extension as a function of the K=(Pm/S)*F*D parameters.

5. Conclusions

In this work, features and performance of a pulsed Nd:YAG laser, with a nominal power of 150 W, in cutting of a 1-mm thick CFRP laminate, are presented. On the basis of the results, the following main conclusions were drawn:

- A pulsed Nd:YAG can be satisfactorily used to cut 1-mm thick CFRP plates, with a cutting speed of about 11 mm/s and a mean power of about 95 W.
- Processing regimes for laser cutting can be established considering the pulse energy and the spot overlapping.
- The kerf width is almost insensitive to variations of the process parameters within the considered range.
- The external thermal damage is very limited. However, consistent damage could occur at the centre of the laminate in correspondence of the unidirectional lamina.
- The HAZ extension is related to the spot overlap through an exponential law. The coefficients of the exponential law depend on the pulse energy range.
- A unique formulation for the prevision of the HAZ was found. The relation binds the HAZ to the following process parameters: Pm, S, F and D.
- In order to obtain a high productivity and high quality cuts, high speed, high pulse energy and a short duration are needed.

Acknowledgements

The authors are particularly grateful to the CIRTIBS Research Centre, for providing the equipments and financial support to develop the present research work.

References

[1] Komanduri R., "Machining fibre reinforced composites", Mechanical Engineering, 27, 1993, pp. 58-64.

[2] Koenig W., Wulff C., Grass P. and Willerscheid H., "Machining of Fiber Reinforced Plastics", Annals of CIRP, 34/2, 1985, pp. 537-548.

[3] Caprino G; Tagliaferri V., "Damage development in drilling glass fibre reinforced plastics", Int. J. Mac. Tools & Manufacture, 35/6, 1995, pp. 817-829.

[4] Tagliaferri V., "Laser cutting of reinforced materials", in: N.P. Cheremisinoff (Ed.), Handbook of Ceramics and Composites, Marcel Dekker, New York, 1990, pp. 451-467.

[5] Tagliaferri V., Visconti Crivelli I., "Machining of fibre reinforced materials with laser beam: cut quality evaluation", Proc. of European Conference on Composite Materials, Elsevier, London, 1, 1987, pp. 190–199.

[6] Cenna A.A., Mathew P., "Analysis and prediction of laser cutting parameters of fibre reinforced plastics (FRP) composite materials", Int. J. of Mac. Tools & Manufacture, 42, 2002, pp. 105-113.

[7] Mathew P., Goswami G.L., Ramakrishnan N., Naik N.K., "Parametric studies on pulsed Nd:YAG laser cutting of carbon fibre reinforced plastic composites", J. of Materials Processing Technology, 89-90, 1999, pp. 198-203.

[8] Caprino G., Tagliaferri V., Covelli L., "Cutting glass fibre reinforced composites using CO_2 laser with multimodal-Gaussian distribution", J. Mac. Tools & Manufacture, 1995, 35/6, pp. 831-840.

[9] Di Ilio A., Tagliaferri V., Veniali F., Machining parameters and cut quality in laser cutting of aramid fibre reinforced plastics, Mat. Manuf. Process. 5/4 1990, pp. 591–608.

[10] Fenoughty K.A., Jawaid A., Pashby I.R., Machining of advanced engineering materials using traditional and laser techniques, J. Mat. Processing Technology, 42, 1994, pp. 391–400.

[11] Lau W.S., Lee W.B., Pulsed Nd:YAG laser cutting of carbon fibre composite materials, Ann. CIRP 39/1, 1990, pp. 179–182.

[12] C. Leone, A. Pagano, V. Lopresto, I. De Iorio, "Solid state Nd:YAG laser cutting of CFRP sheet: influence of process parameters on kerf geometry and HAZ", Proc of 17th Int. Conf. on Composite Materials - ICCM-17, 27 Jul 2009 - 31 Jul 2009, Edinburgh UK, 2009, pp. 1-10, (Complete paper on CD).

Innovative Production Machines and Systems
D.T. Pham, E.E. Eldukhri and A.J. Soroka (eds)

Laser cutting of silicon wafer by pulsed Nd:YAG source

C. Leone[a], V. Lopresto[a], N. Pagano[a], S. Genna[b] I. De Iorio[b]

[a]*Department of Materials and Production Engineering, University of Naples "Federico II", Naples, Italy*
[b]*Department of Aerospace Engineering, University of Naples "Federico II", Naples, Italy*

Abstract

Nowadays, silicon wafer is employed as substrate material for the fabrication of micro-electro and micro-mechanical components. Since silicon is a nominally brittle material, cutting of complex shapes remains a critical topic to study. Laser cutting represents a promising alternative. It does not involve any mechanical cutting force and tool wear. Moreover, thanks to the small spot diameter, laser cutting allows to realize very small and complex shapes too. However, during laser cutting, different kinds of defects can be generated depending on the beam-material interaction phenomena (ablation, melting and brittle cracking).

The aim of this work is to investigate the influence of the process parameters on the kerf geometry and defects in cutting operation of a 0.75 mm thick silicon wafer using a pulsed Nd:YAG, working in the regime of short pulses. Cutting tests were carried out varying mean power, pulse duration, pulse energy and cutting speed.

Experimental results showed that a pulsed Nd:YAG laser, with a nominal power of 150W, can be successfully used to cut 0.75 mm silicon wafer plates obtaining a kerf of about 200 μm at a maximum cutting speed of 15 mm/s (0.9 m/min). However, an accurate selection of the process parameters is necessary in order to obtain a regular kerf geometry and to avoid defects formation.

Keywords: Silicon wafer, laser cutting, kerf geometry, defects.

1. Introduction

Nowadays, silicon wafer is the dominant substrate material for the fabrication of micro-electro and micro-mechanical components. Since silicon is a nominally brittle material, cutting of complex shapes remains a critical topic to study.

Machining of hard materials, like ceramic or silicon, is very difficult with conventional processes. In most cases, the final structure of a part cannot be produced directly. The part has to be additionally processed to ensure the final shape. This holds true especially for small features, such as holes and grooves.

There are only few manufacturing methods available for the precision processing of silicon, such as grinding, lapping, electro discharge machining (EDM), chemo-mechanical polishing (CMP), but they are not suitable for very small and complex structures.

Laser cutting offers new possibilities by selective processing of all kinds of technical ceramics and silicon [1-3]. It does not involve any mechanical cutting force and tool wear. Moreover, thanks to the small spot, laser beam allows to realize very small and complex shapes too.

Laser machining on silicon material is currently focused on nanomachining in the micrometer range. In this field, in order to obtain high accuracies and small geometries, shorter wavelength, e.g., from frequency-tripled Q-Switch Nd:YAG lasers and Excimer lasers characterised by short or ultra short pulse duration (picoseconds and femtoseconds), have to be used [4-9].

However, when machined material volumes become significant, like in microcutting, the low removal rate results in long working time.

On the other hand, the growing interest in this material shown within different high level technical fields is driving towards the development and optimisation of micro and macro laser operations like cutting and drilling on silicon materials.

Nd:YAG laser at fundamental wavelength and at frequency-double wavelength can be an attractive alternative, when the micrometer range is not required [10, 1, 2]. Thanks to the possibility of having higher pulse powers due to the shorter pulse lengths (in the range of ms or ns) offered by the recent sources, operations like microcutting, microdrilling and micromachining have become possible.

For this source, the process parameters will be a compromise solution between cut quality and productivity requirements [11]. In view of that, it is important to choose optimized process parameters like pulse repetition frequency, scanning speed and beam power, according to the removal mechanism, the productivity and the required quality.

The aim of this work is to investigate the influence of the process parameters on the kerf geometry and defects in cutting operation of a 0.75 mm thick silicon wafer using a lamp pumped Nd:YAG, working in the regime of short pulses.

Cutting tests were carried out varying the process parameters. The kerfs sections of all the samples were examined through microscopic analysis by measuring the top and bottom kerf width, the taper angle and the defects.

2. Equipment, materials and experimental procedures

In the experimental tests, an industrial multimode lamp pulsed Nd:YAG (Starcut 150 from Rofin), working at the wavelength of 1064 nm, was used. This laser is characterised by a nominal power of 150 W (safety value declared by the producer), a pulse duration variable in the range of 0.03÷2.5 ms, a pulse frequency up to 3 kHz, and a multimode energy distribution. The laser is controlled by a microprocessor that allows to set the pump lamp voltage (V), pulse frequency (F) and pulse duration (D).

The released mean power (Pm), the pulse energy (E) and the pulse power (Pp) depend on all the three parameters V, F and D. The relation between the selectable parameters (V, F and D) and the laser beam energy characteristics (Pm, E and Pp) is very complex. On the other hand, since the interaction mode is related to the energy release-time dependence, a strategy to decrease the kerf width and the defects could involve the use of a very short interaction time.

For this reason, it was decided to fix the duration at the values of 0.25, 0.5 and 1 ms, and to vary the pulse frequency and the lamp voltage in order to have three values of the pulse energy (about 0.35, 0.50 and 0.65 J). Then, the laser source was characterised and the process parameters were found. The adopted process conditions are reported in table 1.

Table 1
Process conditions adopted for the tests.

D (ms)	E (J)	Pp (kW)	F (Hz)	V (V)	Pm (W)
0.25	0.37	1.48	250	650	92
	0.50	2.00	200	650	100
	0.64	2.55	150	700	96
0.50	0.37	0.74	200	500	74
	0.53	1.06	150	500	80
	0.61	1.22	150	600	92
1.00	0.34	0.34	200	400	66
	0.54	0.54	150	400	81
	0.72	0.72	100	420	72

The laser source was mounted on a CNC machine (StarCut System Universal from Rofin) which allows the generation of the geometric patterns and the selection of the beam speed.

All the cuts were performed using a cutting head (PRECITEC AK YK52) with a focal length of 80 mm and a nozzle of 0.8 mm in diameter. In this configuration the focused beam diameter, indicated by the manufacturer, is about 170 μm. During the tests, air at 10 bar of pressure and 30 l/min of flow was used as assist gas.

For each process condition, the maximum cutting speed was determined. Then, starting from this value, the cutting speed was decreased until a spot overlap value of about 98 % was achieved.

The investigated material was a Ø 200 mm silicon wafer disk, 0.75 mm in thickness, made of two external layers of oriented polycrystalline silicon and a core of non-oriented polycrystalline silicon.

In order to verify the cut quality, the top kerf width (Wt), the bottom kerf width (Wb) and the taper angle (Ta) were measured by microscopic analysis. Furthermore, the inner side of the kerf was observed by SEM microscopy. In figure 1, the section of a kerf together with the analysed parameters are reported.

At least fifteen cuts and profiles were analyzed for each test condition. On each cut, three measurements of the kerf geometry parameters were carried out.

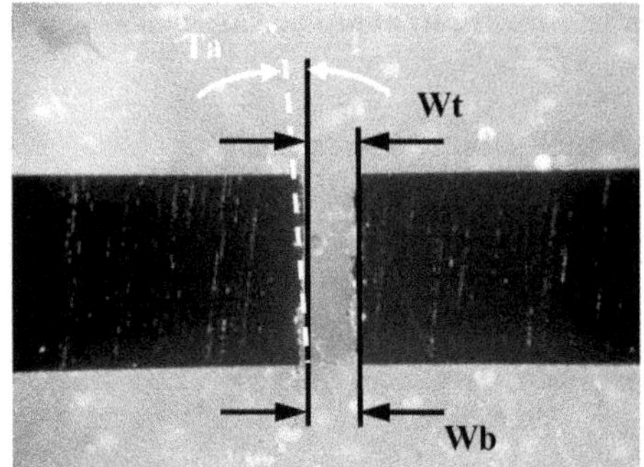

Fig. 1: Examples of kerf geometry. Wt: the top kerf width; Wb: the bottom kerf width; Ta: taper angle.

3. Results and discussion

3.1 Cutting regions maps

In figure 2, a map of the cutting region is reported. It delineates the "cutting region" and the "no-cutting region" for each adopted cutting parameter as a function of cutting speed and beam mean power.

The figure reveals that silicon wafer cutting is possible over a large processing window covering a wide beam mean power-speed range. The only case where no cutting was obtained is for D=0.6 ms, E=0.33, Pm =66W. The maximum cutting speed (red dashed line) depends on the mean power according to an exponential law. The maximum cutting speed is 15 mm/s, obtained with a power of about 100 W.

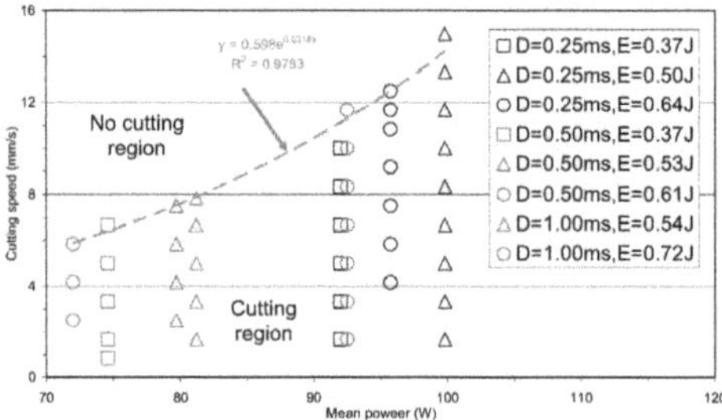

Fig. 2: Cutting regions maps in terms of cutting speed-mean power. The dashed line represents the maximum cutting speed-mean power dependence.

3.2 Kerf geometry

In table 2, the mean values, the maximum, the minimum and the standard deviation of all the measured top and bottom kerf width values as well as of the taper angle values are reported. From the table, it is possible to notice that the top kerf width varies within a very limited range, while the bottom kerf width and the taper angle show a larger variation. So, for the latter, an influence of the process parameters is expected.

Table 2
Mean values, maximum, minimum and standard deviation of the kerf geometry parameters.

	Mean	Max	Min	St. Dev.
Wt (mm)	0.23	0.26	0.20	0.01
Wb (mm)	0.23	0.32	0.14	0.03
Ta (deg.)	0.02	2.87	-3.73	1.06

The cutting process with a pulsed Nd:YAG laser involves the overlapping of a series of individual spots generated by each pulse, as reported in figure 3.

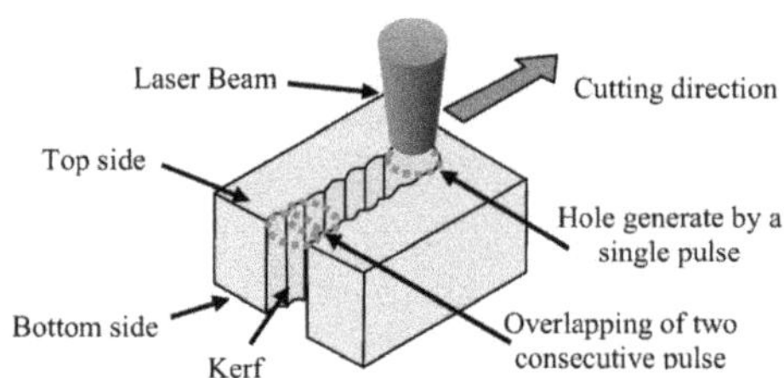

Fig. 3: Scheme of pulsed laser cutting mechanism.

A key parameter in pulsed laser cutting is the spot overlap percentage that represents the overlap between two consecutive pulses. The spot overlap percentage can be calculated by the equation (1):

$$\text{Spot overlap}\,(\%) = R = 100 \cdot \left(1 - \frac{S}{F \cdot W}\right) \qquad (1)$$

where S is the cutting speed, F is the pulse repetition frequency, W is the kerf width and R the spot overlap.

Thus, in the next paragraph, the kerf geometry will be reported as a function of the spot overlap percentage.

In figures 4 and 5, the mean values of fifteen measurements of the top and the bottom kerf width are plotted against the spot overlap. The vertical bars represent the standard deviations. As expected, the top kerf width (figure 4) shows a mean value that is almost constant with limited data scattering. Therefore, it is possible to say that the top kerf width is insensitive to any process parameters variation. On the contrary, as regards the bottom kerf width, figure 5, data scattering appears higher compared to the one observed for the top kerf width. Furthermore, if the data obtained at highest pulse energy are not considered (closed dots), it is possible to say that the bottom kerf width tends to increase with the increase of the spot overlap (as indicate by the arrows in the figure). Furthermore the increase is more evident as the highest is the pulse duration.

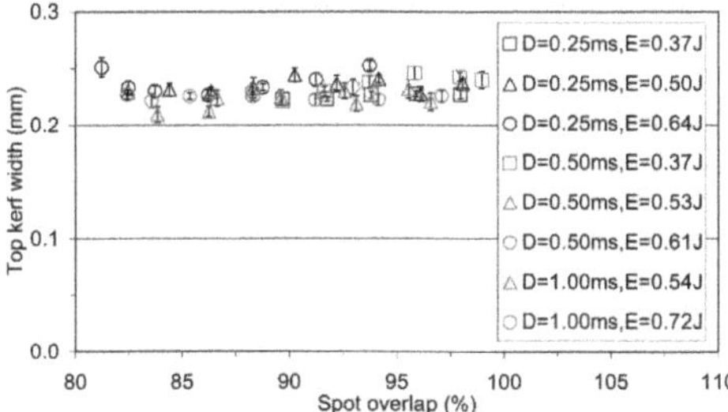

Fig. 4: Top kerf as a function of the spot overlap.

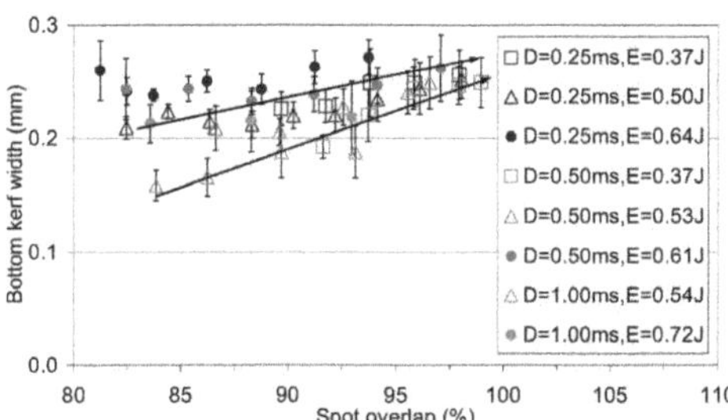

Fig. 5: Bottom kerf width as a function of the spot overlap.

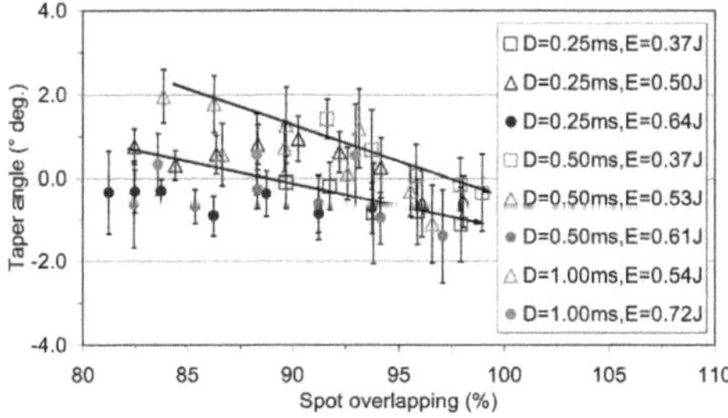

Fig. 6: Taper angle as a function of the spot overlap.

On the basis of the top and the bottom kerf width, the taper angle was obtained and plotted against the spot overlap in figure 6. As expected, also in this case a dependence on the spot overlapping could be observed for the data obtained at lower pulse energy. In fact, the taper angle tends to decrease with the increase of the spot overlap (as indicate by the arrows in the figure).

The above mentioned behaviour can be explained by considering that the cutting process is obtained prevalently by fusion cutting. In this process, the kerf volume is transformed mainly into the molten state and blown out of the kerf by the high pressure of the process gas jet [12]. During the process, part of the melted material re-solidifies on the kerf inner wall, forming a regular series of ridges or striations similar to those shown in figure 7.

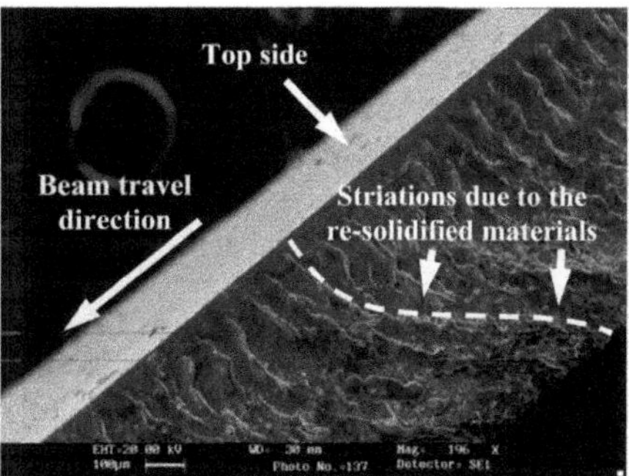

Fig. 7: Examples of the kerf wall aspect. (F=150Hz; E=0,614J; D=0,5ms; S=1.67 mm/s).

The distribution of striations is highly dependent on the spot overlap. At low spot overlap (high cutting speed), the melted material tends to stop on the top of the kerf wall, with a visible recession from the laser beam entry zone. By increasing the spot overlap (or decreasing the cutting speed), the material tends to settle on the bottom wall or to be ejected out of the kerf and a reduction of the recession is visible, figure 8. The extra energy released onto the material is used to enlarge the kerf width at the bottom. So, a more regular surface is obtained. However, in these conditions the presence of drop and bur on the bottom surface was observed. Hence, the effect of the spot overlap or the cutting speed results in a change of the bottom kerf width, figure 9. It is useful to underline that this behaviour is the same as that observable in metal cutting, as reported in [12].

The different behaviour observed for the data obtained at high pulse energy is probably due to the fact that, in this condition, the highest pressure obtained in the plasma phase helps the melted material expulsion. So, a smaller amount of re-solidified material is on the kerf wall. Therefore, the kerf geometry is insensitive to the effect of the speed or the spot overlap.

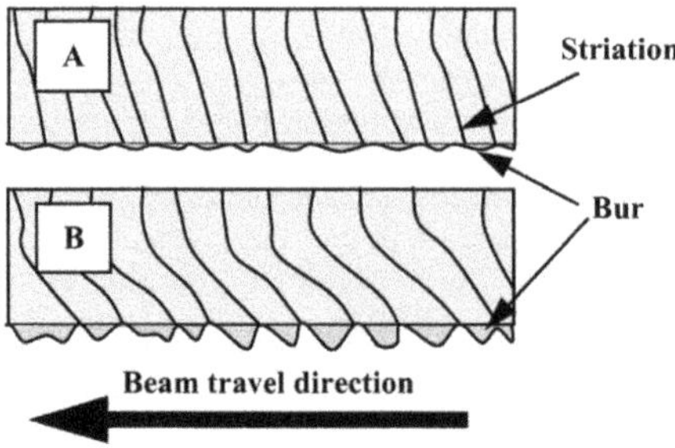

Fig. 8: Effect of the spot overlap (or the cutting speed) on striation geometry. A: high spot overlap or low speed; B) low spot overlap or high speed.

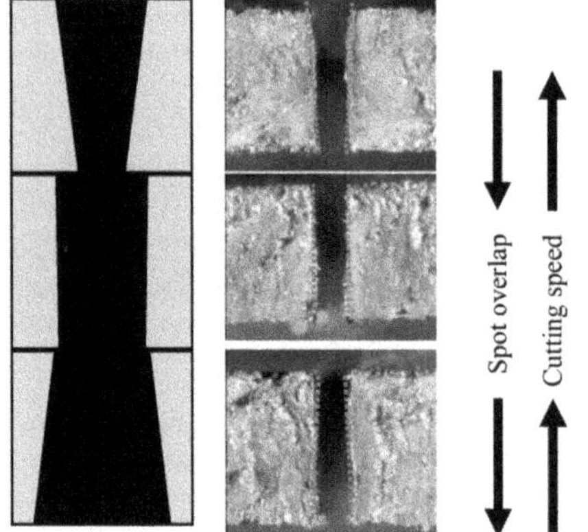

Fig. 9: Effect of the spot overlap or of the speed on the kerf geometry.

3.2 Defects and aspect of the kerf

The SEM analysis showed the presence of different defects: re-solidified material on the kerf wall (figure 7), droplets and burs at the bottom surface (figure 10), micro-cracks and porosity on the re-solidified material (figure 11). The latter defects are very limited and localized only in the re-solidified material, while no cracks were found on the un-worked material.

The contemporary use of high pulse energy and low duration, that means high pulse power values, entails the presence of additional striations close to the upper surface, as visible in figure 12.

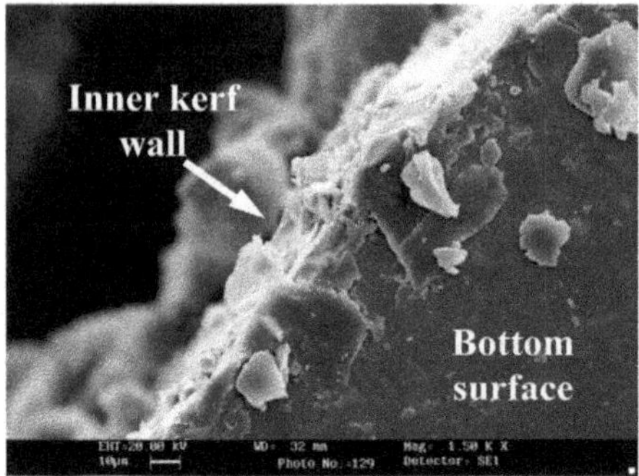

Fig. 10: Bur formation at the bottom surface.

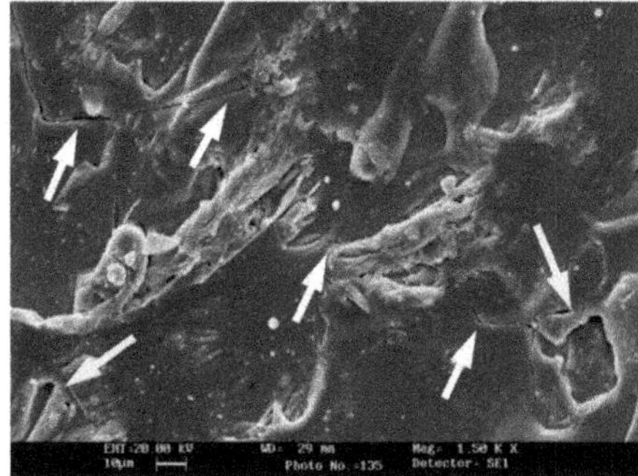

Fig. 11: Presence of micro-crack (white arrows) and porosity (yellow arrows) on the re-solidified material.

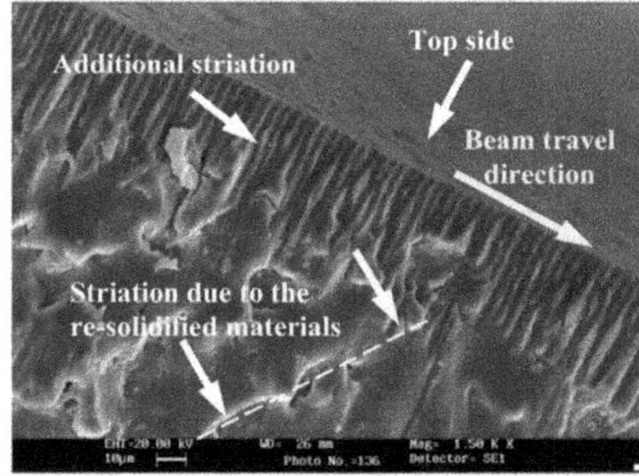

Fig. 12: Aspect of a cut obtained at: Pm=92 W, F=150 Hz; E=0.614 J; D=0.5 ms, Pp=1.22 kW, S=1.67 mm/s.

The latter phenomenon is due to the fact that, as the pulse power increases, a transition from the prevalent melting mechanism to the ablation-melting one occurs. This is clearly visible by comparing figures

12 and 13, where kerf walls obtained at the same cutting speed, similar pulse energy but different pulse power (lower in figure 13) are reported.

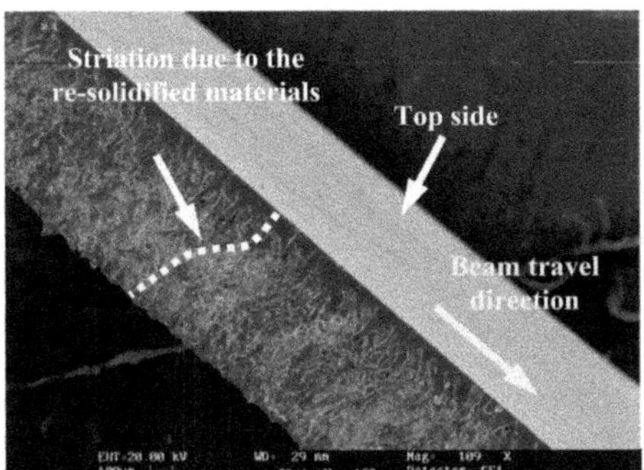

Fig. 13: Aspect of a cut obtained at: Pm=81 W, F=150 Hz; E=0.54 J; D=1.0 ms, Pp=0.54 kW, S=1.67 mm/s.

5. Conclusions

In this work, features and performances of a 100 W lamp pumped pulsed Nd:YAG laser in cutting of a 0.75 mm thick polycrystalline silicon wafer is reported. From the results, the following main conclusions were drawn:

- A pulsed Nd:YAG can be satisfactorily used to cut 1 mm thick CFRP plates, at cutting speed of about 15 mm/s using about 100 W.
- Processing regimes for laser cutting can be established on the basis of mean power and cutting speed, considering that the maximum cutting speed depends on the released mean power according to an exponential law.
- In the range of the adopted process parameters, the top kerf width is about insensitive to any variation of them, while the bottom kerf and the taper angle are sensitive to the spot overlap or the cutting speed.
- The observed defects consist of re-solidified material, droplets and burs at the bottom surface, micro-cracks and porosity on the re-solidified material.

Acknowledgements

The authors are particularly grateful to the CIRTIBS Research Centre, for providing the equipments and financial support.

References

[1] Meijer, J., Du, K., Gillner, A., Hoffmann, D., Kovalenko, V.S.; Masuzawa, T.; Ostendorf, A., "Laser Machining by short and ultrashort pulses, state of the art and new opportunities in the age of the photons", CIRP Annals - Manufacturing Technology, 51/2, 2002, pp.531-550.

[2] Meijer J., "Laser beam machining (LBM), state of the art and new opportunities", J. of Mat. Proc. Tech., 149, Issue: 1-3, June 10, 2004, pp. 2-17.

[3] Gillner A., Holtkamp J., Hartmann C., Olowinsky A., Gedicke J., Klages K., Bosse L., Bayer A., Laser applications in microtechnology, J. Mat. Proc. Tech., 167, 2005, pp. 494–498.

[4] Bärsch N., Körber K., Ostendorf A., Tönshoff K.H., Ablation and cutting of planar silicon devices using femtosecond laser pulses, J. Appl. Physics A 77, 2003, pp. 237–242.

[5] Chen T. C., Darling R. B., Parametric studies on pulsed near ultraviolet frequency tripled Nd:YAG laser micromachining of sapphire and silicon, J. Mat. Proc. Tech., 169, 2005, pp. 214–218.

[6] Matsumura T., Nakatani T., Yagi T., Deep drilling on a silicon plate with a femtosecond laser: experiment and model analysis, Applied Physics A, 86/1, 2007, pp. 107-114

[7] Ninomiya T., Sawada H., Kawahara K., Yokotani A., Kurosawa K., Dicing Process for Thin Silicon Wafer by Using Femtosecond-laser-Effect of Laser Pulse Width on Flexural Strength and Cutting Quality, J. of the Jap. Society for Precision Engineering, 70/12, 2004, pp.1554-1558.

[8] Hopman S., Fell A., Mayer K., Mesec M., Rodofili A., Kray D., Comparison of laser chemical processing and LaserMicroJet for structuring and cutting silicon substrates, Appl. Physics A, 95/3, 2009, pp. 857-866.

[9] Tangwarodomnukun V., Wang J., Laser Micromachining of silicon Substrate, Advanced Materials Research, 76-78, 2009, pp 416-421.

[10] Dobrzanski L.A., Drygala A., Laser processing of multicrystalline silicon for texturization of solar cells, J. Mat. Proc. Tech., 191/1-3, 2007, pp. 228-231.

[11] Leone C., Morace R.E, De Iorio I. (2007). AISI 304 Stainless Steel Engraving by QSwitched Nd:YAG Nanosecond Laser. J. of the Chinese Society of Mechanical Engineers, Vol. 28/2, 2007, pp. 217-224.

[12] Ready J.F. et al, LIA Handbook of Laser Materials Processing, Springer-Verlag, 2001, pp. 425-441. ISBN: 3-540-41770-2, ISBN 13: 978-3-540-41770-5.

Innovative Production Machines and Systems
D.T. Pham, E.E. Eldukhri and A.J. Soroka (eds)

Mechanical characterization of Nd:YAG laser welding on Ti-6Al-4V

G. Serroni [a], C. Bitondo [a], U.Prisco [a], A. Squillace [a], A.Favi [b] and A.Prisco [c]

[a] *Department of Material Engineering and Production, University of Naples "Federico II", Naples, Italy*
[b] *RTM S.p.A., Pomigliano D'Arco (NA)*
[c] *Alenia Aeronautica., Pomigliano D'Arco (NA)*

Abstract

Since 10 years it has been trying to develop an innovative welding technique for titanium. The difficulties encountered are mainly related to the need of developing a joint free from defects (e.g. undercut and porosity). The attention of the researchers passed from the traditional TIG welding to Laser Beam Welding [1]. The laser technology, while achieving a better coupling, requires a great deal of energy, especially the high-power Nd:YAG lasers [2]. Adjusting the parameters of the laser, welds almost free of defects can be obtained, so it would be necessary to plan the experiment using a DoE [3]. This is not possible for laser welding because this type of technology has limitations that do not allow for proper planning. The aim of this study is to provide a welding process that could produce high quality joints. Static tensile tests showed a high resistance joints, fatigue strength is strongly influenced by the presence of undercut, so it is important to work in conditions where such defect is minimized.

Keywords: Laser Welding, titanium, stress test, fatigue test.

1. Introduction

Development of aircraft product is marked by leaps in technology that have allowed its greater use.

The introduction of the technology of composites in all primary structures of aircraft has been the last, in order of time. The criteria for the design, manufacture and operation, were changed to obtain structures lighter and more efficient than the ones made of aluminum. However, the structures in CFRP, although they show greater efficiency for loads in the airplane, require the use of metal structures, especially in areas of great concentration of loads. For highest electrochemical compatibility with the CFRP, titanium structures are preferred, as they are more resistant to corrosion. On the other hand, this implies excessive manufacturing costs related to the cost of raw materials, the high volumes of waste and the complex and expensive finishing. The innovative solutions are designed to manufacture semi-finished products not obtained through solidification of molten and subsequent plastic deformation, but through welding simpler parts. This allows, with the same performance of the final product, to obtain more complex near net-shapes, with a reduction of scrap (up to a ratio of 30:1) and of energy processing.

The considerable advantages, in terms of productivity, and the high quality end-product after the application of lasers in the machining field, encourage the growing use of new technology in the most advanced processing. For these reasons, it was necessary, since its initial application, a detailed study of laser-material interaction, to optimize the welding parameters as a function of the heat treatment of the material to be welded and to get working procedures, able to produce joints responding to the design requirements.

This work was focused on the study of Nd:YAG laser welding of Ti-6Al-4V plates. In particular, we tried to optimize the welding parameters to assure a good penetration [4-5] and the absence of defects and of oxidation. It is provided to check the effectiveness of shielding gas, as has been shown [6] that it can stabilize the welding process. Adjusting the heat input is important to define the final microstructure of the joint and, as a consequence, its mechanical properties [7]. FEM simulations have been carried out recently able to adequately describe the laser welding process [8].

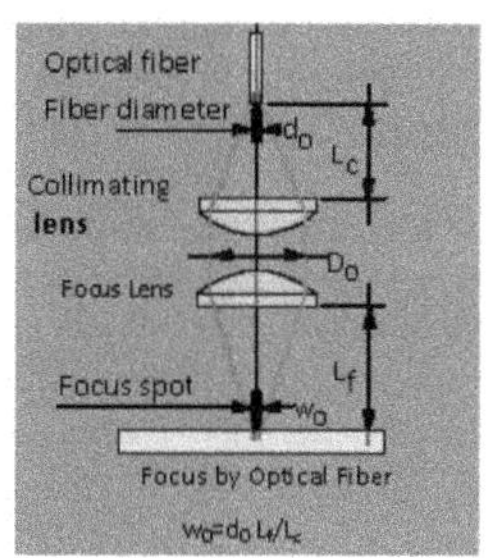

Fig. 1 – focus – collimating lens

Fig. 2 – The color, due to oxidation, is an indication of the quality of the weld

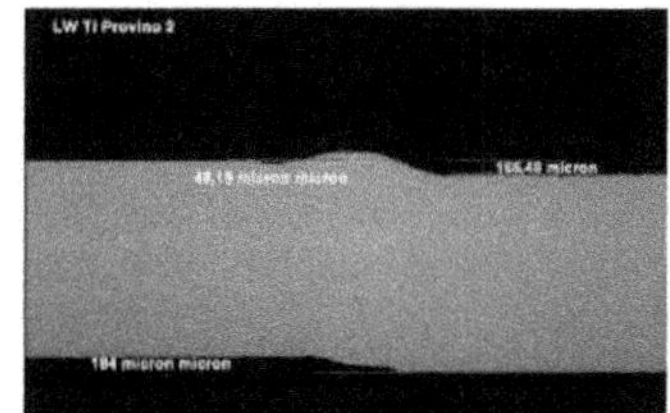
Fig. 3 – Misalignment of weld sheets.

2. Materials and methods

The welding parameters (speed, laser power of phasic, inert gas and the reverse) were selected on the basis of form and dimension of the cord cross-section according to UNI EN ISO 13919-1:1997. To get an overview on the weldability of the alloy, preliminary tests were carried out on sheets of Titanium Grade 5 (Ti-6Al-4V) mm 100x100x2.5. We proceeded, therefore, on welding of titanium with rolled-sized 503x105x2.5 mm in butt-joint configuration.

For the analysis of factors influencing the process, steps were taken to carry out non-destructive testing, such as visual analysis and destructive tests such as tensile test, bending test and fatigue test, and finally microhardness. The welding system consists of a robot, with a maximum power of the laser beam of 2 kW, with a welding head equipped with optic fibers (∅ 0.6 mm). An important trick used was to provide the focusing head of a shoe in which it was applied the supply ducts of the shielding gas. This shoe was fitted into the lower edges of a board made of fire resistant fabric to minimize the loss of shielding gas. The diameter of the fiber has a significant impact on spot size. Reducing the size of the spot is related to the relationship between the focal lengths of re-collimating and focusing lenses, as shown in Figure 1. The instrumentation dimensions are: d_0=0.6mm, L_f=80mm e L_c= 120mm so focus spot was set to 0.4mm.

The system was equipped with two shielding gas: Helium, which because of its high ionization potential, has been used in coaxial and laminar direction so as to create a cloud of protection, and Argon, with low-threshold ionization, which is used in the lower area of the specimen as protective gas in reverse.

In the first phase of the research, some penetration tests have been carried out to define what were the changes to be applied to the process parameters. An index of the goodness of penetration is the colour of the welds. (Fig. 2) The color scale is associated with the following general behaviors:

Table 1
Process Parameter

Spec.	Laser Power [W]	rate. [mm/s]	Coax Gas	Assis . Gas	Rate He [Nl/h]	Prot. Gas
A 1	1700	28	He	none	250	Ar
A 2	1700	28	He	none	250	Ar
A 3	1700	28	He	none	250	Ar
B 1	1700	28	He	none	250	Ar
B 2	1700	28	He	none	250	Ar
B 3	1700	28	He	none	250	Ar
C 1	1500	22	He	He	250	Ar
C 2	1200	18	He	He	250	Ar
C 3	1200	16	He	He	250	Ar
D 1	1500	22	He	He	250	Ar
D 2	1200	18	He	He	250	Ar
D 3	1200	10	He	He	250	Ar
E 1	1200	13	He	He	250	Ar
E 2	1200	13	He	He	750	Ar
E 3	1200	13	He	He	750	Ar
F 1	1500	24	He	He	500	Ar
F 2	1500	20	He	He	500	Ar
F 3	1500	16	He	He	500	Ar
G 1	1500	32	He	He	500	Ar
G 2	1500	28	He	He	500	Ar
G 3	1700	20	He	He	500	Ar
H 1	1700	32	He	He	500	Ar
H 2	1700	28	He	He	500	Ar
H 3	1700	24	He	He	500	Ar
I 1	1700	36	He	He	500	Ar
I 2	1200	16	He	He	500	Ar
I 3	1200	12	He	He	500	Ar
L 1	1200	28	He	He	500	Ar
L 2	1200	24	He	He	500	Ar
L 3	1200	20	He	He	500	Ar

1) shiny silver - indicates a good coverage and a weld is not infected, 2) pale yellow acceptable

welding with reduced ductility, and 3) blue, purple - low ductility, unsuitable for high performance applications , 4) gray - not acceptable.

An overview of the tests is shown in Table 1. 10 samples were used (named with the letters AL) and on each of these three penetration tests were performed (indicated by numbers 1-3). Some tests were carried out with the same parameters to test the repeatability of the process.

Table 2
Geometric Measurements

Spec.	Under-cut [μm]	Exc. Mat. [μm]	Exc. Pen. [μm]	MHV ZTA L [HV]	MHV Bulk [HV]	MHV ZTA R [HV]
I3a	34	88	189	371	324	356
I3b	35	91	222	356	324	343
I2a	26	150	29	371	324	350
I2b	28	186	39	371	324	356
L3a	29	146	29	386	324	365
L3b	29	172	26	386	324	356
L2b	29	117	9	371	324	386
L2a	26	107	19	380	324	350
L1a	23	123	13	371	324	350
L1b	22	134	4	371	324	386
F3a	30	199	62	371	324	350
F3b	30	209	65	371	324	350
F2a	45	186	98	380	324	350
F2b	50	134	166	386	324	350
F1a	39	153	196	371	324	350
F1b	37	150	173	386	324	365
G2a	16	170	27	371	324	350
G2b	25	183	23	380	324	350
G1a	24	173	26	386	324	350
G1b	22	166	26	371	324	350
G3a	44	205	67	386	324	356
G3b	46	245	59	371	324	350
H3a	35	178	163	386	330	343
H3b	37	154	147	396	330	343
H2a	65	208	156	371	330	336
H2b	71	219	163	386	330	343
H1a	34	157	225	386	330	336
H1b	29	183	228	380	330	336
I1a	20	189	55	396	324	356
I1b	16	225	22	371	324	386

After all the penetration tests and optimized the process parameters, were carried out butt welds. For the tests of visual and stereoscopic analysis was used a metallographic microscope Leica DM4000DM.

Two samples of each specimen were encompassed and observed after Kroll attack (designated by the letters a and b in Table 2).

For the microhardness test was used Microdurometer Leitz, equipped with an optical microscope, connected via interface with a digital meter. The load used was 500p for a time of 15 seconds imprinting

The tensile test was carried out by subjecting the material, shape and size appropriate, to a tensile load acting in the direction normal to the straight section of the specimen, ASTM E8-EM8.

The bending tests were conducted in accordance with ASTM E-190/92, choosing a test roll with a diameter of 10mm.

Finally, the cyclic fatigue tests were carried out using ten samples in accordance with ASTM E466.

3. Results and discussions.

In Table 3 are compared the different sections of some representative samples, which have suffered the attack Kroll to highlight the heat-affected zone. It is apparent that the welds made with higher power and lower speeds show a deeper and wider molten zone and HAZ. Table 2 shows the measurements taken on specimens for which was set a level of He flow of 500 Nl/h. All measurements concerning the cross-sectional dimension, such as undercut, excess material in the upper and excessive penetration, falling within the range of values imposed by the UNI EN ISO 13919 1:1997: Whereas the grading more stringent, in fact you can see the values of 0.125 mm to 0.575 mm for undercut and excess material and the excessive penetration.

This reflection allows the direct selection of parameters for subsequent butt welds excluding the samples that do not pass through welds, those with excessive penetration and a welding area too wide. The information obtained from this visual analysis, addressed the choice of the parameters for maximum power, choosing to make the welding parameters of the penetration test H3 (Power 1700 W; Speedr = 24 mm / s) referred to as the weld "BJ H3".

Before proceeding with subsequent tests, was observed under optical microscope incorporated a section of a sample, in order to verify the mismatch between the laminates. The result is a value of

Table 3
Cross Section of weld beads

Rate mm/s	Power [KW]		
	1200	1500	1700
20	L3	F2	G3
24	L2	F1	H3
28	L1	G2	H1

0.184mm, less than the limit imposed by the rule (0.250mm) (Fig.3). Visual inspection of the welding has shown no rust, with a thick cord similar to that of the base material and good quality, although they were found small pieces of molten material, welded, both on the rope to the base material.

Tensile tests have given satisfactory results. All samples under investigation showed a break away from the weld. The appearance of the curves is completely overlapped for four samples (Fig. 5) that had a scratch on the surface that triggered the crack of breaking the last sample has confirmed the trend, however, expressed by the previous (Fig. 5). The average values obtained were: Ultimate Tensile Stress=1035 N/mm^2, Yield Stress = 960 N/mm^2, Maximum Force = 65KN, E=114 KN/mm^2.

The specimens tested in bending were positioned so as to have a top of the weld in tension, resulting in a bending angle of 30 degrees with a fracture in the weld, and the other in order to have the bottom of the weld in tension, resulting in a bending angle of 92°, and most importantly, the initiation of fracture within the bulk material, not inside the cordon.

In this regard it should be noted that the specification SAE AMS 4911L provides, for the specimens prepared with only the base metal, so no welding, a bending angle of 105 degrees (Fig. 8).

The development of cyclic fatigue tests, carried out for load levels ranging between 60% and 90% UTS, is explained in the diagram of Fig. 9: The breaking of all the samples, occurred between the cord and base material, as can be seen from Fig. 10. A more careful analysis of a fracture line denotes that the onset of the crack was caused by the presence of the undercut.

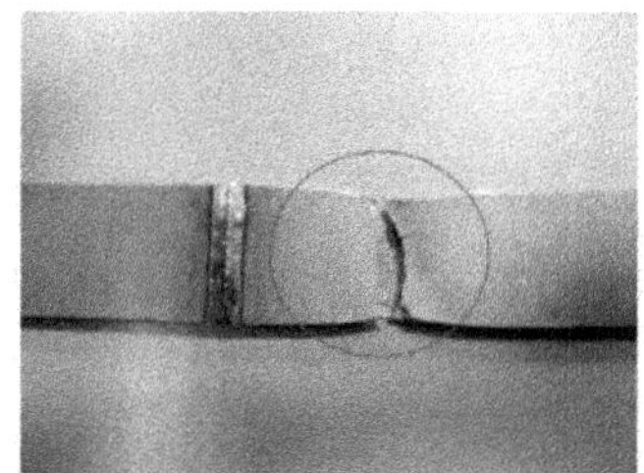
Fig. 4 – Fracture on specimen after Tensile Stress test.

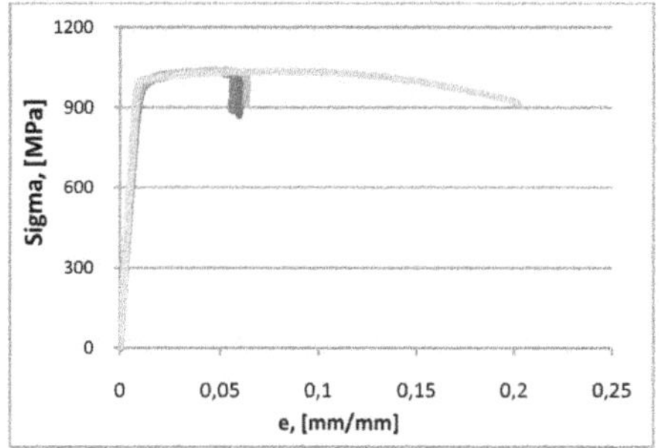

Fig. 5 – Stress Test

Table 4
Hardness Vickers Report

No	x	y	Hardness HV
1	-3	0	324,50
2	-2	0	329,90
3	-1	0	386,40
4	0	0	379,80
5	0	0,5	388,60
6	0	-0,5	382,00
7	1	0	373,50
8	2	0	332,80
9	3	0	324,50
10	0,8	0	369,30
11	-0,8	0	371,40

Table 4 shows the values of microhardness Vickers and the coordinates of the points of measurement (where the source was considered to be the center of the weld,), referring to the sample is not attacked. The microhardness of the bulk material is about 330 HV, and this value increases approaching the HAZ, to reach the maximum of about 390 HV

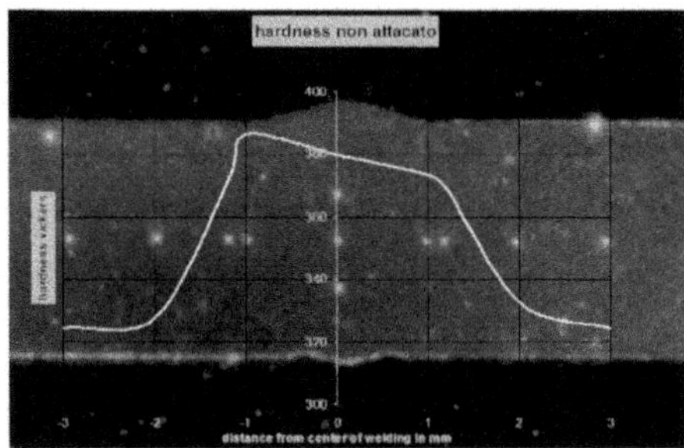
Fig. 6 - Hardness Vickers

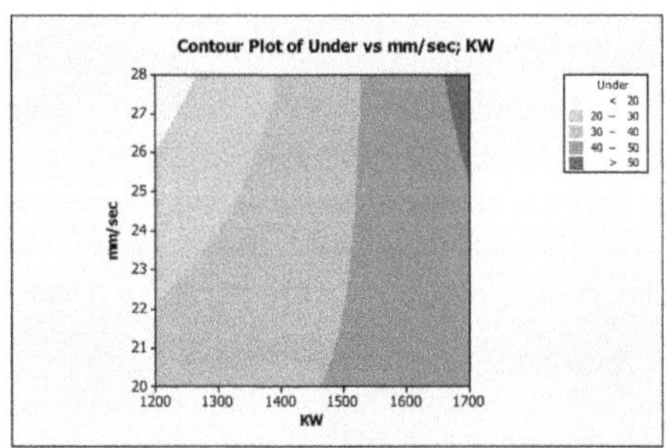

Fig. 7 - Contour Plot of Undercut vs SpeedRate/Power

Fig. 8 – Bending test (a) angle 30° (b-c) angle 92°

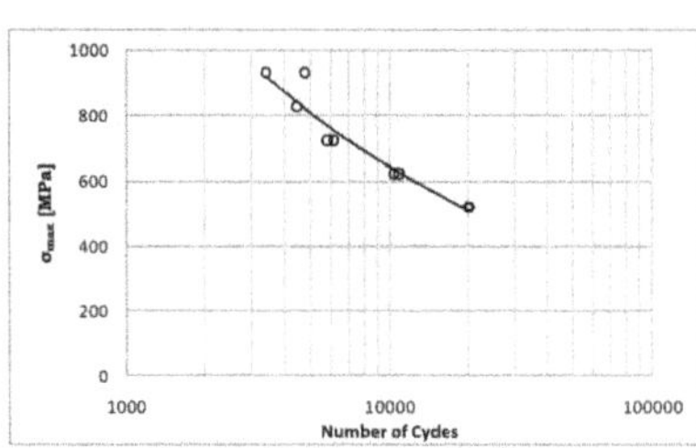

Fig. 9 - Fatigue Cycles Test

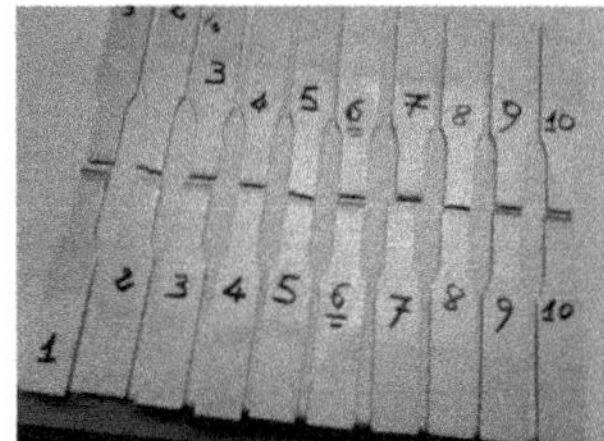
Fig. 10 - Fatigue Cycles Test Fractures

inside the nugget zone. The diagram overlaid on the joint image, also shows how the values, more in the upper zone, maintaining in all measurements, a delta of about 10 HV area between the right and left of the weld. It can be assumed that this effect has a dependency with the process, providing a starting point for further studies.

Table 5
Averages of 3 measurement of UnderCut (micron)

Rate [mm/s]	Power [KW]		
	1200	**1450**	**1700**
20	29	47,5	45
24	27,5	38	36
28	22,5	20,5	68

Table 6
ANOVA Undercut

Source	gdl	Seq SS	Adj MS	F	P
KW	2	1677,78	838,89	260,34	0,000
mm/sec	2	186,11	93,06	28,88	0,000
KW*mm/sec	4	2138,89	534,72	165,95	0,000
Error	9	29,00	3,22		

S = 1,79505 R-Sq = 99,28% R-Sq(adj)= 98,64%

The analysis was completed by a statistical study on the geometry of the UnderCut, investigating the influence of two factors, power and speed. The data analysis are given in Table 5, while Table 6 shows the results for the evaluation of the effects. A model that included the effects of two factors and their interaction was used:

$$x_{ijk}=\mu + \alpha_i + \beta_j + (\alpha\beta)_{ij} + \varepsilon_{ijk} \quad (1)$$

where i=1 to 3 (levels of effect α, *speed rate)*; j=1 to 3 (levels of effect β, *Power of the beam*), k=1 to 2 (number of repetitions)

Fig. 7 allows to confirm that, excluding the technological parameters are not significant, to reduce the formation of undercut which influences the process, the initial choice (speed 24 mm / s, power 1700 W) can be considered a compromise between quality and applicability of the process, ever, given the technological constraints imposed by the plant used laser.

Conclusions

Data from this study may serve as a guide for future work. The purpose was to learn better the interaction of laser welding process of titanium and to wire the quality of the joints to numerical value of process variables.

The tests allow us to observe the excellent protection of the weld . excellent data obtained by tensile tests highlight the potential of the process.

Results from fatigue tests showed that undercut get a decisive role in the beginning of fatigue cracks (due to geometric aspects) as it propagates only in the base material, at the edge of the HAZ.

Acknowledgements

Thanks for Alenia Aeronautica to provide the panel and RTM spa for the welds.

References

[1] Qi Yunlian *, Deng Ju, Hong Quan, Zeng Liying, Electron beam welding, laser beam welding and gas tungsten arc welding of titanium sheet, Materials Science and Engineering A280 (2000) 177–181

[2] Zhang Li,SL Gobbi, I Norris, S. Zolotovsky, KH Richter, Laser welding techniquesfor titanium alloy sheet, J. of Material Processing Technology 65,(1997) 203-208

[3] Avanish Kr. Dubey, Vinod Yadava, Experimental study of Nd:YAG laser beam machining—An overview, journal of materials processing technology 195 (2008) 15–26

[4] O. Perret, M. Bizouard, Ph. Naudy and G. Pascal, D. Nore´ , Y. Horde´ , and Y. Delaisse, Characterization of the keyhole formed during pulsed Nd–YAG laser interaction with a Ti–6Al–4V metallic target, Journal of Applied Physics 90, Number 1, 1 July 2001

[5] R Rai1, JW Elmer2, T A Palmer2 and T DebRoy1, Heat transfer and fluid flow during keyhole mode laser welding of tantalum, Ti–6Al–4V, 304L stainless steel and vanadium, J. Phys. D: Appl. Phys. 40 (2007) 5753–5766

[6] Hong Wang a,b,∗ , Yaowu Shi a, Shuili Gongb, Aiqin Duanb, Effect of assist gas flow on the gas shielding during laser deep penetration welding, Journal of Materials Processing Technology 184 (2007) 379–385

[7] Y. Fan P. Cheng, and Y. L. Yao, Z. Yang and K. Egland, Effect of phase transformations on laser forming of Ti–6Al–4V alloy, Journal of Applied Physics 98, 013518 (2005)

[8] G. Casalino, A.D. Ludovico, Finite element simulation of high speed pulse welding of high specific strength metal alloys, journal of materials processing technology, 197 (2008) 301–305

Innovative Production Machines and Systems
D.T. Pham, E.E. Eldukhri and A.J. Soroka (eds)

Monitoring of cutting tool wear through images analysis

D. D'Addona, R. Teti

Department Materials and Production Engineering, University of Naples Federico II, Piazzale Tecchio 80,80125 Naples, Italy

Abstract

In the manufacturing systems, one of the most important issues is to estimate the rest of cutting tool life under a given cutting conditions as accurately as possible. In order to monitor the tool wear development during machining processes, the interface chosen between the working procedure and the computer was a digital image of the cutting tool detected by an optical sensor (video camera). Images, however, are not homogeneous. Contrast, definition, size and position of the tool vary from one detected image to another and they depend on the particular condition in which they was captured by an optical sensor. This paper is focused on the development of a procedure for the standardization of cutting tool digital images captured during machining tests. These cutting tool images with standard size and pixel density can be utilized for intelligent processing of image data aimed at automatic and real time tool wear characterization.

Keywords: Process monitoring, Tool life, Image processing

1. Introduction

The technology of metal cutting is in permanent evolution and is a field of much interest in terms of computer applications. Cutting tool wear detection and monitoring is a fundamental aspect in the evolution of production techniques. As the quality of the cutting tool is directly related to the quality of the product, the level of tool wear should be kept under control during machining operations.

The recognition of the general conditions of a cutting tool has a major role in the optimisation of machining processes, since the accurate prediction of the exact moment for tool change results in many cases in an effective economy: a longer cutting tool life can be achieved, tolerances can be under control and rejection of pieces by deterioration of the tool conditions can be prevented [1].

Generally, two procedures are followed for cutting tool replacement, depending on the value of the workpiece. If it is a large production where the single workpiece value is low compared to the cost of tool change and tool value, complete tool degradation is reached, then the tool is changed and the workpiece is rejected. On the other hand, if it is a production where the cost of the workpiece is high or the workpiece is in its finishing stage, the tool is replaced before it produces unacceptable results, according to a preventive approach.

Cutting tool state monitoring can allow for the identification of the correct moment to replace the tool, thus protecting the workpiece and/or increasing the tool life.

In order to control the tool wear development, the interface chosen between the real working procedure and the computer was a digital image of the cutting tool detected by an optical sensor.

In this paper, a procedure for the standardisation of cutting tool images detected during machining tests is proposed.

Images with standard size and pixel density were produced by processing tool image files. These standard images of cutting tool can be easily utilised for intelligent processing of image data aimed at automatic and real time tool wear characterisation [2].

2. Machining tests

The work material used during quasi-orthogonal cutting tests is an AISI 1045 steel and the cutting tool is a P3 tungsten carbide insert. The cutting parameters values are shown in Table 1.

Using different parameter combinations, 14 cutting conditions were selected, namely "adh", "aeh", "aei", "afh", "afi", "bdi", "beh", "bfh", "bfi", "ldi", "leh", "lei", "lfh", "lfi".

Every machining test was carried out with a new cutting tool until tool failure. More details of machining tests can be found in [3].

Table 1
Cutting parameters for quasi-orthogonal tests.

Cutting speed (m/min)		Feed rate (mm/rev)		Depth of cut (mm)	
a	200	d	0.06	h	1.0
b	150	e	0.12	i	1.5
l	250	f	0.19		

3. Cutting tool image processing

3.1. Cutting tool image detection

After each minute of cutting process, the machining test was interrupted and two image types of the cutting tool were detected by a video camera: the flank wear (type "l", Fig. 1) and the crater wear (type "c", Fig. 2).

The number of cutting tool images for each cutting test ranges from 6 to 122 depending on each test duration. A total of 717 digital tool images were detected and stored as image files. The distribution of the cutting tool images for the different cutting tests is shown in Table 2. In order to allow for the measurement of the cutting tool wear, the image includes the worn zone of the tool and a reference copper wire with a known thickness of 0.25 mm. Images, however, are not homogeneous. Contrast, definition, size and position of the cutting tool vary from one image to another.

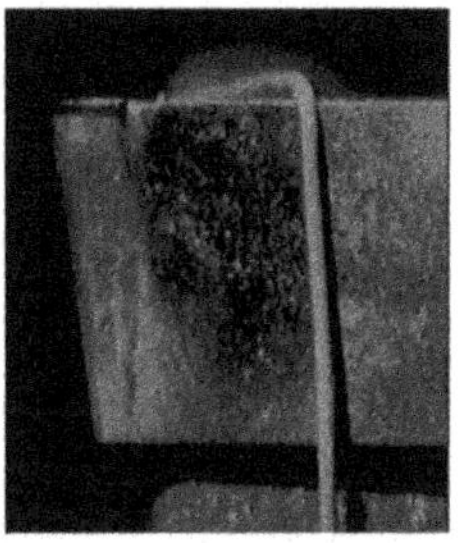

Fig. 1. Raw flank wear image. File: bfh190l.tif.

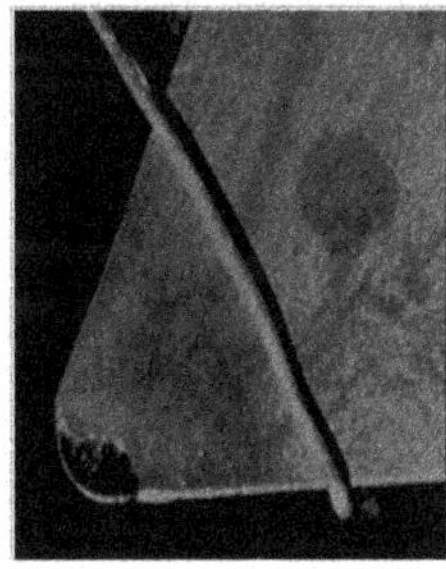

Fig. 2. Raw crater wear image. File: bfh190c.tif.

Table 2
Distribution of images for the different machining tests.

test	crater wear type "c"	flank wear type "l"	total
adh	51	50	101
aeh	27	27	54
aei	15	15	30
afh	17	16	33
afi	11	11	22
bdi	57	56	113
beh	61	61	122
bfh	23	25	48
bfi	33	33	66
ldi	29	28	57
leh	15	15	30
lei	8	8	16
lfh	7	6	13
lfi	3	3	6
Total	357	354	711

Table 3
Format of image file name.

cutting test parameters			minute			type	extension
cutting speed	feed rate	depth of cut					
letter	letter	letter	digit	digit	0	c or l	.tif

3.2 Cutting tool image files

For each tool image detection, an image file in TIFF format was generated and stored on hard disk. The file name has 7 characters and is coded to identify the cutting test, the cutting minute and the type of view that the image reproduces (Table 3).

Tool images from the same machining test are contained in a single test folder with the corresponding test name, for a total of 14 folders. Each test folder contains five sub-folders named from "c0" to "c4". Folder "c0" contains the raw crater wear images; folders "c1" to "c4" contain images obtained after each image processing step. There is also a sixth folder named "l0" containing images of "l" type that are difficult to classify.

The mode in which images were digitised involves the presence of 3 numerical values for each pixel, corresponding to the RGB colours. However, these 3 numbers are identical, which means that images are in B&W, i.e. the colour is identified by a single grey tone value.

3.3 Tool image standardization procedure

For the standardization procedure of tool images, Adobe PhotoShop 3.0 software package was used as image editor. Since these digital images are 2D matrixes of numerical values, Matlab software was employed for matrix transformations. As there were two types of images (cutting tool flank wear and cutting tool crater wear), it was decided to start working with images of cutting tool crater wear.

These images are stored in folder "c0". Although this paper describes only the procedures for crater wear type images, the procedures for flank wear type images are conceptually similar.

3.3.1. Cutting tool orientation

Since the relative orientation and position of the cutting tool varies from one image to another, the first step was to define a common orientation of the tool for all images. Each image was rotated so that the tool cutting edge is vertical and facing the left side of the image (Fig. 3).

This rotation was achieved with PhotoShop command "Image: Rotate: Arbitrary...". To save disk space without losing information, the image mode was changed from RGB to Greyscale, obtaining one numerical value for each pixel, instead of three identical values.

All images submitted to this processing step are stored in folder "c1", keeping their original file name and extension.

3.3.2. Cutting tool position

In order to standardise tool position, the images from folder "c1" were clipped along a horizontal line tangent to the tool tip and a vertical line tangent to the cutting edge. The image clipping was such that the tool became approximately tangent to the upper left borders of the image (Fig. 4). However, image clipping could exclude a part of the reference wire. Therefore, wire measurements were performed using the reference wire before image clipping.

PhotoShop package was utilised for wire measurement with the parameters shown in Table 4.

The average of three wire measurements was calculated for each image using Excel sheet "medidas.xls!misuri". A minimum of 7.67 pixels and a maximum of 21 pixels was obtained.

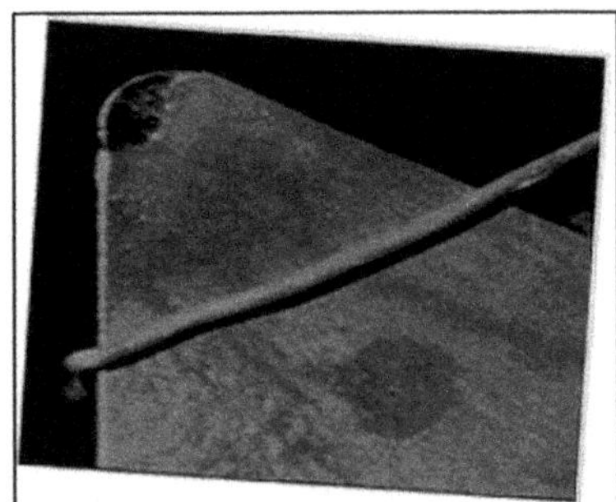

Fig. 3. Standardisation of cutting tool orientation, folder: c1. File: bfh190c.bmp.

Fig. 4. Standardisation of tool position, folder: c2. File: bfh190c.bmp.

Table 4
PhotoShop settings for measuring the reference copper wire.

Property	Value
Marquee options: Shape	Elliptical
Marquee options: Style	Constrained Aspect Ratio
Info Options: Mouse Coordinates	Pixels

A file named "fili.wk1" was created to store the calculated values, so Matlab could use them in subsequent operations. In this way, image operations could be performed without losing size information provided by the reference wire.

After image clipping, the size expressed in pixels of the new image was reported in Excel sheet "medidas.xls!misuri". The file format was changed from TIFF to BMP so Matlab tool could open and read these files as images. Image processing through PhotoShop software was completed at this point.

All subsequent image processing was performed through Matlab package. All images submitted to this processing step are stored in folder "c2", keeping their original file name and extension ".bmp".

3.3.3. Image size and pixel density

To process image information using an intelligent computation approach, images with the same size and the same pixel density must be generated. The standard size to utilise for all images was selected by analysing the size of all images and the maximum crater wear as detected in the last two minutes of each machining test contained in folder "c2".

Using the reference wire measurements and the image sizes reported in Excel sheet "medidas.xls!misuri", the real image size was calculated. Images have width ranging from 0.93 mm to 9.55 mm and height ranging from 2.21 mm to 7.93 mm.

The maximum crater wear as measured using the last but one test images was considered because the last images present a largely damaged tool and crater wear cannot be measured.

Values of 1.70 mm for width and 1.97 mm for height were considered the maximum values for crater wear size, as reported in Excel sheet "medidas.xls!risultati".

By analysing the pixel density that all images would have after the standardisation of image size, values ranging from 31 to 84 pixels were found for width and ranging from 61 to 168 pixels for height.

Based on these observations, two standard image sizes were chosen: standard image size 1 mm (horizontal) x 2 mm (vertical) and standard image size 2 mm x 2 mm.

- Standard image size 1 mm (horizontal) x 2 mm (vertical)

This size was selected because some images had width < 1.70 mm. Even though, for this standard image size 5 images had width < 1 mm. In these cases, the standard size 1 mm x 2 mm was obtained by adding to the image matrix an appropriated number of columns with a predetermined grey tone value, which in our case was 256 = white. Pixel density of these images was standardised to 7225 pixels/mm" generating images of standard pixel size 85 (horizontal) x 170 (vertical) (Table 5).

A procedure through Matlab package for increasing pixel density was used. All images with standard size 1 mm x 2 mm and 85 pixels x 170 pixels (Fig. 5) are stored in folder "c3", keeping their original name and extension ".bmp". Although this standard image size is generally greater than the crater wear size, it is worth noting that a few images, related to the last minutes of cutting, do not show the complete crater wear area. In order to provide all the crater information, a larger standard image size was also adopted.

Table 5
Standard image size in pixel and in mm.

Standard image size		Pixel density
mm	pixel	pixel/mm
1 x 2	85 x 170	7225
2 x 2	170 x 170	7225

- Standard image size 2 mm x 2 mm.

This size was selected to include the complete crater wear information contained in all images. A total of 100 images having width < 2 mm were completed with white columns. Pixel density of these images was standardised to 7225 pixels/mm" generating images of standard pixel size 170 (horizontal) x 170 (vertical) (Table 5).

All images with standard size 2 mm x 2 mm and 170 pixels x 170 pixels (Fig. 6) are stored in folder "c4", keeping their original name and extension ".bmp".

Fig. 5. Standardisation of image size and pixel density: 1 mm x 2 mm (85 pixels x 170 pixels), folder: c3. File: bfh190c.bmp.

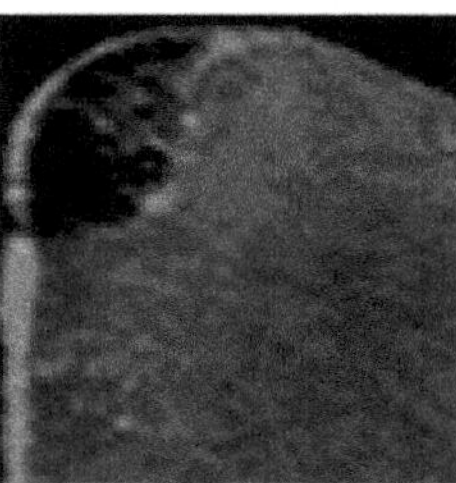

Fig. 6. Standardisation of image size and pixel density: 2 mm x 2 mm (170 pixels x 170 pixels), folder: c4. File: bfh190c.bmp.

4. Cutting tool image standardisation procedure using Matlab package

Each image file was processed in turn through Matlab software according to the following procedure:

- Load image:
  Image is read through BMPREAD and matrixes corresponding to image data are created;
- Identify the image by its file name:
  The machining test and minute are identified by decoding the file name;
- "spess.m" function is used to search in file "fili.wk1" the measure of the reference wire:
  File "fili.wk1" is loaded and a variable "fili" is created. The three machining test letters are converted to a numerical code that identifies the test in variable "fili". The measure of the reference wire is identified;
- Calculate the scale of the image:
  The scale of each image is calculated by dividing the real thickness of the wire by its measure in pixels;
- For some images, "aggius.m" function is used to increase width:
  Images smaller than the selected standard size are completed by adding columns of numerical values to the image matrix.
- "trunca.m" function is used to clip images to size 1 mm x 2 mm or 2 mm x 2 mm;
- Increase pixel density by using "mataum.m" function;

Convert the matrix to an image file and save it in folder "c3" or "c4": An image file given by the calculated matrix and the initial colormap is created and saved on hard disk through command BMPWRITE.

Table 6 reports the 13 files that could not be transformed by the developed Matlab procedure, in fact some images were unreadable while in a few images the reference wire was absent.

Table 6
Files that could not be processed by Matlab procedures for standardisation.

Dimensions	File names				
1 mm x 2 mm	afh050c.bmp	afh100c.bmp	afi050c.bmp	bdi070c.bmp	
	bdi190c.bmp	beh040c.bmp	beh140c.bmp	lfh025c.bmp	
2 mm x 2 mm	beh040c.bmp	afh050c.bmp	afh100c.bmp	afi050c.bmp	lfh025c.bmp

5. Conclusion

In a manufacturing system, machining efficiency is easily influenced by the kind of tool selected at each cutting process. One of the most complex problems for tool selection is that of estimating the life of the tool under a given cutting condition as accurately as possible As the quality of the cutting tool is directly related to the quality of the product, the level of tool wear should be kept under control during machining operations. Automatic cutting tool wear monitoring is an important aspect in the improvement of production systems.

In this paper, a procedure for the standardisation of cutting tool images detected during turning tests is proposed. Images with standard size and pixel density were produced by processing tool image files obtained during machining tests.

A system based on cognitive paradigms to estimate the life and wear type of cutting tools from their image data and cutting conditions will be presented in future research works.

References

[1] Byrne, G., Dornfeld, D., Inasaki, I., Ketteler, G., Koenig, W., Teti, R., Tool Condition Monitoring. (TCM) - The Status of Research and Industrial Application, Annals of the CIRP, vol. 44/2, 1995, pp. 541-567.

[2] TESHIMA, T., Shibasaka, T., Takuma, M., Yamamoto, A., Estimantion of Cutting Tool Life by Processing Tool Image Data with Neural Network, Annals of the CIRP, vol. 42/1/1993, 1993, pp. 59-62.

[3] Teti, R., Sensor Signal Frequency Analysis for Metal Cutting Process Monitoring, 1st AITEM Conf., Ancona, Italy, 22-24 Sept., 1993, pp. 55-64.

Innovative Production Machines and Systems
D.T. Pham, E.E. Eldukhri and A.J. Soroka (eds)

Statistical analysis of fiber laser machining of titanium alloy

F. Tagliaferri[a], N. Pagano[b], C. Leone[b], B. Palumbo[a]

[b]*Department of Aerospace Engineering, University of Naples "Federico II", Naples, Italy*
[a]*Department of Materials and Production Engineering, University of Naples "Federico II", Naples, Italy*

Abstract

In this work a preliminary statistical study on laser machining of titanium alloy is presented. Laser machining tests were carried out on Ti6Al4V alloy sheet, 4 mm thick, using a Q-Switched 30 W Yb:YAG fiber laser. The aim of the paper is to characterize the laser engraving process; that is, to detect which process parameters affect the depth of machined volume and the quality of the machined surface in terms of roughness. The examined parameters were: the lease beam scan speed, the pulse frequency, the distance between the linear patterns of two consecutive laser scans (step), the number of repetitions of the geometric pattern and the scanning strategy. A two-level fractional factorial design and ANalysis Of VAriance (ANOVA) were applied. In addition experimental results have shown that, for each adopted pulse frequency, the machined volume linearly depend on the total amount of released energy. Besides the process map in term of Material Removal Rate (MRR) and roughness has been evaluated and discussed too.

Keywords: Laser machining, fiber laser, Design of Experiment (DOE), ANOVA

1. Introduction

Laser machining is one of the most adopted technologies in rapid prototyping to produce tool and mould operations. A laser beam is used to ablate a solid bulk, following predetermined patterns [1-3]. The sculpture is obtained by repeating this process on each successive thin layer. Compared to traditional machining, this method has some advantages, such as: greater flexibility of use, no mechanical contact with the surface, a reduction in industrial effluents (i.e. no acid, solvent or dielectric oils are required) and a fine accuracy of machining, even with complex forms in injection moulding [4-5]. However, the degree of shape precision, the Material Removal Rate (MRR) and the surface quality during the engraving process strictly depend on the material properties, the laser source characteristics and the process parameters. In order to obtain low roughness and high MRR a process of optimization is required [6-10].

In this paper laser engraving of Ti6Al4V alloy has been studied by using a Q-Switched 30 W Yb:YAG fiber laser. In particular the influence of the process parameters on the depth of machined volume and the roughness of the engraved surface has been evaluated.

This paper involves both statistical and technological aspects. The statistical methodologies applied in this first screening experimental phase, two-level fractional factorial design and ANalysis Of VAriance (ANOVA), are extensively treated in the literature (for example in [11]), and so they will be applied without any explicit introduction or analytical formulation.

In addiction a predictive model of the machined volume was proposed and verified. In practice experimental results have shown that, for each adopted pulse frequency, machined volume linearly depend on the total amount of released energy.

At last the process map in terms of Material Removal Rate (MRR) and roughness has been evaluated and discussed too.

2. Equipment, material and experimental design

2.1. Laser equipment

The engraving tests were performed by using a Q-Switched 30 W Yb:YAG fiber laser (Lasit Fly Fiber 30W), with fundamental wavelength λ = 1064 nm, duration time of 50 ns and pulse frequency variable in the range of 30÷80 kHz, pulse energy up to 1 mJ.

The laser beam is first moved through two galvanometer mirrors, and then it is focused by a "flat field" lens, with a focal length of 160 mm, onto the workpiece. The final focused beam diameter is about 100 μm. The laser system is controlled through a PC, which allows the generation of the geometric patterns and the setting of the following process parameters: the mean beam power (Pm), the pulse frequency and the scan speed.

In Figure 1 the characteristic response of the laser source, in terms of pulse energy and pulse power, is reported as a function of pulse frequency, at different mean power values.

Pulse energy and pulse power play an important role in the laser machining process, by determining the laser beam-material interaction mode and then the amount of machined volume. Their influence will be take into account in this experimental study by changing the pulse frequency.

The adopted experimental conditions, in terms of mean beam power and pulse frequency, are highlighted by circles in Figure 1.

2.2. Material and measurement equipment

The adopted materials is a Alpha/Beta Titanium Alloy (Ti6Al4V), largely applied in medical and aerospace fields. In Table 1 the main thermo-mechanical properties and the chemical composition are reported.

In order to perform the engraving tests four Ti6Al4V alloy plates, 60x60x4 mm^3 in size, were used. A square areas 5x5 mm^2 in plane dimension were machined at nominal maximum mean beam power (30 W).

The depth and the roughness of the engraved cavities were measured by 3D Surface Profilometer Talysurf CLI 2000 by Taylor Hobson. No less than 5 profiles were acquired in the same direction of the scanning speed.

For the analysis of the acquired profiles, a surface analysis software (Taly Map Universal) was adopted, in order to measure the depth of the engraving and the roughness of the engraved surfaces.

In Figure 2 an image of a machined plate, during the measuring process, is reported.

As roughness parameter was used the arithmetic mean of the profile data points, Ra (ISO 1302:2002).

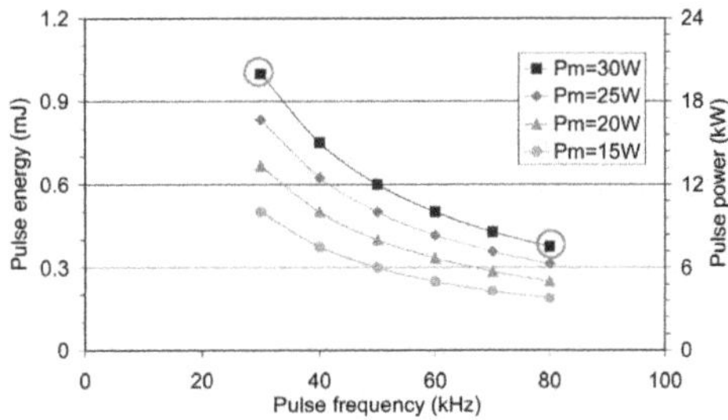

Fig. 1: Pulse energy and pulse power as a function of the pulse frequency at different mean power values (Pm).

Table 1
Main properties and chemical composition of Ti6Al4V alloy.

Properties		Componet (%)	
Density (kg/m^3)	4430	Al	6
Hardness, Brinell	334	Fe	Max 0.25
Ult. Tensile Strength (MPa)	950	O	Max 0.2
Yield Strength, (MPa)	880	Ti	90
Elongation at Break (%)	12	V	4
Young's Modulus (GPa)			113.8
CTE, 20°C (μm/m °C)			8.6
CTE, 500°C (μm/m °C)			9.7
Specific Heat Capacity (J/g °C)			0.5263
Thermal Conductivity (W/m K)			6.7
Melting Point (°C) Solidus			1604
Liquidus			1660
Beta Transition (°C)			980

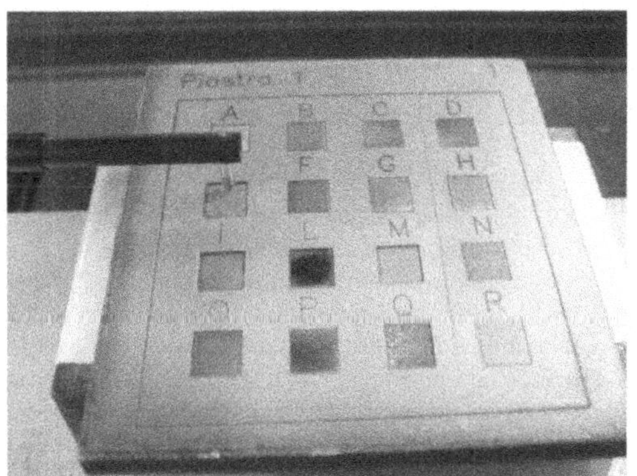

Fig. 2. Example of an engraved sample.

The machined volume was evaluated as product between the mean depth and the plane dimension of the groove [10]. The MRR was quantified as the ratio between the machined volume and the calculated process time.

2.2. Pre-experimental planning, experimental design and set-up

Following the systematic approach to planning designed experiments proposed in [12] and successfully applied in [13], two pre-design sheets (i.e. the main and secondary sheets) were drew up and implemented. These two kinds of sheets catalyze the interaction between statistical and technological competences.

The main sheets contain information about the objective of the experimentation, the relevant background, the response variables and the factors (i.e. control, held-constant and nuisance factors).

On the secondary sheets, for each quantitative control factor, the normal level and range as well as the measurement precision were specified. Moreover the secondary sheets detail the technological relationship between the control factors and the response variables, in terms of the expected main effects and interactions.

In this first experimental phase, the objective was to characterize the laser engraving process; that is, to detect which factors affect the quality of the machining surface in term of depth of machined volume and surface roughness.

For each variable, the normal operating level, the range, the measurement precision and the relationship to the objective were specified on the main sheets. The first step involved listing all factors that came out during team discussions. The second step consisted in classifying each factor as a control, held-constant or nuisance factor [12].

Obviously, different classifications are possible, each strictly related to the specific aims of the experiment.

The following control factors were adopted: scan speed (A), pulse frequency (B), step (C) (distance between two consecutive scan lines), repetition number of the geometric pattern (D) and scanning strategy (E).

The number of repetitions (D) represents the number of the laser beam scanning on the same area. The adopted scanning strategy (E) were two: the first one is based on horizontal lines (//), the other one is based on a sequence of four lines placed at 0°,90° and ±45° (/\).

Factors A, B, C and D are quantitative parameters. Factor E is a qualitative factor that refers to the specific scanning strategy adopted. Figure 3 shows both different scanning strategies.

In this screening experimental phase a 2^{5-1} design was adopted. This design, with I = ABCDE (defining relation), is a resolution V design, so no main effect or two-factor interaction is aliased with other main effects or two-factor interactions, but each main effect is aliased with a four-factor interaction, and each two-factor interaction is aliased with a three-factor interaction.

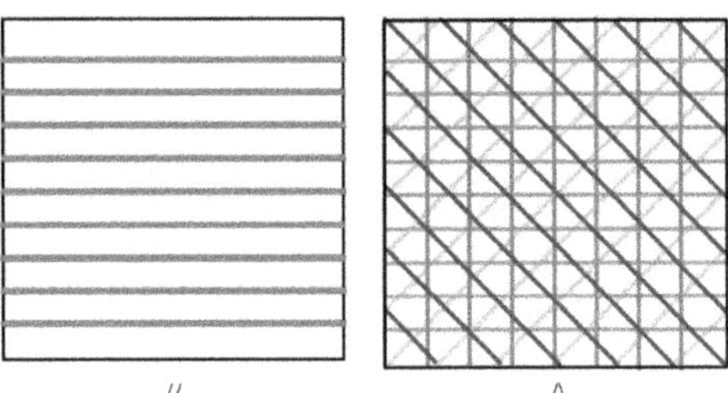

Fig. 3. Scanning strategies.

Table 2
Control factors and their settings.

Control factors	Labels	Low (−)	High (+)	Unit
Scan speed	A	400	1600	mm/s
Frequency	B	30	80	Hz
Step	C	30	60	μm
Repetitions	D	80	120	--
Strategy	E	//	/\	--

Table 3
Matrix for the 2^{5-1} design.

Treatment	A	B	C	D	E=ABCD
a	+	+	-	+	-
b	+	-	+	+	-
c	+	+	+	-	-
d	-	+	+	-	+
e	-	-	+	-	-
f	+	-	-	-	-
g	+	+	+	+	+
h	+	-	-	+	+
i	-	-	-	-	+
l	-	+	-	+	+
m	-	-	-	+	-
n	-	+	+	+	-
o	-	-	+	+	+
p	+	+	-	-	+
q	-	+	-	-	-
r	+	-	+	-	+

Since three-factor (and higher) interactions are negligible, the experimental 2^{5-1} design enables reliable information to be obtained about main effects and two-factor interactions. Table 2 summarizes the levels of control factors and their settings.

Table 3 shows the 2^{5-1} design matrix. For each treatment, 4 replications were executed, giving a total of 64 experimental runs. This number of replications was adopted to provide more consistent response repeatability during this first experimental study.

3. Statistical results and technological considerations

3.1. Statistical analysis of results

The ANOVA method was applied in order to test the statistical significance of the main effects and the two-factor interactions for depth of machined volume and surface roughness. Diagnostic checking was successfully performed via graphical analysis of residuals.

The experimental results for depth of machined volume and surface roughness are shown in Figure 4 using Pareto charts of standardized effects (α = 0.05).

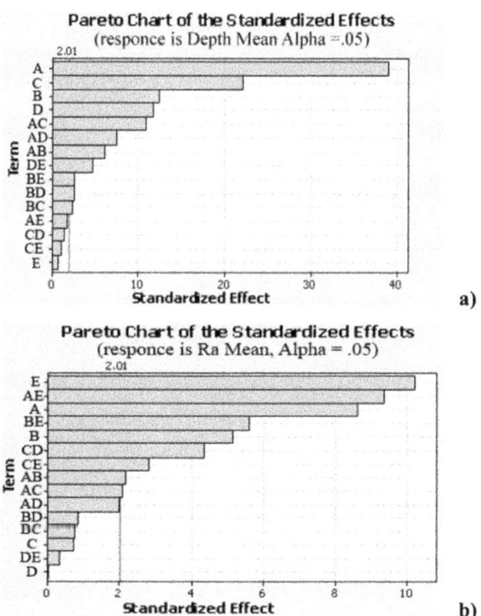

Fig. 4. Pareto charts of standardized effects (α = 0.05): **a)** depth of machined volume material; **b)** surface roughness.

Figures 5 and 6 show the main effects and the interaction effects plots for depth (measured in µm) of machined volume, respectively. The significant effects (α =0.05) are highlighted (red boxes). They are the scan speed (A), the pulse frequencies (B), the steep (C) and the repetition (D).

In terms of mean effects the results are not unexpected since the laser beam-material interaction mode depend on the factors A and B, while the working time depend on the factors A, C and D, as also reported in [5, 7 and 10].

Although strategy (E) is not significant, the interaction between the repetition and the strategy (DE) is significant. This result was not expected and requires further technological investigations.

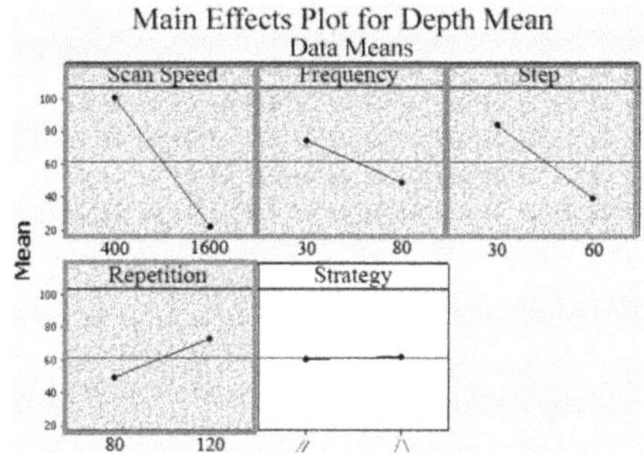

Fig. 5. Main effects plot for depth (in µm) of machined volume.

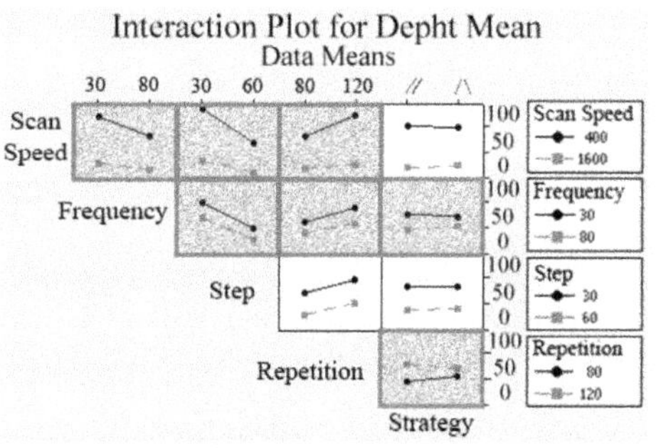

Fig. 6. Interaction effects plot for depth (in µm) of machined volume.

Figures 7 and 8 show the main effects and the interaction effects plots for surface roughness (measured in µm) respectively. The significant effects (α =0.05) are highlighted (red boxes). In terms of main effects of scan speed (A), frequency (B), repetition (D) and strategy (E) the results are in agreement to those observed in [10]. On the contrary the result for step (C) is surprising because, in previously works, its influence was observed [6, 7, 8 and 10].

This contradiction could be first explained taking into account the anti-synergic type interaction between step and repetition (CD) (see Figure 8). It means that the effect of the different level of the repetitions (D) tends to balance the effect of the step (C) [11]. This first interpretation needs further technological investigations.

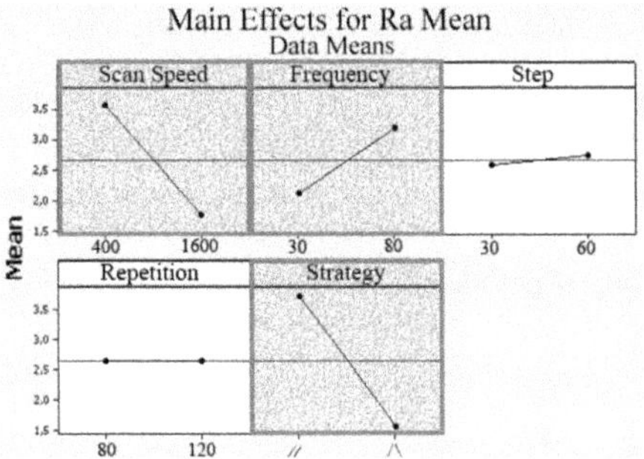

Fig. 7. Main effects plot for surface roughness (μm).

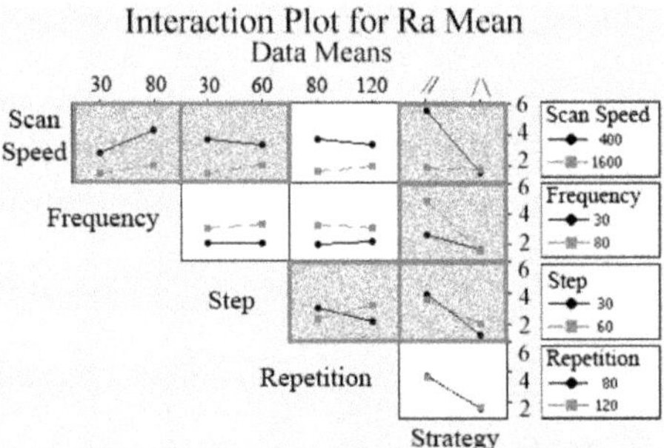

Fig. 8. Interaction effects plot for surface roughness (μm).

3.2. Technological considerations in term of machined volume, roughness and MRR

Since the depth of machined volume strictly depends on the laser beam-material interaction mode and the working time, in order to formulate a predictive model for the machined volume it is possible to use an easy model based on energy consideration, often used for laser machining [7, 14].

The model starts from the equation:

$$V = \alpha \frac{Et}{Q_v} \quad (1)$$

where V is the ablated volume; Et is the total amount of released energy, α is the coefficient of absorption of the materials at the emission wavelength and Q_v is the material vaporization energy for unit volume.

Obviously the model neglects all the energy losses. So, in order to verify the validity of the proposed model, the machined volume was plotted as a function of the total amount of released energy (calculated as product between the mean beam power and the interaction time), as showed in Figure 9.

From Figure 9, it is possible to observe that, as expected, for each adopted pulse frequency (30 and 80 kHz) the machined volume linearly depend on the total amount of released energy.

Furthermore, the different slope of the lines indicates that a different laser beam-material interaction mode occurs changing the pulse frequency as also observed from the ANOVA in terms of depth of machined volume.

In order to obtain indication about both machined volume and surface roughness, the roughness parameter Ra was reported in Figure 10 as a function of the MRR for each treatment shown in Table 3.

In Figure 10 are visible six different groups of treatments, each one indicated by a different symbol (+, ×, ◊, □, ▲, ●), where each treatment is labelled by a letter (see Table 3).

Since high productivity and low roughness are the first requirements, the best process treatments correspond to the ones indicated by the closed dots (b, i and o, also highlighted in Table 3).

On the contrary, the worst working condition corresponds to the process treatments indicated by the squared dots (q and n treatments).

In Figure 10 the images of the obtained surface for the best and the worst cases are reported.

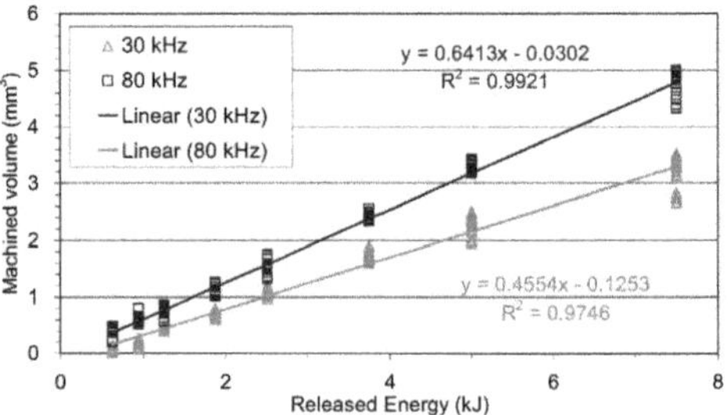

Fig. 9. Machined volume as a function of the total amount of energy. The continuous lines represent the best fitting lines for the two used pulse energy value.

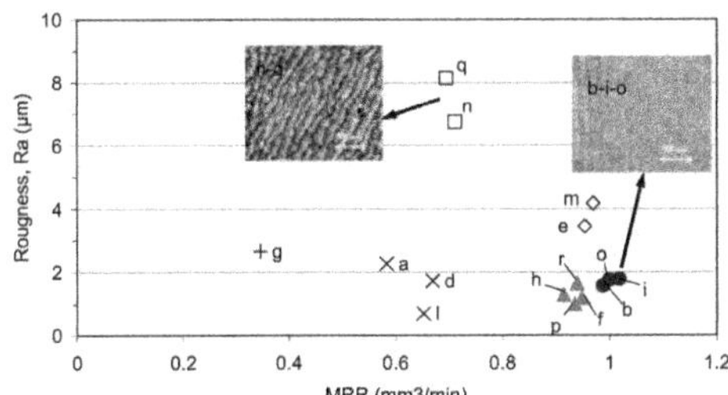

Fig. 10. Roughness, Ra, as a function of the MRR for the different process treatments.

5. Conclusions

This first experimental study concerning the laser engraving of the Ti6Al4V alloy by means of a Q-Switched 30W Yb:YAG fiber laser has sown the strategic role that a systematic approach to planning for a design of experiments plays in technological process innovation.

The results obtained in this first screening experimental phase enabled the factors that affect the depth of the machined volume and the quality of the machined surface in terms of roughness.

In addition, experimental results have shown that, for each adopted pulse frequency, the machined volume linearly depend on the total amount of released energy. At last the best treatments in terms of roughness Ra and MRR have been evaluated too.

Acknowledgements

The authors are particularly grateful to the CIRTIBS Research Centre, for providing the equipments and financial support to develop the present research work.

References

[1] Chryssolouris G., Laser Machining - Theory and Practice. Mechanical Engineering Series, Springer-Verlag, New York Inc., NewYork, 1991.

[2] Meijer J., Laser beam machining (LBM), state of the art and new opportunities. J. of Materials Processing Technology, 149, 2004, pp. 2–17.

[3] Dubey K., Yadava V.. Laser beam machining - A review. I. J. of Machine Tools & Manufacture, 2008, 48, pp. 609–628.

[4] Kaldos, H.J. Pieper, E. Wolf, M. Krause. Laser machining in die making—a modern rapid tooling process. J. of Materials Processing Technology, 155–156, 2004, pp. 1815–1820.

[5] Leone C., Quadrini F., Santo L., Tagliaferri V., Trovalusci F. Nd-YAG laser sculpture of WC punches for micro-sheet forming. Key Engineering Materials, 344, 2007, pp. 783-789.

[6] Leone C., Morace R. E., De Iorio I.. AISI 304 Stainless Steel Engraving by Q-Switched Nd:YAG Nanosecond Laser. J. of the Chinese Society of Mech. Eng., 28 No.2: 217-224, 2007.

[7] Leone C., Lopresto V. and De Iorio I..Wood engraving by Q-switched diode pumped frequency doubled Nd:YAG green laser. Optics and Lasers in Engineering, 47/1: 161-168, 2009.

[8] Lorenz A., Emmelmann C.. Methodical investigations for increasing process capability of short pulse laser ablation in high precision die and mould industry, High-Power Laser Ablation VI. In: Proc. of SPIE, Vol. 6261, pp. 626122-13, 2006.

[9] Cicală E., Soveja A., Sallamand P., Grevey D., Jouvard J.M., The application of the random balance method in laser machining of metals. J. of Mat. Proc Technology , 196:393–401, 2008.

[10] Genna S., Leone C., Lopresto V., Santo L., Trovalusci F., Study of fiber laser machining of C45 steel: influence of process parameters on material removal rate and roughness, Int. J. of Material Forming, 3/1, 2010, pp. 1115-1118.

[11] Montgomery D.C., Design and Analysis of Experiments. Wiley, New York, 2001.

[12] Coleman D.E., Montgomery D.C.: A systematic approach to planning for a designed industrial experiment. Technometrics. 35(1), 1993, pp. 1–12

[13] Palumbo B., De Chiara G., and Marrone R., "Technological Process Innovation via Engineering and Statistical Knowledge Integration" edit by Pasquale Erto, Statistics for Innovation, Springer, 2009.

[14] Duley W.W., CO2 Laser: Effect and applications, Academy Press, New York (1977).

Innovative Production Machines and Systems
D.T. Pham, E.E. Eldukhri and A.J. Soroka (eds)

Achieving CO2 Emission Targets for Energy Consumption at Canadian Manufacturing and beyond; using Hybrid Optimisation Model

A. Marzi[a], H. Marzi[b], E. Marzi[b]

[a] Dr. John Hugh Gillis Regional High School, Antigonish, Nova Scotia B2G 1L3, Canada
[b] Department of Information Systems, St. Francis Xavier University, Antigonish, Nova Scotia B2G 2W5, Canada

Abstract

With increased attention directed towards the climate change crisis, international leaders have generated the Copenhagen Accord with national greenhouse gas reduction targets. The presented paper proposes an intelligent hybrid system, a composite of the bees optimisation algorithm and back-propagation neural network, to calculate optimal minimisation of oil, natural gas and coal emissions for Canada's manufacturing sector to meet the sector's quota of the Copenhagen emission targets by the 2020 deadline.

Keywords: Bees algorithm, Artificial neural networks, Optimisation, Emission reduction, Sensitivity analysis

1. Introduction

For the past several decades, scientist have indicated that the Earth is faced with an increasing threat from global warming which stems from greenhouse gases (carbon dioxide, methane, sulphur oxides, etc). These global changes, in turn, endanger and damage numerous ecosystems, which then affects the industries, such as fisheries, lumber, mineral, which then causes damage to the world market and points to the inevitable collapse of world trade and more importantly nature. Attempts are being made by world leaders to reduce society's carbon footprint. These attempts include the ratification of the Kyoto Accord and the latest global climate conference held in Copenhagen, Denmark, where new international carbon dioxide targets have been discussed for the year 2020.

This study focuses on the application of an artificial neural network (ANN) based optimisation algorithm (OA) to determine a practical plan for minimising oil, coal, natural gas and fuel consumption in the manufacturing sector. The goal is to meet the portion of the manufacturing sector's emission quota as determined in Copenhagen for Canada by 2020. The Copenhagen Conference declared that Canada should reduce its carbon dioxide emissions by 17% of our 2005 levels by 2020; at the same time, the portion of Canada's carbon dioxide that is produced by the manufacturing industry is coincidentally 17% on average [1]. The plan will consist of; firstly, calculating on an annual basis, between the years 2010 and 2020, the idealistic total emission per year, enabling the manufacturing industry to accomplish the reduction in its quota by 2020. Secondly, the OA will be applied, using an embedded ANN, which will check the fitness of the OA's solutions. When the criteria of the OA are met, optimal solutions, or optimal reduction coefficients for oil, coal, natural gas and energy consumption will be given. These optimal reduction coefficients when applied to the 2005 fuel source emission levels will indicate the necessary changes that need to be implemented in order to meet the proposed Copenhagen targets.

2. Hybrid Optimisation Model

2.1. *Artificial Neural Networks*

An artificial neural network is an arrangement of neurons, where each neuron functions as an abstraction of the biological neuron. An ANN does not yet have the ability to mimic the complete intelligence of a human. It allows simulated learning processes to occur such that the network is able to produce formulaic or non-formulaic connections and trends between presented data sets. The most common fields of problem solving for an ANN are pattern recognition, data clustering and curve fitting [22].

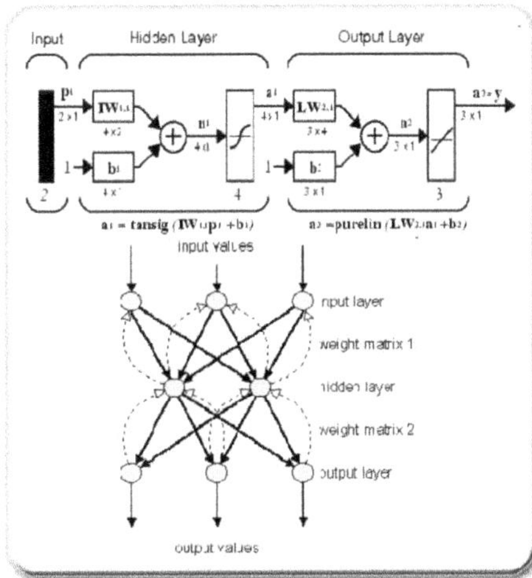

Fig. 1. The architecture of the utilised artificial neural network [23]

Various types of ANNs exist that differ due to the variety of activation functions and procedures used to develop the connections and relationships between the input data set and the target data set. A multi-input Back-Propagation neural network using tan-sigmoid and linear activation functions was used in this case study. There are three main stages in the processing of ANNs, the user is able to determine how much of their data is used for each particular stage. In this study, 60% of the data dedicated to the training and 20% in validation and 20% for testing. The main and most crucial stage is the training stage where by using a series of alternating weights (based on connection strength) and biases (usually constant of one) are applied to each input value in the form of:

$$N_j = w_{ij} I_i + b \tag{1}$$

where;

N is the hidden neuron j of which the product of the equation is assigned,
i is the input neuron,
j is the hidden neuron,
w are the weights,
b are the biases [22].

The summation of all the hidden neurons is given as the argument of the first activation function. Then a secondary set of weights and biases are applied to the product of the activation function and summation is done of

$$w_{jT} P_j + b \tag{2}$$

where T is the target neuron elements, and P is product of the activation function. The result of the summation is given as the argument of the second activation function. The product of the second activation function is then compared to the target data set. To determine how well the network performed, the mean squared error (MSE) is found by comparing the network's output and the target data set. After each iteration, or epoch, is completed, the ANN will check to see if the MSE has improved, or reduced. If the MSE is not improving, the network readjusts the weights and biases accordingly, and repeats the process mentioned above until the MSE meets the desired criteria; at which point the network is said to be trained.

2.2. *Bees Algorithm*

The Bees Algorithm (BA) was developed to evaluate optimisation problems using the real bees' forging method. In the BA there are several variables and parameters that have to be initialised, variables like: (n), the total number of scout bees; (m), the selected sites out of n visited sites; (e), number of best sites out of m selected sites; (n_1), number of bees recruited for best e sites; (n_2), number of recruited bees for the other for selected sites; (*ngh*), initial size of the sites or patches [21].

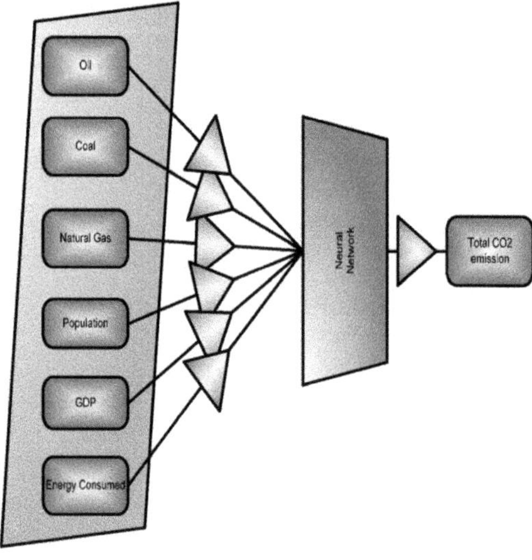

Fig. 2. Specific artificial neural network inputs and target

The BA procedure is presented in the following pseudo code and later in a flowchart in Fig.3.

1. Initialise population with random solutions.
2. Evaluate fitness of the population.
3. While (stopping criterion not met)
//Forming new population.
4. Select sites for neighbourhood search.
5. Recruit bees for selected sites (more bees for best e sites) and evaluate fitnesses.
6. Select the fittest bee from each patch.
7. Assign remaining bees to search randomly and evaluate their fitnesses.
8. End While.

The nature of swarm optimisation algorithms, like the BA, is to obtain a set of random solutions and then to find improved solutions by deploying more members of the swarm to search in the neighbourhood patches of the best solutions. This characteristic of the BA plays a crucial part in shaping how much the manufacturing industries have to reduce their emission. It is vital for BA to produce accurate solutions, as opposed to extreme reduction coefficients to produce drastic reduction of emissions. To produce accurate solutions the fitness function of the algorithm was embedded with an ANN to ensure that coefficient produced by the BA's random search are effective and have minimal error in comparison to the idealistic yearly emissions.

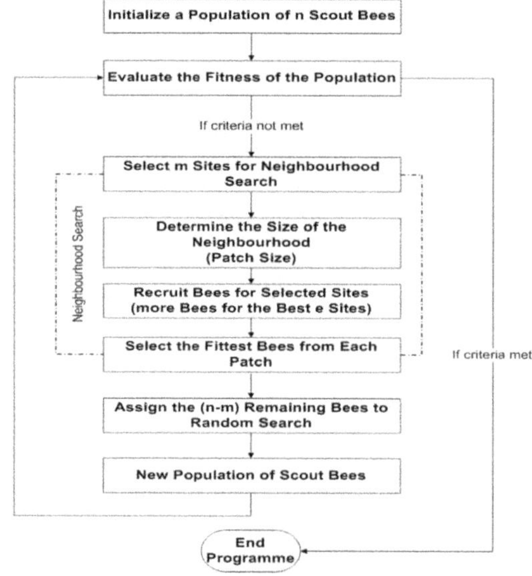

Fig. 3. Flowchart of BA procedure [20].

3. Data selection

Data was mainly used from the Canadian Industrial Energy End-use Data and Analysis Centre [2], located at Simon Fraser University, and from http://datafinder.worldbank.org/ [8], a database of international statistics. Other comparative sources include [4], [5], and [7]

The inputs of the ANN include emissions from oil, coal and natural gas based fuel sources, the population, the gross domestic product (GDP), and the total energy consumed from all manufacturing industries of Canada. The data ranges from 1990 to 2008. The target data set was comprised of the total emissions from the fuel sources in the manufacturing sector per year during the training, validation and testing of the network. The input and target data sets are entered into the networks as matrices; the input matrix consists of the rows representing each of the different factors (i.e. fuel source emissions, population, etc) and the columns representing years 1990 to 2008. Similarly, for the target data set, the single row represents the total emissions of the manufacturing industry, and the columns representing 1990 to 2008.

Table 1. Fuel source emissions data

Year	Oil (kT)	Coal (kT)	Natural Gas (kT)	Energy Consumed (J)
1990	20893.25	15563.00	39148.00	2508355
1991	19503.79	16358.00	35122.00	2322165
1992	17543.00	16267.00	35733.00	2346993
1993	18422.00	15504.00	37489.00	2381873
1994	17836.00	14727.00	39561.00	2544161
1995	18247.47	15532.00	40306.00	2639741
1996	18324.00	15759.00	41332.00	2638585
1997	17912.00	15820.00	42657.00	2678331
1998	18256.00	15403.00	41283.00	2657271
1999	17986.00	15722.00	40815.00	2704487
2000	17698.00	16224.00	40746.00	2736450
2001	18463.00	15475.00	37666.00	2647809
2002	17244.00	15177.00	37741.00	2648201
2003	19452.63	15716.00	34994.00	2663219
2004	20510.01	16529.00	36084.00	2753338
2005	17685.61	16442.00	34289.00	2644609
2006	16073.24	17381.00	32363.00	2548445
2007	15917.65	17367.00	33379.00	2561559
2008	13223.00	16908.00	32131.00	2444290

Using presented methodology, idealistic emission targets are initialised, mainly, years 2010 to 2020 as years 2005-09 are considered 'missed' targets since planning was not accounted for those years that passed. After sensitivity analysis is done, these desired targets are given to NN and input values are altered such to meet the desired targets. The idealistic yearly CO_2 emission targets are created for 2010 until 2020 using the procedure described in Eq. 2 to Eq. 4.

3.1. Determining idealistic and practical annual CO_2 emission targets

Eq. 3 corresponds to the Copenhagen Protocol's expectation of Canada's manufacturing industry.

$$(CO_2)_{2020} = (CO_2)_{2005} \times (1 - 0.17) \quad (3)$$

Therefore, the difference, with respect to the annual CO_2 emissions, can be expressed in the form of Eq, 5. Practical annual CO_2 emission targets are set by dividing the amount of CO_2 emissions that is being reduced over a fifteen year period; 2005 to 2020. However, years 2006 to 2009 are considered not to be applicable since they have already passed

$$\Delta(CO2)_{2005-2020} = 0.17(CO_2)_{2005} \quad (4)$$

Eq. 5 produces the practical annual CO_2 emission targets for 2010 to 2020.

$$(CO_2)_{Annual\ Decrease} = \left[1 - \frac{(year) \times 0.17}{15}\right](CO_2)_{2005} \quad (5)$$

Table 2. Reduction Factors

Year	Reduction factor	Total Yearly reduction
2010	0.7708	61658604.4
2011	0.725	57994925.0
2012	0.6791	54323246.3
2013	0.6333	50659566.9
2014	0.5874	46987888.2
2015	0.5416	43324208.8
2016	0.4957	39652530.1
2017	0.4499	35988850.7
2018	0.4041	32325171.3
2019	0.3583	28661491.9
2020	0.3124	24989813.2

4. Sensitivity analysis

With the intention of identifying the influence of each variable in the input data set, all of the values in the input and target matrices had to be scaled or normalised [10]. Normalisation is done by taking the largest value in the data set and dividing all the data set values by it; this produces a threshold of values between zero and one. Therefore, when performing a sensitivity analysis of a variable in the input matrix, it can be initialised as one, whereas all other values would consequently be initialised as zero. The purpose of the binary matrix is to isolate each individual variable in a particular column. A column is also added to represent when all of the variables are engaged, thus all variables are initialised as ones; therefore there are seven columns. Six rows are present, each row corresponding to a variable. The tabular version of the binary matrix is displayed below in Table 1.

To confirm the sensitivity output when the binary matrix is given as input to the trained network, a second matrix is created with a reverse isolation scheme. The validity of the sensitivity output of the ANN is confirmed because the output of the second sensitivity matrix returns very similar results. These outputs now represent the influence of each value of the input matrix. The ANN's output, using the binary matrix in Table 1, is used to produce common ratios between the oil, natural gas and energy consumption based on coal's sensitivity percentage.

Table 3 Binary matrix used for sensitivity analysis

	Oil	Coal	N gas	P O p	G D P	Energy consu-mption	All
Oil	1	0	0	0	0	0	1
Coal	0	1	0	0	0	0	1
Natural gas	0	0	1	0	0	0	1
Populati on	0	0	0	1	0	0	1
GDP	0	0	0	0	1	0	1
Energy	0	0	0	0	0	1	1

Variables, namely population and GDP, demonstrated an influence on the output; however, it would not be practical to reduce them. Therefore, due to the nature of those two variables, identifying a growth pattern would be more practical. By using MatLab's curve fitting toolbox, a polynomial expression was identified for both GDP and population to determine an approximate rate of growth for each variable for years between 2010 and 2020.

5. Conclusion and future work

The case study was successful in meeting its objectives. The first objective was to calculate annual carbon dioxide goals for years 2010 to 2020. The second objective was to utilise the hybrid algorithm, to obtain optimal reduction coefficients. These coefficients, in turn, forecast how much Canada's manufacturing sector must reduce their fuel emissions and energy consumption to meet annual Copenhagen targets. The simulations of the hybrid algorithm did not have cost constraints or limitations on the magnitude of fuel source reduction. In the future research, there is the potential to introduce limitations on the slope of fuel source emission reduction based on a particular company or industry's abilities and restrictions.

Acknowledgements

The research work presented in this article was presented, by the first author Arash Marzi, at the Canada Wide Science Fair (CWSF) in 2010, where it earned the gold medal in the Computing and Information Technology category.

References

[1.] http://en.cop15.dk, Retrieved Febuary 15th 2010.

[2.] http://www.cieedac.sfu.ca/CIEEDACweb/index.php?menu=1 Retrieved December 15th 2009.

[3.] http://www.canada.com/business/Canada+target+back+headed+into+Copenhagen+summit/2309960/story.html Retrieved December 16th.

[4.] http://www.iaea.org/inisnkm/nkm/aws/eedrb/data/CA-enemc.html, Retrieved December 16th.

[5.] http://www.smh.com.au/interactive/2009/world/carbon-emissions/index.html, Retrieved December 16th.

[6.] http://www.conferenceboard.ca/hcp/details/environment/greenhouse-gas- Retrieved December 16th, emissions.aspx Retrieved December 10th, 2009.

[7.] http://cdiac.ornl.gov/trends/emis/can.html.

[8.] http://datafinder.worldbank.org/co2, - Retrieved November 16th, emissions.

[9.] http://www.geo.upm.es/postgrado/CarlosLopez/materiales/cursos/www.ncgia.ucsb.edu/giscc/units/u188/u188.html.

[10.] Bevilacqua, Vitoantonio; Intini, Francesca; Kühtz, Silvana; *A model of Artificial Neural Network for the analysis of climate change*; ISF, 2008.

[11.] Rong,Aiying; Lahdelma, Risto; *CO2 emissions trading planning in combined heat and power production via multi-period stochastic optimisation*; European Journal of Operational Research; 2005.

[12.] Takahashi, Takuya; Nakamura, Masao; van Kooten, G. Cornelis; Vertinsky, Ilan; R*ising to the Kyoto challenge: Is the response of Canadian industry adequate?*; Journal of Environmental Management; 2001.

[13.] Elkamel, A.; Ba-Shammakh, M.; Douglas, P.; Croiset, E; *An Optimisation Approach for Integrating Planning and CO2 Emission Reduction in the Petroleum Refining Industry*; Ind. Eng. Chem. Res; 2008.

[14.] Wang, Jing-Min; Wang, Li-Ping; *A new method for short-term electricity load forecasting*; Transactions of the Institute of Measurement and Control; 2008.

[15.] Sheta, Alaa F.; and De Jong, Kenneth; *Time-series forecasting using GA-tuned radial basis functions*; Information Sciences; 2001.

[16.] Crone, Sven F.; *Forecasting with Artificial Neural Networks*; EVIC 2005 Tutorial; December 15, 2005

[17.] Razmi-Rad, Elhame; Ghanbarzadeh, Babak; Rashmekarim, J.; *An Artificial Neural Network for Prediction of Zeleny Sedimentation Volume of Wheat Flour*; International Journal of Agriculture &Biology; 2008.

[18.] Abido M. A.: *A Novel Multi-objective Evolutionary Algorithm for Environmental/Economic Power Dispatch.* Electric Power Systems Research, Vol. 65 2003.

[19.] Ah King R. T. F., Rughooputh H. C. S.: *Elitist Multi-objective Evolutionary Algorithm for Environmental/Economic Dispatch. IEEE Congress on Evolutionary Computation*, Canberra, Australia, Vol. 2, 2003.

[20.] Pham D.T., Ghanbarzadeh A., Koç E., Otri S., Rahim S., Zaidi M.: *The Bees Algorithm – A Novel Tool for Complex Optimisation Problems*. Cardiff University, Manufacturing Engineering Centre, 2006.

[21.] Marzi Arash; *A Bees Inspired Multi-Objective Optimisation Algorithm applied to the Environmental/Economic Dispatch Problem*; Moncton, Canada, 70th MCDA Conference; 2009.

[22.] Neural Network Design; Hagan, Martin T.; Demuth, Howard B.; Beale, Mark; PWS Publishing Company; 1996.

[23.] MathWork; MatLab r2008b; 2008.

[24.] Siddharth Sharma; MatLab Technical Support Assistant; Re: Extrapolation using Neural Network Toolbox; March 12, 2010.

Innovative Production Machines and Systems
D.T. Pham, E.E. Eldukhri and A.J. Soroka (eds)

A Neighbourhood Expansion Technique in Improving the Bees Algorithm

D.T. Pham*, S.A. Ahmad* and M. Castellani**

** School of Engineering, Cardiff University*
Queen's Building, The Parade, Newport Road, Cardiff CF24 3AA, UK.

*** Modelling Group, Department of Biology, University of Bergen, Postboks 7803, 5020 Bergen, Norway*

Abstract

The Bees Algorithm mimics the foraging behaviour of honey bees in nature. This work presents a modification of the local search procedure that improves the performance of the algorithm. The proposed modification consists in manipulating the neighbourhood size, expanding it dynamically as the optimization progresses. The Bees Algorithm with neighbourhood expansion (BA-NE) is tested on eight benchmark test functions of various shape and dimensionality. Thanks to this new neighbourhood expansion technique, the modified Bees Algorithm outperforms the standard Bees Algorithm (BA) in terms of accuracy and number of iterations.

Keywords: Bees Algorithm, expansion

1. Introduction

Many phenomena in Nature have influenced scientists to develop solutions to problems in daily life. The interaction and collaboration of bees, for instance, has inspired the development of a number of optimization algorithms that are widely used in engineering and economy. Using a decentralized decision-making system, bees are able to perform complex tasks such as to optimise the food foraging process, select a tree for dwelling, or repel predators. The achievements of bee societies have inspired researchers to devise models that mimic their bees.

There are many optimization procedures based on biological concepts in the literature. Some of them use iterative methods and combine local and random search such as Genetic Algorithms (GA) [1], Particle Swarm Optimization (PSO) [2], Ant Colony Optimization (ACO) [3], and a number of recently developed algorithms inspired by different behaviours of honey bees. Abbass [4] created the first bees-inspired optimization algorithm taking inspiration from the mating-flight behaviour of bees. Pham et al. [5] implemented an algorithm based on the foraging behaviour of honey bees. A similar algorithm was developed also by Karaboga and his co-workers [6]. Pham's and Karaboga's algorithms differ in terms of the implementation of the search mechanism and the number of parameters. In addition, Algorithms based on the foraging behaviour of bees are also proposed in [7-10].

This work presents the Bees Algorithm with Neighbourhood Expansion (BA-NE), a modification of Pham's BA [4]. The paper is organized as follows: Section 2 describes the BA and the proposed BA-NE algorithm. Section 3 describes the experimental tests, and presents the results obtained by the BA-NE procedure. Section 4 analyses the sensitivity of the BA to the changes of the neighbourhood size and Section 5 summarises the achievements.

2. The Bees Algorithm

2.1. Standard procedure

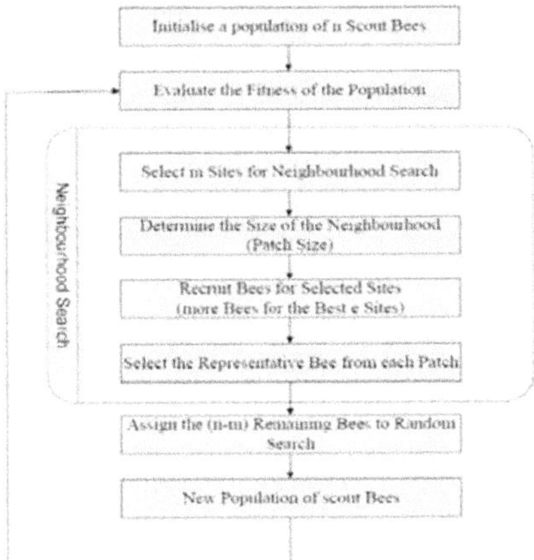

Fig.1: Flowchart of the Bees Algorithm

The BA requires six parameters, namely the number of scout bees (*n*), number of selected sites (*m*), number of bees recruited for each selected site (*nsp*), number of top-ranking (elite) sites among the *m* selected sites (*e*), number of bees recruited for each elite site (*nep*), and the neighbourhood size (*ngh*). The flowchart of the Bees Algorithm is shown in Fig. 1 [11]. The optimisation process starts with *n* scout bees randomly spread across the solution space. Each bee is associated to the site (solution) visited. The algorithm enters then the main loop.

The main loop is iterated until the stopping criterion is met. It comprises the selection (waggle dance), local and global search procedures. The visited sites are evaluated and ranked in descending order of the fitness, and the best *m* locations are selected for local search. In the local search step, more forager bees are sent in the neighbourhood of the elite (*e*) bees, and fewer bees around the remaining *m-e* bees. According to this strategy, the foraging effort is concentrated on the very best (i.e. elite) solutions. That is, *nep* bees are sent to forage around the elite *e* sites. The area around the remaining *m-e* locations is searched by *nsp* (*nsp*<*nep*) forager bees. Within the given neighbourhood area *ngh*, some of the newly generated solutions are expected to be better than that found by the scout bee. In the global search procedure, *n-m* scout bees are used to explore at random the solution space. At the end of each cycle, a new list of scout bees is formed, including the fittest solutions from each neighbourhood (local search results), and the new randomly generated solutions (global search results). This list will be sorted in the next iteration, and used for a new phase of local and global optimisation. These steps are repeated until either the maximum number of iterations (*imax*=50000) is reached, or the global optimum is located with a precision better than 0.001.For every test function, the values of all parameters are optimised by trial and error since there is no documented procedure to guide the user in selecting the most suitable parameters [9]. In [10], the distance of flight is a key factor and it significantly affects the time to reach the global optimum. Similarly in this work, the size of neighbourhood area greatly influences the performance of the algorithm. Therefore, the right value of *ngh* must be carefully chosen.

2.2 BA-NE

This work builds on the BA [5], and proposes a new adaptive neighbourhood expansion procedure to modify the current algorithm.

Like the BA, the proposed algorithm requires six parameters and employs the same initialisation procedure. At the present, when the local search procedure fails to improve the current solution, the size of a flower patch is gradually shrunk until either a better solution is produced, or the patch is abandoned [11]. Contrarily to the standard practice [5, 11], the idea proposed in this paper is to expand the neighbourhood size when the local search procedure progresses along the fitness landscape, and keep it unchanged when local search brings no improvement. The proposed neighbourhood expansion technique increases the range of the local search around the most promising solutions, aiming to speed up the optimisation of the fitness landscape.
According to the proposed procedure, all the sides of the local neighbourhood are expanded of a small amount. For each variable, the side of

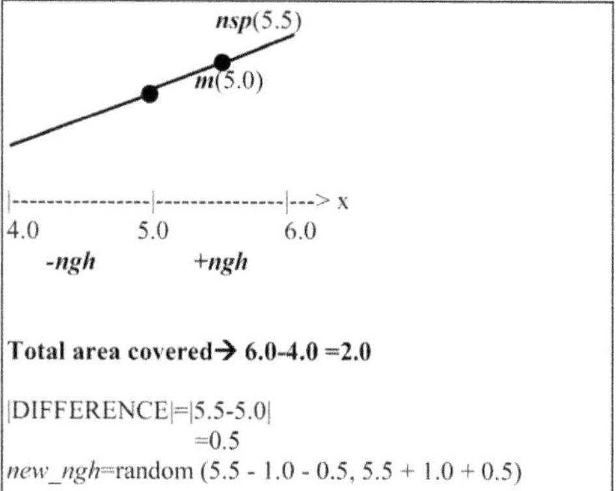

Fig. 2a. The position of *m* and *nsp* in the current iteration

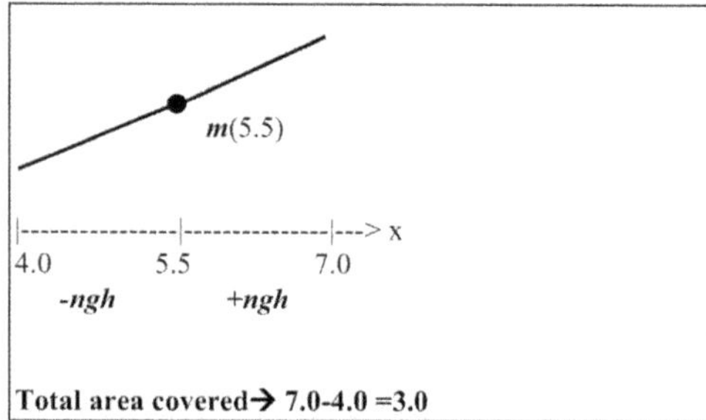

Fig. 2b. The position of *m* in the second iteration

the expanded neighbourhood is calculated according to Equation (1):

new_ngh=

random(position_*m*-*ngh*-DIFFERENCE, position_*m*+*ngh*+DIFFERENCE) (1)

where,

DIFFERENCE=

|position of best recruited bees, *nsp*-position of selected bees, *m*|

Fig. 2a and 2b illustrate an example of the application of the proposed method to the local search procedure in a one-dimensional flower patch, where *ngh*=1.0, *m* is at position (5.0), *nsp* lands at (5.5), and the optimisation problem is a maximisation one. Fig. 2a describes the search process at iteration *i*, while Fig. 2b describes the situation at iteration *i*+1. Since the local search procedure deploys and evaluates the forager bees one by one, rather than simultaneously, Fig. 1a represents a step of the neighbourhood search process around a given *m* (*e*) local best solution. Fig. 2b shows the case when one of the *nsp* (*nep*) foragers finds a fitter solution than *m* (*e*).

In the example of Fig. 2a and 2b, the total neighbourhood size (area covered by the local search procedure) is increased from 2.0 to 3.0. That is, the neighbourhood size is expanded from 1.0 to 1.5 on the right and left sides of the local optimum. In the event that no improvement takes place, the neighbourhood size is kept equal to *ngh*.

The advantage of using a dynamically varying patch size is that it allows the bees to adjust adaptively the neighbourhood area, with the aim to speed up the search process.

Fig. 3 shows the pseudo code of the BA-NE.

---

1. Initialize population with random solutions.
2. Evaluate and rank of the fitness of the population.
3. While (stopping criterion not met).
// Forming new population.
4. Select the *m* of the solution.
5. Recruit bees and evaluate their fitness.
6. Select the fittest bee from each patch.
7. Update new neighbourhood size to be used in the next iteration.
8. Assign bees to search randomly and evaluate their fitness.
9. End While.

Fig. 3. Pseudo code of the Bees Algorithm with neighbourhood expansion technique

Whilst the stopping criterion is not satisfied, the reach of the local search procedure is individually adjusted for each of the selected solutions. If a new local optimum is found, the neighbourhood size is expanded proportionally to the distance between the new best and the previously known one.

3. Experiment and result

For every function, a set of 100 trials were run. Table1 shows the eight test functions used [12, 13], and Table2 presents the parameter setting. The average and standard deviation of the accuracy and number of iterations are shown in Table3. It shows

Table1. Test Functions

No.	Function	Interval	Equation	Minimum
1	Goldstein & Price (2D)	[-2,2]	$A(x_1,x_2)=1+(x_1+x_2+1)^2(19-14x_1+3x_1^2-14x_2+6x_1x_2+3x_2^2)$ $B(x_1x_2)=30+(2x_1-3x_2)^2(18\text{-}32x_1+12x_1^2+48x_2-36x_1x_2+27x_2^2)$ $f(x_1,x_2)=AB$	$\vec{x}=(0,-1)$ $f(\vec{x})=3$
2	Branin (2D)	[-5,10]	$f(x_1,x_2)=a(x_2-bx_1^2+cx_1-d)^2+e(1-f)\cos(x_1)+e$ $a=1, b=\frac{5.1}{4\pi^2}, c=\frac{5}{\pi}, d=6, e=10, f=\frac{1}{8\pi}$	$\vec{x}=(-\pi,12.275),(\pi,2.275),(9.4286,2.475)$ $f(\vec{x})=0.3977272$
3	Schwefel (2D)	[-500,500]	$f(x_1,x_2)=-x_1\sin(\sqrt{\lvert x_1\rvert})-x_2\sin(\sqrt{\lvert x_2\rvert})$	$\vec{x}=(420.97,420.97)$ $f(\vec{x})=-837.97$
4	Schaffer (2D)	[-100,100]	$f(x_1,x_2)=0.5+\frac{(\sin\sqrt{x_1^2+x_2^2})^2-0.5}{[1.0+0.001(x_1^2+x_2^2)]^2}$	$\vec{x}=(0,0)$ $f(\vec{x})=0$
5	Rosenbrock (4D)	[-1.2,1.2]	$f(\vec{x})=\sum_{i=1}^{4}100(x_{i+1}-x_i^2)^2+(1-x_i)^2$	$\vec{x}=(\vec{1})$ $f(\vec{x})=0$
6	Hypersphere (10D)	[-5.12,5.12]	$f(\vec{x})=\sum_{i=0}^{10}x_i^2$	$\vec{x}=(\vec{0})$ $f(\vec{x})=0$
7	Ackley (10D)	[-32,32]	$f(\vec{x})=20-20e^{-0.2\sqrt{(1/10)\sum_{i=1}^{10}x_i^2}}-e^{(1/10)\sum_{i=0}^{10}\cos(2\pi x_i)}+e$	$\vec{x}=(\vec{0})$ $f(\vec{x})=0$
8	Rastrigin (10D)	[-5.12,5.12]	$f(\vec{x})=\sum_{i=1}^{i=10}[(x_i)^2-10\cos(2\pi x_i)+10]$	$\vec{x}=(\vec{0})$ $f(\vec{x})=0$

Table2. Parameters used

No.	Function	*n*	*m*	*nsp*	*e*	*nep*	*ngh*
1	Goldstein & Price (2D)	10	3	2	1	13	0.005
2	Branin (2D)	10	3	3	2	4	0.05
3	Schwefel (2D)	10	2	5	1	6	0.5
4	Schaffer (2D)	100	4	10	2	30	3
5	Rosenbrock (4D)	10	7	10	1	30	0.0015
6	Hypersphere (10D)	10	7	20	1	30	0.05
7	Ackley (10D)	100	8	10	1	20	0.7
8	Rastrigin (10D)	100	3	20	1	40	0.01

that the global optimum was found in all functions within the allocated *imax* (50000), except for the Ackley and Rastrigin benchmarks.

4. Sensitivity test

The expansion of the best *m* at the first axis in the final iteration was recorded. After 100 runs, the average of the expansion is calculated as in Table4 and this value is assumed as the 'optimum' *ngh*. The new set of the BA parameters including the 'optimum' *ngh* is presented in Table5. An experiment on the BA by using this fixed and 'optimum' ngh is then carried out to see if the

Table3. Results

No.	Function	Average of accuracy	Std. Dev.	Average number of iterations	Std. Dev.
1	Goldstein & Price (2D)	3.0005	0.0003	14.86	7.33
2	Branin (2D)	0.3982	0.0003	16.14	7.26
3	Schwefel (2D)	837.9653	0.0005	49.61	22.88
4	Schaffer (2D)	0.0005	0.0003	913.47	1407.92
5	Rosenbrock (4D)	0.0009	0.0002	401.61	148.35
6	Hypersphere (10D)	0.0007	0.0002	810.30	1208.40
7	Ackley (10D)	0.9355	0.1962	50000	0.00
8	Rastrigin (10D)	20.1264	7.2673	50000	0.00

Table4. Expansion of neighbourhood

No.	Function	Original *ngh*	avg. expansion (100 runs)	Ori. *ngh*+avg. expansion (100 runs)
1	Goldstein & Price(2D)	0.005	0.0027	0.0077
2	Branin (2D)	0.05	0.0165	0.0665
3	Schwefel (2D)	0.5	0.1076	0.6076
4	Schaffer (2D)	3.0	0.0005	3.0005
5	Rosenbrock (4D)	0.0015	0.0004	0.0019
6	Hypersphere (10D)	0.05	0.0106	0.0606
7	Ackley (10D)	0.7	0.1166	0.8166
8	Rastrigin (10D)	0.01	0.0017	0.0117

Table5. New parameter set

No.	Function	*n*	*m*	*nsp*	*e*	*nep*	new ***ngh***
1	Goldstein & Price (2D)	10	3	2	1	13	0.0077
2	Branin (2D)	10	3	3	2	4	0.0665
3	Schwefel (2D)	10	2	5	1	6	0.6076
4	Schaffer (2D)	100	4	10	2	30	3.0005
5	Rosenbrock (4D)	10	7	10	1	30	0.0019
6	Hypersphere (10D)	10	7	20	1	30	0.0606
7	Ackley (10D)	100	8	10	1	20	0.8166
8	Rastrigin (10D)	100	30	20	10	40	0.0117

adaptive expansion procedure works better than expanding the neighbourhood of a fixed amount.

This test gives the reader an insight if the adaptive neighbourhood is beneficial to the solution of any of the benchmark problems. The significance of the differences in the accuracy and the number of iterations is justified using Student's t-test, at 0.05(5%) level of significance. Significant differences are represented by 'S', and not significant differences as 'NS'. The superiority of the algorithm is indicated by using hierarchical method, where the accuracy has the primacy over the number of iterations.

According to Table6 and Table7, the Rosenbrock, Ackley, Branin, Schwefel and Hypersphere produced better performance with the BA-NE. It is because this method promotes the algorithm to converge faster and hence the bees do not require more iteration to capture the global optimum.

Table6. Results of the expanding the neighbourhood of a fixed amount and comparison with the BA-NE

(Accuracy)

No.	Function	The BA with fixed and 'optimum' *ngh*		student's t-test	
		Average of Accuracy	Std. Dev.		
1	Goldstein & Price (2D)	3.0005	0.0003	0.282	NS
2	Branin (2D)	0.3982	0.0003	0.559	NS
3	Schwefel (2D)	837.9653	0.0005	0.615	NS
4	Schaffer (2D)	0.0004	0.0003	1.030	NS
5	Rosenbrock (4D)	0.0010	8.24E-05	3.070	S
6	Hypersphere (10D)	*0.0008	0.0002	0.331	NS
7	Ackley (10D)	1.0564	0.2752	3.580	S
8	Rastrigin (10D)	20.8889	7.5155	0.729	NS

Table7. Results of the expanding the neighbourhood of a fixed amount and comparison with the BA-NE (Number of iterations)

No.	Function	The BA with fixed and 'optimum' *ngh*		student's t-test	
		Average of number of iterations	Std. Dev.		
1	Goldstein & Price (2D)	16.36	8.06	1.380	NS
2	Branin (2D)	19.51	12.18	2.380	S
3	Schwefel (2D)	62.63	34.40	3.150	S
4	Schaffer (2D)	521.79	800.16	2.420	S
6	Hypersphere (10D)	*3854.90	5945.55	5.010	S
8	Rastrigin (10D)	50000	0	0	NS

Table8. Percentage of neighbourhood size

No.	Function	limits	*ngh*	*x*=(*ngh*+*ngh*)^dim	search space=(upper limit-lower limit)^dim	% of *ngh*=(*x*/search space)*100%
1	Goldstein & Price (2D)	x_i=[-2,2]	0.005	0.0001	16	0.000625
2	Branin (2D)	x_1=[-5,10],x_2=[0,15]	0.05	0.01	225	0.004444444
3	Schwefel (2D)	x_i=[-500,500]	0.5	1	1000000	0.0001
4	Schaffer (2D)	x_i=[-100,100]	3.0	36	40000	0.09
5	Rosenbrock (4D)	x_i=[-1.2,1.2]	0.0015	8.1E-11	33.1776	2.44141E-10
6	Hypersphere (10D)	x_i=[-5.12,5.12]	0.05	1E-10	1.268E+10	7.88861E-19
7	Ackley (10D)	x_i=[-32,32]	0.7	28.9254655	1.153E+18	2.50888E-15
8	Rastrigin (10D)	x_i=[-5.12,5.12]	0.01	1.024E-17	1.268E+10	8.07794E-26

On the other hand, when the neighbourhood is expanded at a fixed amount, the

Hypersphere only achieved 93% of success. Meanwhile, the Schaffer is better when expanding the neighbourhood of a fixed amount is used. However, the Goldstein & Price and Rastrigin did not give any difference between these two methods. Therefore, it can be concluded that the performance of these functions does not depend on the method and they are not sensitive to the changes of the neighbourhood size in the algorithm.

The Schaffer and Rastrigin are well-known for featuring a numerous local optima. The *ngh* value for these functions should be carefully chosen since a too large *ngh* may cause the forager bees to overshoot the global optimum, whereas a too small *ngh* would increase the chance that the bees get trapped into a local optimum. In this work, a neighbourhood *ngh*=3.0 is used for the Schaffer function, and it is considerably small relatively to total search space. Considering that the search space is [-100,100], the neighbourhood area only constitutes as little as $([(3+3)/(100-(-100))]^2)$, that is 0.09% of the search space. The percentage of size of the neighbourhood of every function is shown in Table8.

The smallness of the neighbourhood area compared to the whole search space may imply that local search is restricted to a too small area, and thus may never overcome a local obstacle (e.g a ridge or plateau) to find an area of better fitness. In the case that the chosen neighbourhood had been too small, it would have also slowed down the progress towards the global optimum. As a result, the bees would never find the global optimum, but they would have stagnated at the same spots producing similar solutions over the remaining iterations.

Therefore, it might be postulated that the algorithms have a difficulty to set a suitable neighbourhood size for the local search process [13], especially when the test function has numerous local optima.

There is no performance difference on the Goldstein & Price functions. It means that the role of adaptive and fixed neighbourhood is not distinct for this function. Any method could locate the global optimum for the Goldstein & Price as it is a fairly easy optimization problem.

5. Conclusion

Expanding and varying the neighbourhood size on all sides could capture the global optimum quickly and the results are better than expanding the neighbourhood of a fixed amount on majority of the cases. The size of neighbourhood is sometimes sensitive to the algorithm. For example, by changing it from 0.05 to 0.0606, it would result non-100% success on the Hypersphere benchmark.

The landscape of the surface highly influences the performance of the algorithm. A smooth surface encourages bees to capture the global optimum in a short period, while rough surface would require more time in finding the global optimum.

The proposed new technique still preserves the main structure of the original Bees Algorithm metaphor, such as exploration, exploitation and recruitment. However, the adaptive expansion of the neighbourhood size allows to maximize the performance of the algorithm without time consuming trial and error optimisation.

References

[1] David, E.G., *Genetic Algorithms in Search, Optimization and Machine Learning*. 1989: Addison-Wesley Longman Publishing Co., Inc. 372.

[2] Eberhart, R. and J. Kennedy. *New optimizer using particle swarm theory*. in *Proceedings of the International Symposium on Micro Machine and Human Science*. 1995.

[3] Dorigo, M., V. Maniezzo, and A. Colorni, *Ant system: optimization by a colony of cooperating agents*. Systems, Man, and Cybernetics, Part B, IEEE Transactions on, 1996. **26**(1): p. 29-41.

[4] Abbass, H.A. *MBO: marriage in honey bees optimization-a Haplometrosis polygynous swarming approach*. in *Evolutionary Computation, 2001. Proceedings of the 2001 Congress on*. 2001.

[5] Pham, D.T., A. Ghanbarzadeh, E. Koc, S. Otri, S. Rahim and M. Zaidi. *The Bees Algorithm, A Novel Tool for Complex Optimisation Problems*. in *Proc Proc 2nd Int Virtual Conf on Intelligent Production Machines and Systems (IPROMS 2006)*. 2006: Oxford: Elsevier.

[6] Karaboga, D. and B. Basturk, *A Powerful and Efficient Algorithm for Numerical Function Optimization: Artificial Bee Colony (ABC) Algorithm*. Journal of Global Optimization, 2007. **39**(3): p. 459-

471.

[7] Akbari, R., A. Mohammadi, and K. Ziarati, *A novel bee swarm optimization algorithm for numerical function optimization.* Communications in Nonlinear Science and Numerical Simulation, 2010. **15**(10): p. 3142-3155.

[8] Banharnsakun, A., T. Achalakul, and B. Sirinaovakul, *The best-so-far selection in Artificial Bee Colony algorithm.* Applied Soft Computing. **11**(2): p. 2888-2901.

[9] Dereli, T. and G.S. Das, *A hybrid 'bee(s) algorithm' for solving container loading problems.* Applied Soft Computing Journal. **11**(2): p. 2854-2862.

[10] Sundareswaran, K. and V.T. Sreedevi. *Development of novel optimization procedure based on honey bee foraging behavior*. 2008.

[11] Ghanbarzadeh, A., *The Bees Algorithm: A Novel Optimisation Tool,* in *Manufacturing Engineering Centre*. 2007, Cardiff University.

[12] Mathur, M., S. B. Karale, S. Priye, V. K. Jayaraman and B. D. Kulkarni. *Ant colony approach to continuous function optimization.* Industrial and Engineering Chemistry Research, 2000. **39**(10): p. 3814-3822.

[13] Pham, D.T. and M. Castellani, *The Bees Algorithm: modelling foraging behaviour to solve continuous optimization problems.* Proceedings of the Institution of Mechanical Engineers, Part C: Journal of Mechanical Engineering Science, 2009. **223**(12): p. 2919-2938.

Innovative Production Machines and Systems
D.T. Pham, E.E. Eldukhri and A.J. Soroka (eds)

Enhancing multi-stage deep-drawing processes through the novel use of the Bees-Algorithm

F.J. Ramírez[a], J.Y.Lee[b], M.S. Packianather[b], D.T. Pham[b]

[a] *School of Industrial Engineering, University of Castilla-La Mancha, Albacete, 02071, Spain*
[b] *Manufacturing Engineering Centre, Cardiff University, Cardiff CF24 3AA, UK*

Abstract

The paper describes the application of the Bees Algorithm to multi-stage deep-drawing and ironing processes for improving the manufacturing time and cost of these type of axisymmetric components. The algorithm is applied to a scientific model which provides an initial solution. This initial solution is optimized by means of objective functions and constraints. The resolution of the optimization process is carried out firstly by means of an algorithm named "Deep-drawing tool" (DDT) based on a recursive function to search for the best solution in order to minimize the total time and cost of the global process. Secondly, the Bees Algorithm (BA) is applied. The results show that the enhanced BA achieving a significant reduction in time and cost of the process compared to the initial solution and the DDT algorithm. The experimental results show the reliability of the model and the robust performance of the Bees Algorithm.

Keywords: Bees Algorithm, Optimization, Deep-drawing, Ironing.

1. Introduction

The manufacturing industry requires more efficient processes, and when it comes to deep-drawing procedures it is important to achieve optimal process times and costs. Thus, it is necessary to devise specific algorithms that will satisfy these demands. Such algorithms should be based on technological and scientific basis that will provide solutions which can be easily adopted by the industries. The deep-drawing process which consists of drawing, redrawing and ironing stages has been analysed with this perspective.

This paper describes the application of the Bees Algorithm(BA) [1,2,3] to the deep-drawing process represented by its scientific model [4,5,6]. The model provides a comprehensive analysis of the phenomena occurring during the multi-stage process of an axisymmetric geometry work-piece. The scientific development departs from the work of Leu [7] and Sonis et al. [8] which provides limiting drawing ratio (LDR) solutions based on normal anisotropy value, strain hardering exponent and others, applied to the drawing and redrawing stages. The technological model developed by the authors has a scientific foundation based on the literature regarding the plastic deformation processes and, in particular about drawing [9, 10], redrawing [11] and ironing stages. This study focuses on analyzing geometries of axisymmetric components manufactured by a multi-stage deep-drawing process.

The model developed by the authors allows for a quantitative and integrated analysis to be carried out in a multi-stage deep-drawing process based on a scientific solution and real measurements. The model permits the optimisation of the initial solution for the total process time and the manufacturing cost. This optimisation is based on the resolution of an algorithm, namely, "Deep-drawing tool" (DDT) by means of recursive functions, which explores all the possibilities of the process and selects the most adequate solution. The algorithm is supported by a software tool that provides the final solution [4].

The Bees Algorithm (BA) has been applied to various optimization problems and has produced good results [1-3, 12]. This paper describes the first application of the BA to the deep-drawing process. The results are compared with the DDT solution and the experimental results show the validity of the proposed optimisation method.

2. Deep-drawing optimization problem

The optimization of the deep-drawing process is based on two models. Model A describes the optimization of the total process time and Model B describes the optimization of the manufacturing cost. The aim of both models is the minimization of both variables: time and cost.

2.1. Model A. Process time optimization.

This model is based on the minimization of the total process time according to the following objective function:

$$F_o = \min t_e = \min \sum_{i=1}^{i=n} t_i \quad (1)$$

where:
- F_o is the objective function
- t_e is the total process time
- t_i is the process time in stage i

The process time of stage i is given by the expression:

$$t_i = \frac{1.1 \cdot l_{i-1}}{v_{ai}} + \frac{L_{ui} + l_i}{v_{ei}} + \frac{1.1 \cdot l_{i-1} + L_{ui} + l_i}{v_{si}} \quad (2)$$

where:
- l_{i-1} is the part length in stage $i-1$
- L_{ui} is the punch length in stage i
- l_i is part length in stage i
- v_{ai} is the approach velocity of the press in the stage i
- v_{ei} is the ironing velocity in the stage i
- v_{si} is the recovery velocity of the press in the stage i

Using the previous expressions, the objective function can be rewritten as:

$$F_o = \min t_e = \min \sum_{i=1}^{i=n} t_i = = \min \sum_{i=1}^{i=n} \left(\frac{1.1 \cdot l_{i-1}}{v_{ai}} + \frac{L_{ui} + l_i}{v_{ei}} + \frac{1.1 \cdot l_{i-1} + L_{ui} + l_i}{v_{si}} \right) \quad (3)$$

2.2. Model B. Process cost optimization.

Model B describes the optimization of the total process cost. The model is based on the following objetive funcion:

$$F_o = \min C_t = \min(C_{mf} + C_E) = = \min \sum_{i=2}^{n} \left(c_{lb} t_i + \frac{c_e}{3.6 * 10^6} F_{e,i} l_i \right) \quad (4)$$

where:
- C_t is the total process cost
- C_{mf} is the total manufacturing cost
- C_E is the total energy cost in the process
- C_{lb} is the labour cost
- c_e is the energy cost
- l_i is part length at stage i (Variable)
- $F_{e,i}$ is the operation load at the ironing stage i

The operation load at the ironing stages is given by the expression:

$$F_{e,i} = \frac{\pi}{4}(d_{i-1}^2 - d_i^2) n_e S_u \quad (5)$$

where:
- d_{i-1} is the diameter at the stage $i-1$
- d_i is the diameter at the stage i
- n_e is an ironing coefficient
- S_u is the Ultimate Tensile Strenght of the material

(2)

2.3. Constraints

The objective functions defined on Model A and Model B have a group of constraints defined by the following expression:

$$d_n \le D_n \le d_n(1 + \Delta DR_e) \quad (6)$$

where, ΔDR_e is the drawing surplus ratio between the stages n and m, defined by the expression:

$$\Delta DR_e = DR_m - DR_n \quad (7)$$

The optimization requires a distribution, between the stages 1 to n, of the drawing ratio that has not been used, ΔDR_e. This distribution is conditioned to the minimization of the total process time as the sum of the time taken at each stage. Thus, an improvement is achieved in each drawing stage. On this manner the optimized drawing conditions are independent of the maximum limits fixed by the initial drawing conditions. The optimization is resolved using firstly the DDT algorithm and after the BA.

3. Optimization using the DDT algorithm

The DDT algorithm uses a recursive function to resolve the optimization problem. The resolution of the recursive function is carried out according to values of all the possible diameters between d_n and $d_n(1+\Delta DR_e)$. The algorithm identifies which combination is the most suitable, and it selects the optimised diameter D_n in each stage, with the condition of satisfying the objective function: the combination of diameters must give as final result the minimum time of the process.

The algorithm permits to modify the velocity parameters in each press required in the ironing process: approach velocity of the punch towards part (v_{an}), ironing velocity (v_{en}) and recovery velocity of the punch in the direction of the initial position (v_{sn}). Accordingly, it is possible to realize a more adequate distribution of available presses in the facility.

Once the parameters are calculated and the optimized solution has been found, it is possible to determine the process cost, considering the costs of workforce and electricity.

The algorithm performs the resolution in four phases:

- *Phase 1*.- Definition of incremental factor t, given by the following expression:

$$t = \frac{\Delta DR_e}{x_e} \tag{7}$$

where:

ΔDRe	is the surplus ratio of drawing
x_e	is the number of time that the drawing surplus ratios is fractioned

- *Phase 2*.- Progressive increment of diameter of each process stage, from d_n to $d_n(1+kt)$, where k is a parameter that varies from 0 to factor x_e.

- *Phase 3.-* the resolution of the recursive function is based on obtaining all possible substages, by means of the modification of each stage diameters, from d_n to $d_n(1+\Delta DRe)$.

-*Phase 4.-* Once defined all the possible diameters of each stage, the algorithm selects the arrangement that allows for the minimizing of the total process time.

4. Optimization using the Bees Algorithm

The Bees Algorithm is an optimisation algorithm inspired by the natural foraging behaviour of honey bees [1-3]. A colony of honey bees fly over long distances in multiple directions in order to exploit a large number of food sources at the same time. While searching, flower patches with plentiful amounts of nectar or pollen that can be collected with less effort are visited by more bees, whereas patches with less nectar or pollen receive fewer bees. The foraging process begins in a colony by sending scout bees randomly to search for promising flower patches. Those scout bees that found a patch which is rated above a certain threshold go to the "dance floor" to perform a dance known as the "waggle dance". This dance enables the colony to evaluate the relative merit of different patches and recruit worker bees to harvest it. This allows the colony of bees to gather food quickly and efficiently.

The BA which mimics the behaviour of honey bees requires a number of parameters to be set as given in Table 1, with the maximum number of generations set to 60. The search space is defined according Model A and B, applied to all stages of the multi-stage deep-drawing process.

Table 1: Parameters of the Bees Algorithm for the deep-drawing process

Bees algorithm parameters	Symbol	Value
ScoutBees	n	50
SelectedBees	m	8
EliteBees	e	3
SendingBeesToElite	me	10
SendingBeesToSelect	mm	5
NeighbourhoodSize	mm	0.01

5. Analysis and experimentation of real cases

In order to show the comparative results obtained with the DDT algorithm and the Bees Algorithm, the resolution of three experimental cases are presented here. The resolution of these cases requires the combination of different manufacturing processes (drawing, redrawing and ironing), highlighting the potential of the proposed model. Table 2 shows the dimensions of the three cases.

Table 2: Dimensions of real parts used in the experiment

Dimensions (mm)	Case I	Case II	Case III
External diameter d_n	98.5	134.4	102.7
Length, l_n	620	650	1,480
Bottom thickness, s_n	13.5	13.8	15
Wall thickness, e_n	2	1.45	0.95
Material	UNS C26000	UNS C26000	UNS C26000

5.1. DDT Algorithm. Experimental results

Figures 1 and 2 shows the experimental results obtained from the application of the Deep-drawing Tool Algorithm to the cases I, II and III.

The figures show the improvements on process time and manufacturing cost compared to the initial solution, in function of the number the iterations.

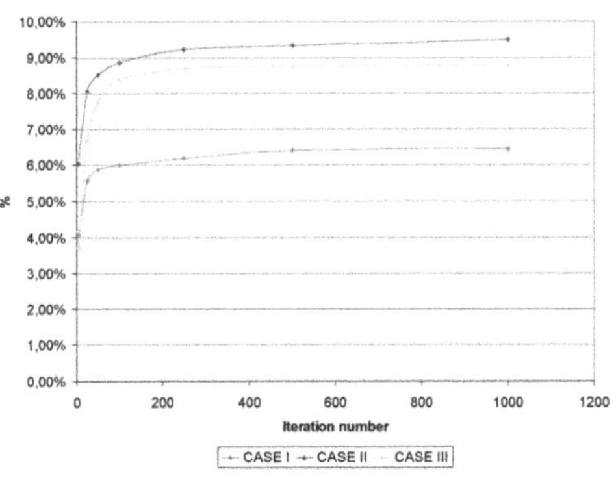

Fig. 1. DDT algorithm. Process time improvements.

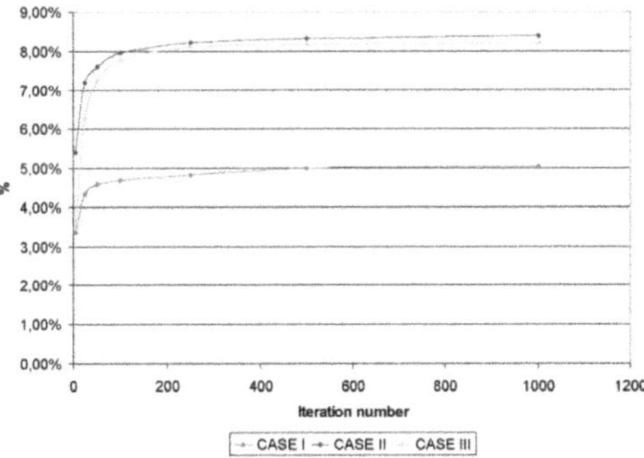

Fig. 2. DDT algorithm. Process cost improvements.

5.2. Bees Algorithm. Experimental results

On the same manner, the Bees Algorithm is applied to the cases I, II and III. The results obtained by this optimization tool in the figure 3 show the evolution of the lowest time and cost in each iteration for Case I. It can be seen that the objective function decreases very quickly in the first iterations and then rapidly converges to the minimum time and cost. Similar results are obtained for the Case II and Case III, as is showed in the figures 4 and 5.

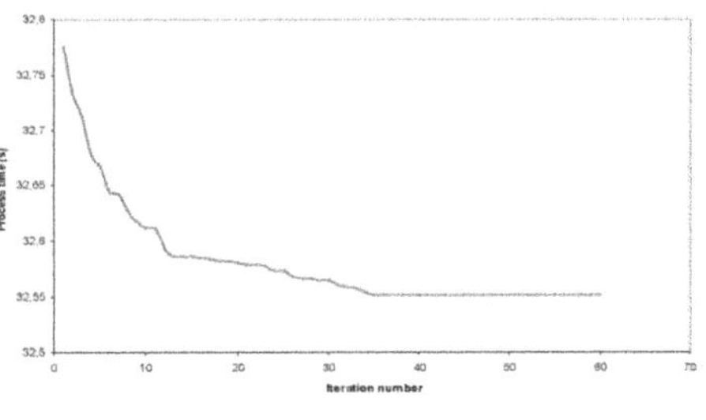

a) Process time

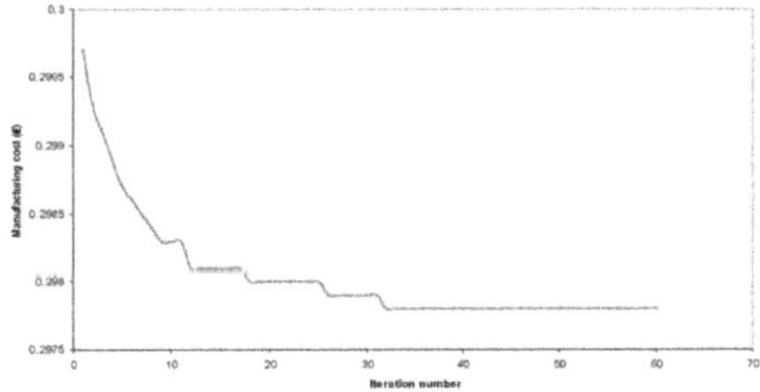

b) Manufacturing cost

Fig. 3. Bees Algorithm Results. Case I

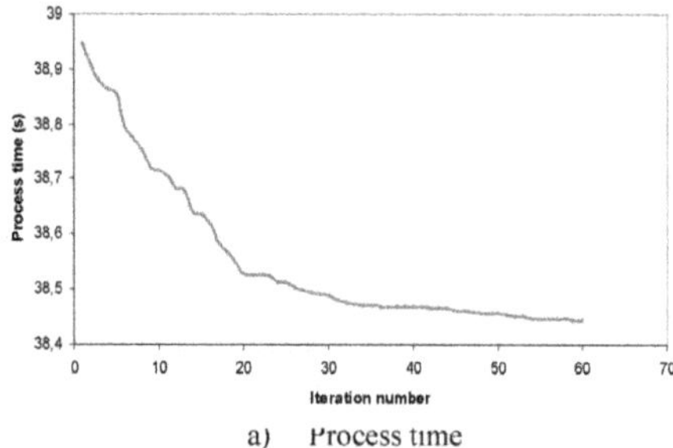

a) Process time

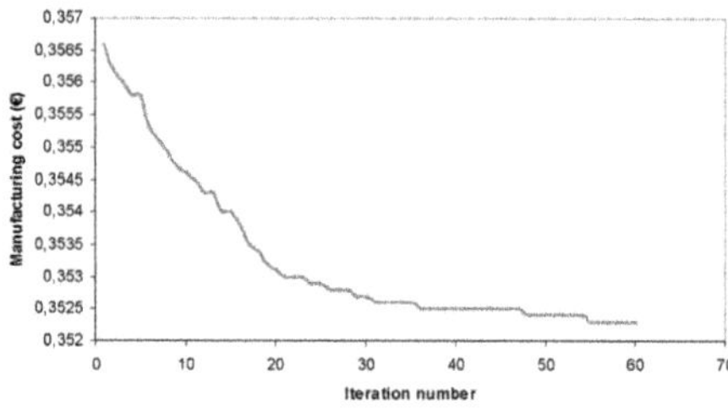

b) Manufacturing cost

Fig. 4. Bees Algorithm Results. Case II

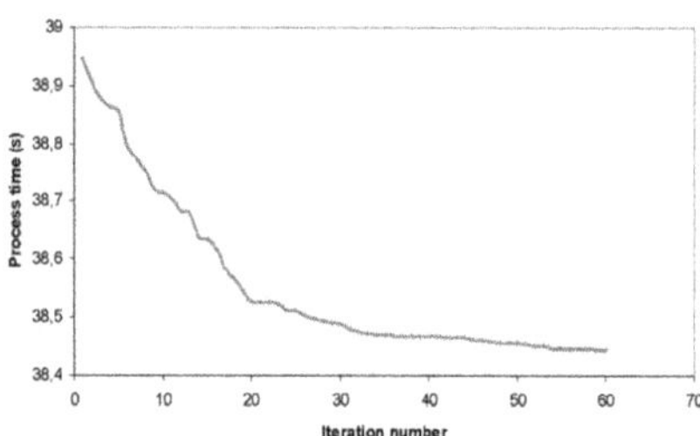

a) Process time

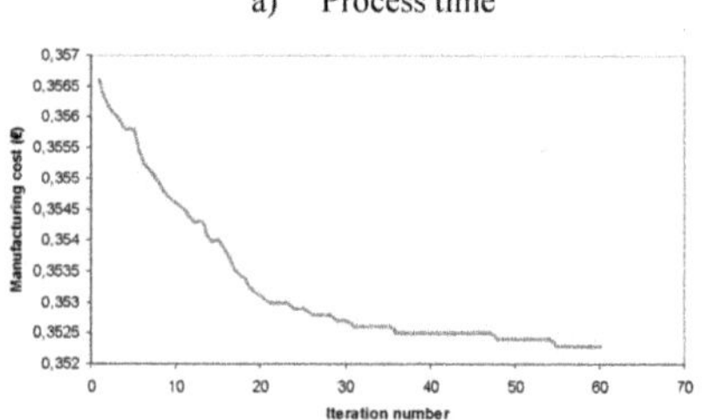

b) Manufacturing cost

Fig. 5. Bees Algorithm Results. Case III

6. Discussion of results

The resolution of the Industrial Cases I, II and III by means of the DDT Algorithm and the Bees Algorithm presents important advantages.

Table 3 presents the final results for the variable "Process time" and "Manufacturing cost" applied to the three cases under study. According to the results, significant improvements have been obtained for the three cases, with the best improvements achieved by the Bees Algorithm.

Table 3: Comparative results – CASES I, II and III

	Process time (s)			Manufacturing cost (€)		
	I	II	III	I	II	III
IS	34.83	42.4	76.58	0.318	0.398	0.688
DDT	33.03	38.93	70.95	0.304	0.367	0.639
BA	32.55	38.44	69.82	0.297	0.352	0.627
	Improvement (%)			Improvement (%)		
DDT vs IS	5.17	8.18	7.35	4.40	7.78	7.12
BA vs IS	6.55	9.34	8.83	6.60	11.55	8.88
BA vs DDT	1.45	1.26	1.59	2.3	4.08	1.87

As shown in the table 3 the DDT Algorithm produces improvements between 5.17% and 8.18 % compared with the initial solution for the process time variable, and between 4.40% and 7.78% for the manufacturing cost.

On the same manner, the application of the Bees Algorithm presents an improvement between 6.55 % and 9.34 % on the process time and from 6.60 % to 11.55 % for the manufacturing cost.

Comparing both optimization processes the best results are obtained for the Bees Algorithm that presents improvements between 1.26 % to 1.59 % for the process time variable and from 1.87% to 4.08% for the manufacturing cost, both compared with the Deep-drawing Tool solution.

7. Conclusions

In this paper two optimization methods have been presented which were applied to deep-drawing process. The models have been resolved by means of two novel algorithms, an algorithm named DDT based on recursive functions and the Bees Algorithm. Three industrial cases have been resolved using both techniques. The results show the reliability of both methods to resolve this type of deep-drawing optimization processes. The algorithms have been shown to converge to the optimum solution, obtaining the minimum time and cost for the process.

The results further show that the best improvements have been obtained by the novel technique of the Bees Algorithm. The application of the BA to the deep-drawing tool process confirms the efficiency and the robust performance of this algorithm to optimize this type of processes.

References

[1] Pham D.T. and Castellani M., The Bees Algorithm: modelling foraging behaviour to solve continuous optimization problems. Proc. ImechE Part C: J. Mechanical Engineering Science, 2009, 223: 2919-2938.

[2] Pham D.T., Ghanbarzadeh A., Koç E., Otri S., Rahim S. and Zaidi M. The Bees Algorithm. Cardiff University Technical Report – MEC 0501, Cardiff Manufacturing Engineering Centre, 2005.

[3] Pham D.T., Ghanbarzadeh A., Koç E., Otri S., Rahim S. and Zaidi M. The Bees Algorithm. A novel tool for complex optimisation problems. In Proceedings of the 2nd International Virtual Conference on *Intelligent production machines and systems* (IPROMS 2006), 2006, pp.454-459 (Elsevier, Oxford), available from http://www.iproms.org/

[4] Ramirez FJ, Domingo R, Sebastian MA (2009) "Design of an Aided System to Optimise Times and Costs in Deep Drawing Process", Proc. 2nd IPROMS International Researchers Symposium, Vol 1, pp. 191-196.

[5] Ramirez, F.J., Domingo, R., 2010. Application of an aided system to multi-step deep drawing process in the brass pieces manufacturing. In: AIP Conf. Proc. 1181, 370-379, Alcoy, Spain.

[6]Ramírez, F.J., Domingo, R., Packianather, M.S., Sebastian, M.A., 2010. Design and optimization of multi-step deep-drawing processes of axisymmetric cups by means of a scientific model. In: Proceedings of the 2nd International on Process Machine Interactions, Vancouver, Canada.

[7] Leu DK. The limiting drawing ratio for plastic instability of the cup drawing process. Journal of Materials Processing Technology 1999; 86: 168-176.

[8] Sonis P, Venkata N, Lal GK. On multistage deep drawing of axisymmetric components. Journal of Manufacturing Science and Engineering 2003; 125: 352-362.

[9] Lange K. Handbook of Metal Forming. McGraw-Hill Book Company, 1985.

[10] Kalpakjian S, Schmid SR. Manufacturing, Engineering and Technology. Prentice-Hall, 2007.

[11] Chung SY, Swift SH. An experimental investigation into the re-drawing of cylindrical shells. Proceedings Institute of Mechanical Engineering 1975, 1B: 437-447.

[12] Pham D.T., Ghanbarzadeh A., Otri S. and Koç E. Optimal design of mechanical components using the Bees Algorithm. Proc. ImechE Part C: J. Mechanical Engineering Science, 2009, 223: 1051-1056.

Innovative Production Machines and Systems
D.T. Pham, E.E. Eldukhri and A.J. Soroka (eds)

Integration of Agile Ontology Mapping towards NLP Search in I-SOAS

Z. Ahmed[a,b], I. Tacheva[b,c]

[a] *University of Wuerzburg, Germany*
[b] *Vienna University of Technology, Austria*
[c] *Technical University Sofia, Bulgaria*

Abstract

In this research paper we address the importance of Product Data Management (PDM) with respect to its contributions in industry. Moreover we also present some currently available major challenges to PDM communities and targeting some of these challenges we present an approach i.e. I-SOAS, and briefly discuss how this approach can be helpful in solving the PDM community's faced problems. Furthermore, limiting the scope of this research to one challenge, we focus on the implementation of a semantic based search mechanism in PDM Systems. Going into the details, at first we describe the respective field i.e. Language Technology (LT), contributing towards natural language processing, to take advantage in implementing a search engine capable of understanding the semantic out of natural language based search queries. Then we discuss how can we practically take advantage of LT by implementing its concepts in the form of software application with the use of semantic web technology i.e. Ontology. Later, in the end of this research paper, we briefly present a prototype application developed with the use of concepts of LT towards semantic based search.

General Terms: NLP, Search, I-SOAS

Keywords: Natural Language Processing, Lexer, Parser, Ontology

1. Introduction

Product data management (PDM) is the computer based system which electronically maintains the organizational technical and managerial data to take advantage in maintaining and improving the quality of products and followed development processes [11]. Major objectives of product data management are to improve the quality of products, improve team coordination, deliver products at the time, reduce engineering environment based problems, provide better and secure access to the configuration based information, prevent error creation and propagation by increasing customization of products, efficiently managing the large volumes of engineering data in reusable form generated by computer based systems and reusing design information.

In 1980s, in the begging of PDM proposition, the concept was new and not very welcomed by the industry of that time but with the passage of time now PDM is becoming more famous and widely in use of many multinational companies. Now a days, most of the PDM based applications are contributing in industry by providing engineering information management module control to access, store, integrate, secure, recover and manage information in data warehouse, distributed networked computer environment based infrastructure, resource management, information structure management, workflow control and system administration [1]. There are many PDM based application developers like Metaphase (SDRC) [7], SherpaWorks (Inso), Enovia (IBM), CMS (WTC), Windchill (PTC) [8], and Smarteam (Smart Solutions) [10].

Like many other communities of different fields the community of PDM System development is also struggling in solving one of currently available challenges i.e. Static and Unintelligent Search. In every PDM System a search mechanism is required and implemented to locate the user's needed information. But unfortunately still there is no such intelligent search mechanism is available which can process user's natural language based queries and can extract the most optimized results in minimum possible time in return. Keeping eyes on discussed problem in PDM System development, we can say, right now PDM community needs a new approach which can be very helpful in implementing the concepts of Natural Language Processing and capable of intelligently handling user's structured and unstructured natural language based requests, process and model the information for fast, optimized and efficient information retrieval mechanism.

2. I-SOAS

To take advantage in solving the problem of implementing of an intelligent human machine interface consisting of intelligent user system communication, meta data extraction out of unstructured data, semantic oriented information modeling, fast managed data extraction and final user end data representation an approach Intelligent Semantic Oriented Agent Based Search (I-SOAS) [1] has been proposed. The proposed conceptual architecture of I-SOAS is consists of four main sequential iterative components i.e. Intelligent User Interface (IUI), Information Processing (IP) [6], Data Management (DM) and Data Representation (DR), as shown in Figure 1.

IUI is responsible for the intelligent user system communication. IUI is proposed as a flexible graphical user interface capable of first analyzing the source of input, forwarding inputted data for further processing and responding back to the user with end results. Moreover IUI is supposed to be flexible enough so then it can be learned in shortest possible time and redesigned by user itself according to his need and wish. To implement the IUI as shown in Figure.1, IUI is divided into two main sub-categories i.e.Graphical User Interface and Communication Sources. Graphical User Interface is consists of the concept of three more sub-categories i.e. Intelligent, Flexible and Agent to intelligently handle the user's unstructured requests, provide multiple options to redesign the graphical user interface according to the ease of the user by user itself and perform internal architectural component's agent based communication. Where as in Communication Sources, first the corresponding user is supposed to be identified to enable the correct communication mode, if it is a digital system then electronic data communication mode will be enabled and if it is natural system then natural language based communication mode will be enabled [12].

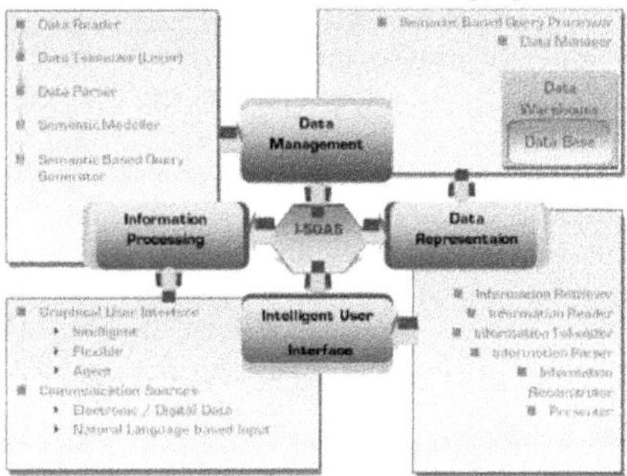

Figure. 1. I-SOAS [12]

IP unit is the most important component of the I-SOAS. The quality of performance of I-SOAS depends upon the accuracy in the results produced by this component. The overall job of IP is divided into five main iterative sequential steps i.e.Data reading, Tokenization, Parsing, Semantic Modeling, Semantic based query generation. The main concept behind the organization of these five steps is to first understand the semantic hidden in the context of natural language based set of instructions and generate a semantic information processable model for the system's own understanding and information processing [5]. In the first step, Data Reader is supposed to read and organized inputted data from IUI into initial prioritized instructions list. Then in the second step Data Tokenizer is supposed to tokenize instruction one by one, which are then treated in the third step by Data Parser for parsing and semantic evaluation with respect to the grammar of used natural language. Then in the fourth step Semantic Modeler is supposed to first filter the irrelevant semantic less data and then generate Meta data based semantic model. Then in the last and fifth step Semantic Based Query Generator is supposed to generate a new query used for further data storage and extraction of desired result.

DM is responsible for two main functions i.e. Semantic based Query Processing and Database Management. Semantic based query built in IP is treated by Semantic based Query Processor to generate SQL query to run in to database to store and extract the required information. The job of Data Manager is to manage the processes of SQL query building, data extraction and creation of new indexes and storage based on newly retrieved information [12]. DR is responsible for responding back to the user with finalized end results. This component consists of six sub components i.e. Information Retriever, Information Reader, Information Tokenizer, Information Parser, Information Reconstructor and Presenter. The job of this component is somehow similar to the job of IP, but major difference is of handling data and information. IP treats data to process but DR treats information. Required extracted and managed information from DM is passed to DR using Information Retriever, which simply read and organized by Information Reader without performing any analytical action except the prioritization of informative statements. Then using Information Tokenizer and Information Parser statements are tokenized and parsed, then using Information Reconstructor finalized formatted information is supposed to be built in user's used natural language based grammar. Finally Presenter presents the resultant information to IUI to respond back to the user [12].

The presented research in this paper, has been conducted for the implementation of one of the main modules i.e. Information Processing of a research project I-SOAS. The scope of this research is limited with respect to the construction of grammar rules. In section 3 of this research paper we give a description of the Natural Language Processing (NLP), in section 4 the ANTLR tool is briefly discussed, which later is used to write natural language grammar. In section 5 we present our newly constructed grammar and in section 6 we discuss mapping of Grammar into Ontology. In section 7 we briefly present the prototype of I-SOAS and conclude the discussion in section 8 of this research paper.

3. Natural Language Processing (NLP)

Today one of the most targeted problems in the field of artificial intelligence (computer science) is to make machine this much intelligent so then it can almost behave like a human being. Some of the behaviours of human beings have been accomplished during machine implementation e.g. now days machines can hear with the use of microphone, speak by producing sound, see with the use of cameras, smell with sensors but still there are some areas where this machine development is not completely successful and some of them are to understand natural language, learning from experience and making autonomous decisions in real time environment etc. In this research paper we will not discuss the moral problems which arise when it comes to machine development with intellectual capacities that rivals human beings but we will handle with the problems of how to make the machine understand the human languages. Natural Language Processing is one of the major step towards the development of field Artificial Intelligence as it deals with the propositions consisting of the production ability to implement an intelligent system which can not only process information but also can understand the user instructions in natural language e.g. English. Natural Language Processing performs two major tasks i.e. Natural language understanding (NLU) and Natural language generation (NLG) as shown in figure 2, to full fill the set goals of human machine communication implementation.

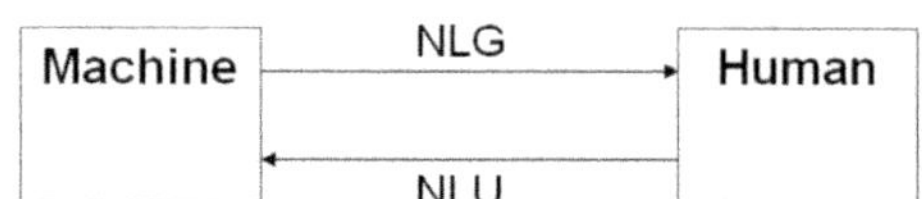

Figure.2. Human Maschine Communications

These tasks are implemented in two ways i.e. processes of converting the information from natural language to machine language and then vice versa converting the machine readable data to natural language based instructions. The main problem exists in between this communication is the structuring and restructuring of information with respect to the grammar of natural language and programming language involved.

Deep down at basic (low) level computers can only understand binary instructions to process with but here as we are dealing with the problem to make machine understand natural language, we need a natural language processor to translate languages from natural language to binary language and then back from binary to natural. To take advantage in the implementation of our targeted solution, we have used some programming language's concepts, whose compilers with the inclusion of assemblers are already doing the job of converting the information to binary level and then returning back the information after converting back to natural language statement (strings) from binary. To meet aforementioned goals, compiler consists of two main components i.e. Lexer and Parser. Lexer is basically a lexical analyser, based on the dictionary of the defined tokens of concerned language. Where as the Parser is the semantic analyzer based on the rules defined in the grammar.

NLP is one of the subfield of artificial intelligence, claiming the jobs of mainly analyzing, understanding and generating human (natural) languages, following the three step language conversion procedure as shown in figure 3. Likewise a programming language's compiler/interpreter, NLP consists of mainly three components i.e. Lexical Analyzer, Semantic Analyzer, and Translator. The lexical analyzer is also known as lexer, defining symbols or separate groups of symbols from the phrase [5]. It works like a filter program, which searches for certain characters by breaking actual character based statement(s) into tokens e.g. symbols, letters, digits, constants, reserved words, whitespace and comments etc., to perform certain defined tasks [4]. Furthermore it also regroups the input as series of characters with group significance i.e. tokens. Token is a symbol or group of symbols, don't have much sense at this level but they acquire meaning in the next step namely parser. The meaning of the tokens is given from the lexer rules. The stream of tokens generated by the lexer is received by the Semantic Analyzer to further process with.

Figure.3. Language Conversion

The Semantic Analyzer is known as a Parser which receives the input source program and breaks these instructions into parts. It groups the tokens received from the lexer according to the given language grammar. If the grammar doesn't contains this group of tokens, the parser denies the successful processing otherwise proceed with.

There are two types of parser i.e. LL and LR parsers. The difference between them is in the derivations. The LL parsers construct a leftmost derivation of the input and LR constructs the rightmost derivation. This means that the LL parser replaces the left-most non terminal first and LR replaces the right most non terminal first, but both of them parses from left to right. Furthermore parser creates the sequences of tokens to put them into an Abstract Syntax Tree (AST) and makes one or many tables with information about the tokens or group of tokens i.e. symbol table, used to validate the types of the data.

4. ANTLR

ANTLR is a tool, developed in 1983 [3] by Professor Terence Parr and his colleagues to write grammar for both lexer and parser. It is implemented in Java but it can generate source code in Java, C, C++, C#, Objective C, Python and Ruby. ANTLR use EBNF (Extended Backus-Naur Form) for the grammars, which is very formal way to describe the grammar. ANTLR provides a standard editor for grammar writing and generating lexer and parser. Till now this tool has been used for programming language's grammar writing but we are considering it for natural language processing by writing natural language's grammar and generating lexer and parser to make the machine understand it.

ANTLR has many belonging applications and opportunities to extensibilities. One of the biggest benefits is the grammar syntax; it is in EBNF form, which is a Meta syntax notation. Each EBNF rule has a left-hand side (LHS) which gives the name of the rule and a right-hand side (RHS) which gives the exact definition of the rule. Between the LHS and RHS there is the symbol ":" (colon), which separates the left from the right side and means "is defined as". One another benefit is the graphical grammar editor and debugger called ANTLRWorks, written by Jean Bovet and gives us the possibility to edit, visualize, interpret and debug any ANTLR grammar.

5. I-SOAS NLP Grammar

We have written and designed the grammatical view for the I-SOAS Lexer and I-SOAS Parser. This grammar is based on only English (Natural Language). At the moment, according to the scope of our research, the proposed grammar has been divided into three main categories i.e. A, B and C. A category is representing all English words belonging to the User (requesting for some results), B category is representing the English grammar structure and C is representing the main Object user is looking for. This grammar consists of 15 different lexer rules and 11 different parser rules in total. Please find generated grammar in section 9 Appendix of this research paper.

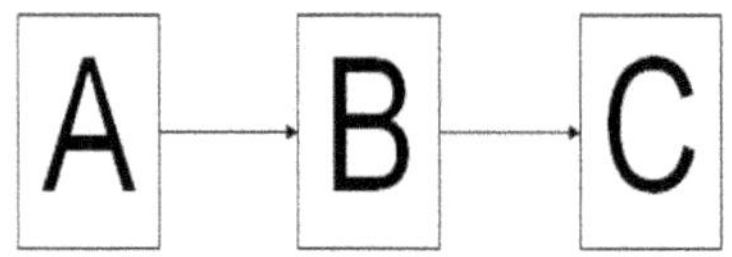

Figure.4. I-SOAS Lexer and Parser Grammar

6. Mapping Grammar to Ontology

One of the steps from Information Processing part in I-SOAS is the Semantic Modeler. I-SOAS Modeler makes a semantic model of the grammar and saves it into the Database (this is the connection between IP and DM). After the information is forwarded to the Resolver, which puts the relationships between the statements in the semantic model and also saves this into the Database. To make a conceptual model of the grammar we should use the ontology theory, which will describe its structure. Ontology clarifies the semantics of a conceptual modeling grammar [7]. In computer science the word ontology means to show one area of knowledge with conceptual scheme. The most used definition of ontology is from Gruber

"Ontology is a formal, explicit specification of a shared conceptualization" Gruber (1993)

Formal means that the specification should be machine processable, explicit means that the elements should be simply defined and specification is an abstract model. According to Gruber the ontology is such a representation of a domain, where a set of objects and their relationships is described by a vocabulary [9]. There are different kinds of ontologies but we will use Natural Language Ontology (NLO) to provide the relationships between the different statements in our grammar [8]. There is also Domain Ontology (based on the knowledge of a particular domain) and Ontology Instance (which generates automatic object based web pages), all are connected to extract hidden semantic out of data.

This Ontology scheme is made from fields with the data and connections between the different objects, which declare the rules in the knowledge area. We can make an ontology scheme for our grammar to show how rules can be constructed by connecting the words. It includes a dictionary and logical relations between the terms and their meanings, the treatment of one term with others, different variations between them and a semantic model of the terms.

The ontology construction process begins with the parsed text from the natural language i.e. Words. The nouns in the ontology are called classes and the verbs are the properties, having different values for the different relationships. The first thing to do is making an oriented graph with all the classes and properties with the semantic relationships in between, then the ontology can be made with some of the ontology supporting languages like XML (eXtensible Mark-up Language), RDF (Resource Description Framework) and OWL (Web Ontology Language) [8] etc.

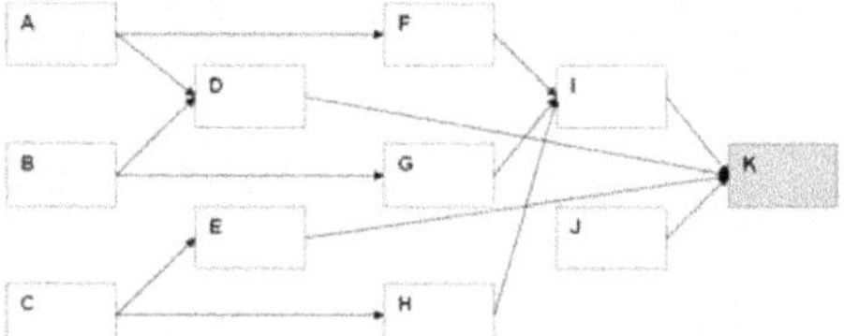

Figure.5. NLP Grammar Ontology Mapping

The map of our ontology should explain the relationships between the words in the grammar, what word comes after another. To answer the question "why", we will explain the English grammar. There are two types of verb tense that are used in our grammar i.e. Present Simple and Present Continuous. With them we have also two different persons and two different figures i.e singular and plural. This explains the statements from A to I i.e. here we are answering the questions: "Who", "What is he doing" or "What does he do" etc. Of course in the real search we can not use the personal pronoun because we can just say the action. That is why we have statements A D K, B D K, C E K, but also statements without A, B and C (which are the pronouns). D K, E, K is the example for Present Simple Tense, but we have it also for Present Continuous Tens. We have also statement J, which never uses a pronoun with itself (* J K).

In the picture of our ontology you can see the exact connections between the statements. There are in way just form left to right and there is no need to start always from A, we can start any sentence from where we want but follow the directions. The only mandatory statement is K, a key word, which defines what we want to search. In the following map scheme you can see the grammar with the connections between the words. As shown in figure 5, to produce ontology based search we have mapped the used grammatical view in Lexer and Parser in to an ontological view as shown in figure. There three main entities i.e. A, B and C. There are five main properties of A i.e. I, We, He, She and They, six properties of B i.e.am, are, is, looking/searching for, need(s)/want(s) and unknown and C has only one property K.

6. I-SOAS Prototype

I-SOAS Desktop Application as shown in figure 6. is capable of running as a stable application, taking natural language based instructions as input from user in the form of text file, and then lexing and parsing input text to build semantic based SQL queries to extract stored results from attached repositories.

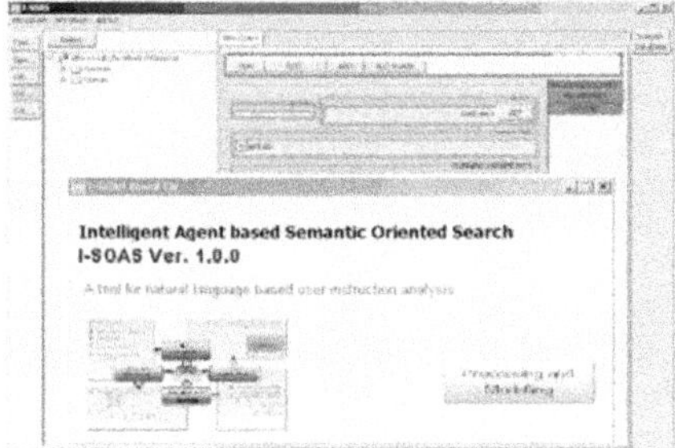

Figure.6. I-SOAS Desktop Prototype

6. Conclusion

In this research paper we have briefly addressed the importance of Product Data Management (PDM) along with currently available major challenge of natural language based search towards the Product Data Management System Development. Targeting the challenge, in this research paper, we have discussed a way for the implementation of a semantic based search by writing a natural language based grammar with the use of ANTLR and later mapped that grammar into ontology. As this is an ongoing research, we are hoping for a web based product data management system development in future, with the use of approach discussed in this research paper.

Acknowledgements

We are thankful to Erasmus for funding Ina Tacheva for working on this research project. We are thankful to Prof. Dr. Detlef Gerhard for his supervision during this research and pay our gratitude to Prof. Dr. Thomas Dandekar for his generous support.

References

[1] Ahmed. Z, D. Gerhard: "Semantic Oriented Agent based Approach towards Engineering Data Management, Web Information Retrieval and User System Communication Problems", In the proceedings of 3rd International Conference for Internet Technology and Secured Transactions, ICITST 08, IEEE - CST, June 23–28 2008

[2] Ahmed. Z. 2009. Intelligent human machine interface design for advanced product life cycle management systems. In Proceedings of the 6th international Conference on Frontiers of information Technology (Abbottabad, Pakistan, December 16 - 18, 2009). FIT '09. ACM, New York, NY, 1-4. DOI= http://doi.acm.org/10.1145/1838002.1838058

[3] Mark Volkmann, "ANTLR 3", Object Computing, Inc., 2008

[4] Programming language,last reviewed 22 June 2010, http://en.wikipedia.org/wiki/Programming_language

[5] ANTLR, Ashley J.S Mills, Last reviewed 27.06.2009, < http://supportweb.cs.bham.ac.uk/docs/tutorials/docsystem/build/tutorials/antlr/antlr.html>

[6]. Ahmed. Z, S. Majeed, T. Dandekar: "Towards Design and Implementation of a Language Technology based Information Processor for PDM Systems", in IST Transactions of Information Technology- Theory and Applications, Vol. 1, No. 1 (2), ISSN 1913-8822, pp. 1-7, Date Published: July 2010

[7] Hong Zhang, Rajiv Kishore, Ram Ramesh, "Semantics of the MibML Conceptual Modelling Grammar: An Ontological Analysis Using the Bunge-Wand-Weber Framework", Journal of Database Management, Vol. 18, Issue 1, edited by K. Siau, pp. 1-19, 2007.

[8] Ahmed. Z D. Gerhard: "How does Ontology Contribute in Semantic Web Development?", In the proceedings of 6th CIIT Workshop on Research in Computing, (CWRC Fall 2007), P 3, 27 October 2007

[9] Karin Koogan Breitman, Marco Antonio Casanova and Walter Truszkowski, "Semantic Web: Concepts, Technologies and Applications", ISBN 978-1-84628-581-3 (print), Springer-Verlag London Limited

[10] M.Y. Huang, Y.J. Lin, Hu Xu, "A framework for web-based product data management using J2EE", In Springer-Verlag London Limited,Received: 8 October 2002 / Accepted: 13 March 2003 / Published online: 27 October 2004

[11] Ahmed. Z, D. Gerhard, " Contributions of PDM Systems in Organizational Technical Data Management ", In the proceedings of The First International Conference On Computer, Control & Communication, Karachi Pakistan, 12 13 November 2007,

[12] Ahmed, Z. 2009. Intelligent semantic oriented agent based search (I-SOAS). In Proceedings of the 6th international Conference on Frontiers of information Technology (Abbottabad, Pakistan, December 16 - 18, 2009). FIT '09. ACM, New York, NY, 1-4. DOI= http://doi.acm.org/10.1145/1838002.1838065

Innovative Production Machines and Systems
D.T. Pham, E.E. Eldukhri and A.J. Soroka (eds)

Transfer prices for two-enterprize supply chain optimization by DCA

Le Thi Hoai An, Tran Duc Quynh

Laboratory of Theoretical and Applied Computer Science LITA EA 3097, Univresity of Paul Verlaine - Metz, France

Abstract

In this paper we consider the problem of fair transfer price and inventory holding policies in two-enterprise supply chains. It was formulated as a mixed integer nonlinear program (MINLP) by Gjerdrum et al. in 2002. Existing global optimization methods to solve this problem are computationally expensive. In this work, we propose a continuous approach based on DC (Diffrence of Convex functions) programming and DCA (DC Algorithm) for solving this combinatorial optimization problem. The preliminary numerical results show the effectiveness and the efficiency of our algorithm: it furnishes an good approximate solution in a short time.

Keywords: Supply chain, Nash bargaining, DC programming, DCA, Mixed integer nonlinear program.

1 Introduction

An important aim of multi-enterprize supply chains is to determine the policies that optimize the performance of the supply chain while ensuring adequate rewards for each participant. In [2], the authors proposed a provider-partnership model based on profit sharing for two partners in the supply chain where the objective is to maximize the whole profits of both partners and the demands of the customers depend on the selling price. This model concerns only one product and a single period. In practice, the firms usually want to optimize their production plan in a time horizon. So, we need to consider a multi-period model for this problem. On the other hand, the price transfer is used by a large majority of firms (refer [5],[6], [7] for an extensive view of transfer price). Moreover, it has an important impact for more fairly distributing the profit between the divisional organizations. This fact explains the necessity of the transfer price in the multi-enterprize supply chain model.

The multi-period problem of fair transfer price and inventory holding policies in two enterprize supply chains with determined demand was formulated by Gjerdrum et al. [1]. For each product, they introduce transfer price levels and select the one which maximize the performance of the supply chain. The authors proposed three approaches for this problem:

- The single approach focuses on maximizing the whole profit of both companies. This approach leads to solve a mixed integer linear problem (MILP). But, the profit distribution between two companies is unfair.

- The Nash approach is considered to avoid the unfair situation. The idea is based on the Nash bargaining function [3], which maximizes the product of two elements where each element is the difference of the achievable profit and the minimum profit level of each organization. The mathematical model is a mixed integer non linear problem (MINLP).

- The hybrid approach mixes the single approach and the Nash approach. One maximizes the whole profits such that the minimum profit requirements are the profits provided by the Nash approach.

The computational results in [1] based on the so called spatial and binary variable branch and bound algorithm show that the Nash approach furnishes the whole profit very close to the one furnished by the single approach and the profit distribution is much fairer. However, this algorithm ([1]) is computationally expensive. In this paper, we directly tackle the mixed integer nonlinear problem according to the Nash approach. Our method is based on DC (difference of convex functions) programming and DCA (DC algorithm). DCA was first introduced by Pham Dinh Tao in 1985 and has been extensively developed since 1994 by Le Thi Hoai An and Pham Dinh Tao in their common woks. It has been successfully applied to many large-scale (smooth or nonsmooth) nonconvex programs in various domains of applied sciences, and has become now classic and popular (see e.g [8],[9],[10], [12], [13] and references therein). This motivates us using DCA in the present paper.

For applying DC programming and DCA on the above problem, we first reformulate it as a DC program by using an exact penalty technique. Afterward, DCA is developed to solve the resulting problem. The computational results with random data show the effectiveness and the efficiency of the proposed method.

The paper is organized as follows. In Section 2, we introduce the mathematical model of the considered problem. Section 3 presents the solution method via DC programming and DCA. The computational results are shown in Section 4 while Section 5 is reserved for the conclusion.

2 Mathematical Model

The mathematical model of fair transfer price and inventory holding policies in two enterprize supply chains was introduced by Gjerdrum et al. [1]. In this section, we improve this model by eliminating some unnecessary variables and constraints. We first recall the problem for clarity of presentation.

We consider two organizations: Plant Corporation (A) and Distribution & Sales Corporation (B). The company A produces and delivers the goods to B. The company B then sells the goods to the external customers with a deterministic profile at each time during period $[0, T]$. Let us suppose that A manufactures I products by using R resources. The price for each product A sells to B is chosen from K transfer price levels. The minimum inventory levels for each product are specified. The costs at A include the manufacturing cost (include raw material cost), the administrative cost, the inventory cost and the transportation cost. The company B suffers the administrative cost, the inventory cost, the transportation cost and the buying cost. The revenue for A is the material purchased in B and the B's revenue is determined by the market price. The problem is to find production plan, inventory policies, transportation plan, transfer price levels such that the customer demands are fulfilled and the profit distribution between A and B is fair.

Let us note the parameters and the variables used in this paper as follows:
UR_{ir}: demand of resource r for the product i,
R_{rt}: capacity of resource r during period t,
MFC_i: manufacturing cost of product i,
TC_i^A: A's transportation cost of product i,
TC_i^B: B's transportation cost of product i,
H_i^A: A's administrative cost of product i,
H_i^B: B's administrative cost of product i,
$\hat{P}_{ik}$: k^{th} transfer price levels of product i,
D_{it}: quantity demand of customer of product i during period t,
σ_i^A, σ_i^B: minimum inventory levels of product i in A and B,
ρ_i: market price of the product i,
r: inventory holding cost constant,
Π_A^L: minimum profit requirement of A,
Π_B^L: minimum profit requirement of B.

The decision variables X_{ik} are defined as

$$X_{ik} = \begin{cases} 1 \text{ if } k^{th} \text{ price is chosen for the product } i, \\ 0, \text{otherwise}, \end{cases} \tag{1}$$

and the continuous variable are given by:
Π: profit,
P_i: transfer price of the product i between A and B,
M_{it}: production level of the product i during period t,
I_{it}: inventory level of the product i during period t,
$\overline{XI}_{ikt}$: linearized inventory of the product i during period t, k^{th} level price,
F_{it}: amount of the product i shipped from A to B during period t,
$\overline{FX}_{ikt}$: linearized shipment quantity of the product i during period t, k^{th} level price.

The resources used at each time period can not exceed the available levels. Mathematically, these constraints are expressed as:

$$\sum_i UR_{i,r}M_{i,t} \leq R_{r,t} \quad \forall r, t. \tag{2}$$

The relation between the inventory level, the production level, the amount of product shipped from A to B and the demand of the customer can be stated as follows:

$$I^A_{i,t} = I^A_{i,t-1} + M_{i,t} - F_{i,t} \quad \forall i, t \geq 2, \tag{3}$$

$$I^A_{i,1} = M_{i,1} - F_{i,1} \quad \forall i. \tag{4}$$

$$I^B_{i,t} = I^B_{i,t-1} + F_{i,t} - D_{i,t} \quad \forall r, t \geq 2, \tag{5}$$

$$I^B_{i,1} = F_{i,1} - D_{i,1} \quad \forall i. \tag{6}$$

The minimum inventory levels at A and B are expressed as follows:

$$I^A_{i,t} \geq \sigma^A_i \quad \forall i, t, \tag{7}$$

$$I^B_{i,t} \geq \sigma^B_i \quad \forall i, t. \tag{8}$$

Let us introduce K different transfer price levels for each product i. Only one of them is chosen as the price of product i when A ships it to B. The transfer price level constraints are so expressed by

$$P_i = \sum_k \hat{P}_{i,k}X_{i,k} \quad \forall i, \tag{9}$$

$$\sum_k X_{i,k} = 1 \quad \forall i, \tag{10}$$

The profit of A is equal to the income of A minus it's expenditure. A's expenditure is the total of the manufacturing cost, the administrative cost, the inventory cost and the transportation cost. A's income is determined by the the total of $P_iF_{i,t}$. Note that the inventory cost on each product depends on the outgoing price and the inventory holding cost constant r. The profit of B is calculated by the same way.

$$\Pi_A = \sum_i \sum_t \Big[P_iF_{i,t} - MFC_iM_{i,t} - r.P_iI^A_{i,t} - H^A_i(M_{i,t} + F_{i,t}) - TC^A_iF_{i,t}\Big],$$

$$\Pi_B = \sum_i \sum_t \Big[D_{i,t}\rho_i - P_iF_{i,t} - r.\rho_iI^A_{i,t} - H^B_i(D_{i,t} + F_{i,t}) - TC^B_iD_{i,t}\Big].$$

In the above equation, the term $P_iF_{i,t}$ and $P_iI^A_{i,t}$ are non linear. We apply a linearization technique to linearize these terms (refer [4]).

$$\overline{FX}_{i,k,t} \leq X_{i,k}.MAX \quad \forall i, k, t,$$

$$F_{i,t} = \sum_k \overline{FX}_{i,k,t} \quad \forall i, k,$$

$$P_iF_{i,t} = \sum_k \hat{P}_{i,k}\overline{FX}_{i,k,t},$$

$$\overline{XI}_{i,k,t} \leq X_{i,k}.MAX \quad \forall i, k, t,$$

$$I^A_{i,t} = \sum_k \overline{XI}_{i,k,t} \quad \forall i, k,$$

$$P_iI^A_{i,t} = \sum_k \hat{P}_{i,k}\overline{XI}_{i,k,t},$$

where MAX is a sufficiently large number. The profit constraints are so expressed as:

$$\Pi_A = \sum_i \sum_t \Big[\sum_k \hat{P}_{i,k}\overline{FX}_{i,k,t} - MFC_iM_{i,t} - r.\sum_k \hat{P}_{i,k}\overline{XI}_{i,k,t} - H^A_i(M_{i,t} + F_{i,t}) - TC^A_iF_{i,t}\Big], \tag{11}$$

$$\Pi_B = \sum_i \sum_t \Big[D_{i,t}\rho_i - P_iF_{i,t} - r.\rho_iI^A_{i,t} - H^B_i(D_{i,t} + F_{i,t}) - TC^B_iD_{i,t}\Big], \tag{12}$$

$$\overline{FX}_{i,k,t} \leq X_{i,k}.MAX \quad \forall i, k, t, \tag{13}$$

$$F_{i,t} = \sum_k \overline{FX}_{i,k,t} \quad \forall i, k, \tag{14}$$

$$\overline{XI}_{i,k,t} \leq X_{i,k}.MAX \quad \forall i, k, t, \tag{15}$$

$$I^A_{i,t} = \sum_k \overline{XI}_{i,k,t} \quad \forall i, k, \tag{16}$$

The constraints on minimum profit of A and B can be stated as:

$$\Pi_A \geq \Pi^L_A, \tag{17}$$

$$\Pi_B \geq \Pi^L_B. \tag{18}$$

Finally, the non negative constraints are applied for all variables.

$$\overline{FX}_{i,k,t}, \overline{XI}_{i,k,t}, M_{i,t}, F_{i,t}, I^A_{i,t}, I^B_{i,t}, P_i \geq 0 \quad \forall i, k, t. \tag{19}$$

According to the Nash approach, we want to maximize the Nash bargaining function $(\Pi_A - \Pi^L_A)(\Pi_B - \Pi^L_B)$. Hence the mathematical model of the considered problem is

$$\text{(P)} \quad \max\{f(\zeta) = (\Pi_A - \Pi^L_A)(\Pi_B - \Pi^L_B) : \zeta \in S\},$$

where S is the feasible set which contains the points ζ satisfying (1)-(19).

This problem is a mixed integer nonlinear program for which we investigate an approach based on DC programming and DCA.

3 Solution Method based on DCA

DC Programming and DCA constitute the backbone of smooth/nonsmooth nonconvex programming and global optimization. They address the problem of minimizing a function f which is difference of convex functions on the whole space $\mathbb{R}^p$ or on a convex set $C \subset \mathbb{R}^p$. Generally speaking, a DC program takes the form

$$\alpha = \inf\{f(x) := g(x) - h(x) : x \in \mathbb{R}^p\} \quad (P_{dc}) \tag{20}$$

where g, h are lower semicontinuous proper convex functions on $\mathbb{R}^p$. Such a function f is called DC function, and $g - h$, DC decomposition of f while g and h are DC components of f. The convex constraint $x \in C$ can be incorporated in the objective function of (P_{dc}) by using the indicator function on C, denoted χ_C, which is defined by $\chi_C(x) = 0$ if $x \in C$, $+\infty$ otherwise. Let

$$g^*(y) := \sup\{\langle x, y\rangle - g(x) : x \in \mathbb{R}^p\}$$

be the conjugate function of g where $\langle . \rangle$ denotes a scalar product. Then, the following program is called the dual program of (P_{dc}):

$$\alpha_D = \inf\{h^*(y) - g^*(y) : y \in \mathbb{R}^p\}. \quad (D_{dc}) \tag{21}$$

One can prove that $\alpha = \alpha_D$, and there is the perfect symmetry between primal and dual DC programs: the dual to (D_{dc}) is exactly (P_{dc}).
For a convex function θ, the subdifferential of θ at $x_0 \in \text{dom}\ \theta := \{x_0 \in \mathbb{R}^p : \theta(x_0) < +\infty\}$, denoted $\partial\theta(x_0)$, is defined by

$$\partial\theta(x_0) := \{y : \theta(x) \geq \theta(x_0) + \langle x - x_0, y\rangle, \forall x \in \mathbb{R}^p\}. \tag{22}$$

The subdifferential $\partial\theta(x_0)$ generalizes the derivative in the sense that θ is differentiable at x_0 if and only if $\partial\theta(x_0) \equiv \{\theta'(x_0)\}$.
The necessary local optimality condition for the primal DC program, (P_{dc}), is

$$\partial h(x^*) \subset \partial g(x^*). \tag{23}$$

A point that x^* verifies the generalized Kuhn-Tucker condition

$$\partial h(x^*) \cap \partial g(x^*) \neq \emptyset \tag{24}$$

is called a critical point of $g - h$.

The idea of DCA is quite simple: each iteration of DCA approximates the concave part $-h$ by one of its affinne majorization defined by $y^k \in \partial h(x^k)$ and minimizes the resulting convex function (i.e., computing $x^{k+1} \in \partial g^*(y^k)$).

Input:
Let $x^0 \in \mathbb{R}^p$, $0 \leftarrow k$.
Repeat:
Calculate $y^k \in \partial h(x^k)$.
Calculate
$x^{k+1} \in \arg\min\{g(x) - h(x^k) - \langle x - x^k, y^k\rangle\}$.
$k + 1 \leftarrow k$.
Until: convergence of x^k.

It has been proved that (see e.g. [8]) if the optimal value of problem (P_{dc}) is finite and the infinite sequence $\{x^k\}$ is bounded, then every limit point $\tilde{x}$ of the sequence $\{x^k\}$ is a critical point of $g - h$. For a complete study of DCA we refer the reader to [8], [10], [12], [13].

We show now how to apply DCA to solve the problem(P). The idea is to reformulate (P) into a DC program by using a penalty technique given in [11].

Theorem 1. *[11] Let Ω be a nonempty bounded polyhedral convex set, f be a finite DC function on Ω and p be a finite nonnegative concave function on Ω. Then there exists $\eta_0 \geq 0$ such that for $\eta > \eta_0$ the following problems have the same optimal value and the same solution set*

$$(P_\eta) \qquad \alpha(\eta) = \min\{f(x) + \eta.p(x) : x \in \Omega\},$$

$$(P) \qquad \alpha = \min\{f(x) : x \in \Omega, p(x) \leq 0\}.$$

Proof: see [11].

Denote by L the variable number of the problem (P), $L = 2 + 2IKT + 4IT + I + IK$ and $\zeta = (\Pi_A, \Pi_B, X, \overline{XI}, \overline{FX}, M, F, I^A, I^B, P) \in \mathbb{R}^L$.

Let Ω be the linear relaxation domain of S, say

$$\Omega := \{\zeta \in \mathbb{R}^L \text{ s.t. } (2-19), X_{i,k} \in [0,1]\}.$$

We consider the function $p : \mathbb{R}^L \to \mathbb{R}$ defined by

$$p(\zeta) = \sum_{i=1}^{I}\sum_{k=1}^{K} X_{i,k}(1 - X_{i,k}).$$

It is clear that $p(\zeta)$ is concave and non negative on Ω. By using the penalty theorem 1, we obtain, for a sufficiently large number η, the following equivalent DC program of (P)

$$\min\Big\{ - (\Pi_A - \Pi_A^L)(\Pi_B - \Pi_B^L) + \eta p(\zeta) : \zeta \in \Omega\Big\}.$$

This is a DC program of the form

$$\min\{f_\eta(\zeta) := g(\zeta) - h(\zeta) : \zeta \in \Omega\} \qquad (25)$$

where

$$g(\zeta) := 0.25\Pi_A^2 - 0.5\Pi_A\Pi_B + 0.25\Pi_B^2,$$

$h(\zeta) := 0.25\Pi_A^2 + 0.5\Pi_A\Pi_B + 0.25\Pi_B^2 - \Pi_B^L\Pi_A - \Pi_A^L\Pi_B + \eta\sum_{i=1}^{I}\sum_{k=1}^{K} X_{i,k}(X_{i,k}-1) + \Pi_A^L\Pi_B^L$ (g and h are clearly convex functions).

From the definition of h, we see that h is differentiable and $u = \nabla h(\zeta)$ is computed as:

$$\begin{cases} u_1 = 0.5(\zeta_1 + \zeta_2) - \Pi_B^L, \\ u_2 = 0.5(\zeta_1 + \zeta_2) - \Pi_A^L, \\ u_{2+(i-1)I+k} = 2\eta X_{i,k} - \eta, \\ u_j = 0, \quad \text{otherwise}. \end{cases} \qquad (26)$$

DCA applied to the DC program (25) consists of computing, at each iteration ℓ, the two sequences $\{u^\ell\}$ and $\{\zeta^\ell\}$ such that $u^\ell = \nabla h(\zeta^\ell)$ and $\zeta^{\ell+1}$ is an optimal solution of the next convex quadratic program

$$\min\{g(\zeta) - \langle \zeta, u^\ell\rangle : \zeta \in \Omega\}. \qquad (27)$$

Finally, DCA applied to (25) can be described as follows:

DCA-SC

Initialization

Let ϵ be a sufficiently small positive number. Set $\ell = 0$ and $\zeta^0 \in \mathbb{R}^L$.

Repeat

Calculate $u^\ell = \nabla h(\zeta^\ell)$ via (26).

Solve the convex quadratic problem (27) to obtain $\zeta^{\ell+1}$.

$\ell \longleftarrow \ell + 1$

Until $\|\zeta^{\ell+1} - \zeta^\ell\| \leq \epsilon$ or $\|f_\eta(\zeta^{\ell+1}) - f_\eta(\zeta^\ell)\| \leq \epsilon$.

Theorem 2. *(Convergence properties of DCA)*

i) **DCA-SC** *generates the sequence* $\{\zeta^\ell\}$ *such that the sequence* $\{f_\eta(\zeta^\ell)\}$ *is decreasing.*

ii) The sequence $\{\zeta^\ell\}$ *converges to* ζ^* *which satisfies the necessary local optimality condition (23) to* (P_η).

Proof: These properties are consequences of the convergence properties of DC program and the fact that h is differentiable.

4 Numerical experiments

We test the performance of our algorithm on random data with different sizes. In all the instances, the manufacturing costs MFC_i is randomly generated in $[1.0, 10.0]$. The market prices ρ_i is randomly generated in $[1.2MFC_i, 1.4MFC_i]$. The administrative costs, the transportation costs are respectively determined by 1% and 2% of the manufacturing costs. For each product, 10 transfer price levels are chosen from 1.0 to 10.0. The minimum inventory level is selected as 10 for all products. The inventory holding cost constant r is fixed as 0.05. We consider 5 cases. For each case, we test 20 problems.

Case 1 (50 binary variables): $I = 5$, $K = 10$, $T = 12$, there is no limit of resources. The demands of customers $D_{i,t}$ are randomly generated in $[100, 300]$. The minimum profit requirement of A and B are 100.

Case 2 (100 binary variables): $I = 10$, $K = 10$, $T = 12$, there is no limit of resources. The demands of customers $D_{i,t}$ are respectively randomly generated in $[300, 500]$. The minimum profit requirement of A is 1500 and that of B is 1000.

Case 3 (150 binary variables): $I = 15$, $K = 10$, $T = 12$, $R = 6$, the demands of customers $D_{i,t}$ and the demands of resources $UR_{i,r}$ are respectively randomly generated in $[300, 500]$ and $[0, 1.0]$ while the resource levels $R_{r,t}$ are randomly generated in $[3000, 5000]$. The minimum profit requirement of A is 15000 and that of B is 5000.

Case 4 (200 binary variables): $I = 20$, $K = 10$, $T = 12$, $R = 6$, the resource levels $R_{r,t}$ are randomly generated in $[4000, 6000]$. The minimum profit requirement of A is 20000 and that of B is 15000. The other parameters are generated by the same way in case 3. *Case 5* (300 binary variables): $I = 30$, $K = 10$, $T = 12$, $R = 6$, the resource levels $R_{r,t}$ are randomly generated in $[6000, 8000]$. The minimum profit requirement of A is 15000 and that of B is 30000. The other parameters are generated by the same way in case 3.

To evaluate the quality of the solution furnished by DCA, we compare the objective function value given by DCA with a upper bound. It is easy to see that

$(\Pi_A - \Pi_A^L)(\Pi_B - \Pi_B^L) \leq \frac{(\Pi_A + \Pi_B - \Pi_A^L - \Pi_B^L)^2}{4}$.

Hence, $(\Pi_A - \Pi_A^L)(\Pi_B - \Pi_B^L) \leq \frac{(\beta - \Pi_A^L - \Pi_B^L)^2}{4}$ where

$\beta = \max\left\{\Pi_A + \Pi_B \ s.t. \ (1) - (19)\right\}$. Therefore, $\gamma := \frac{(\beta - \Pi_A^L - \Pi_B^L)^2}{4}$ provides a upper bound of the objective function value. The results obtained by DCA and the upper bounds are shown in Table 1. We use some notations:
C: Case; $AvOpt$: the average (on 20 test problems) of the objective function values; $AvUB$: the average of the upper bounds; AvT: the average of the CPU times; $AvG = \frac{AvUB - AvOpt}{AvUB}.100\%$.
Table 1.
Results obtained by DCA

C	$AvOpt$	$AvUB$	AvT	AvG
1	18338717.0	20235626.2	2.40	8.53
2	404180075.6	447998114.7	3.94	9.58
3	758556797.0	827592010.4	20.92	8.64
4	1216387736.9	1309220576.2	17.35	6.92
5	2751028609.4	3006820399.2	47.37	7.80

We observe, from the results, that:

- DCA always furnishes an integer solution although it works on a continuous domain.

- The solution furnished by DCA is good. The quality of the solution is proved by the average of Gap which is about 8%. Hence, it is interesting to use DCA for finding an approximation solution for the considered problem.

5 Conclusion

In this work, we have presented an efficient algorithm for solving the multi-period problem of fair transfer prices and inventory holding policies in two-enterprize supply chains. The results proved that the proposed algorithm is quite efficient. In a future work, we will combine DCA with Branch and Bound algorithms for globally solving the problem.

References

[1] J. Gjerdrum, N. Shah and L.G. Papageogiou *Fair transfer price and inventory holding policies in two-enterprize supply chains,* European Journal of Operational Research **143** (2002), 582-599.

[2] S.S. Chauhan and J.M. Proth, *Analysis of a supply chain partnership with revenue sharing,* International Journal of Production Economics **97**, (2005), 44-51.

[3] J. Nash, *The bargaining problem,* Econometrica **18**, (1950), 155-160.

[4] F. Glover, *Improved linear integer programming formulation of non linear integer problem,* Management Science **22(4)**, (1975), 455-460.

[5] T. Pfeiffer, *Transfer pricing and decentralized lot sizing in multistage, multiproduct production processes,* European Journal of Operational Research **116**, (1999), 319-330.

[6] I. Waysman, *A model of negotiated transfer pricing,* Journal of Accounting and Economics **25**, (1998), 349-384.

[7] W.G. Howe, J.F. Cox, *Enterprise implications of transfer pricing,* Production and Inventory Management Journal **35(2)** (1994), 35-38.

[8] H.A. Le Thi, *Contribution à l'optimisation non convex and l'optimisation globale: Théorie, Algorithmes et Applications, Habilitation à Diriger des recherches,* Université de Rouen, (1997).

[9] H.A. Le Thi and T. Pham Dinh, *A Continuous approach for globally solving linearly constrained quadratic zero-one programming problem* , Optimization **50(1-2)** (2001), 93-120.

[10] H.A. Le Thi and T. Pham Dinh, *The DC(difference of convex functions) Programming and DCA revisited with DC models of real world non convex optimization problems,* Annals of Operations Research **133** (2005), 23-46.

[11] H.A. Le Thi, T. Pham Dinh and V.N. Huynh, *Exact penalty techniques in DC programming,* Research Report, LMI, National Institute for Applied Sciences-Rouen, France, July,(2005).

[12] T. Pham Dinh and H.A. Le Thi, *Convex analysis approach to d.c Programming: Theory, Algorithms and Applications,* Acta Mathematica Vietnamica, dedicated to Professor Hoang Tuy on the occasion of his 70th birthday, **22(1)** (1997), 289-355.

[13] T. Pham Dinh and H.A. Le Thi, *DC optimization algorihms for solving the trust region subproblem,* SIAM J.Optimization **8** (1997), 476-505.

[14] C.J. Vidal, M. Goetschalckx, *Strategic Production-distribution models: A critical review with emphasis on global supply chain models* European Journal of Operational Research **98** (1997), 1-18.

Innovative Production Machines and Systems
D.T. Pham, E.E. Eldukhri and A.J. Soroka (eds)

Using pheromone marking as method for implementing memory in the Bees Algorithm

D.T. Pham, A.J. Soroka

Manufacturing Engineering Centre, Cardiff School of Engineering Cardiff University, Cardiff, CF24 3AA UK

Abstract

This paper presents a potential approach for implementing memory within the Bees Algorithm (BA). This population-based search algorithm mimics the natural foraging behaviour of swarms of bees including the pheromone marking of sites to remove unnecessary visits. In its basic version, the algorithm performs a kind of neighbourhood search combined with random search. However, due to random elements within the algorithm it is possible that a search may take place in a location more than once. The paper describe how a stack inspired structure can be used to replicate the pheromone marking left by a bee an its degradation.

Keywords: Bees algorithm, PSO, memory, pheromone

1. Introduction

A lot of complex multi-variable optimisation problems cannot be solved exactly within polynomially bounded computation times. This has generated interest in search algorithms that find near-optimal solutions in reasonable running times. The algorithm described in this paper is a search algorithm capable of locating good solutions efficiently. It is inspired by the food foraging behaviour of honey bees and could be regarded as belonging to the category of "intelligent" optimisation tools [1].

This paper proposes a method to allow the Bees Algorithm[1] to remember the locations where searches have been performed.

The following sections introduce the algorithm and the principle of using a stack to simulate pheromone markers and hence provide memory.

2. Intelligent Swarm Based Optimisation

Swarm-based optimisation algorithms (SOAs) mimic nature's methods to drive a search towards the optimal solution. A key difference between SOAs and direct search algorithms such as hill climbing and random walk is that SOAs use a population of solutions for every iteration instead of a single solution. As a population of solutions is processed in an iteration, the outcome of each iteration is also a population of solutions. If an optimisation problem has a single optimum, SOA population members can be expected to converge to that optimum solution. However, if a problem has multiple optimal solutions, an SOA can be used to capture them in its final population. Optimisation algorithms include the Ant Colony Optimisation (ACO) algorithm [2], the Genetic Algorithm (GA) [3] and the Particle Swarm Optimisation (PSO) algorithm [4].

Common to all population-based search methods is a strategy that generates variations of the solution being sought. Some search methods use a greedy criterion to decide which generated solution to retain. Such a criterion would mean accepting a new solution if and only if it increases the value of the objective function (assuming the given optimisation problem is one of optimisation). A very successful non-greedy population-based algorithm is the ACO algorithm which emulates the behaviour of real ants. Ants are capable of finding the shortest path from the food source to their nest using a chemical substance called pheromone to guide their search. The pheromone is deposited on the ground as the ants move and the probability that a passing stray ant will follow this trail depends on the quantity of pheromone laid. ACO was first used for functional optimisation by Bilchev [5] and further attempts were reported in [5, 6].

The Genetic Algorithm is based on natural selection and genetic recombination. The algorithm works by choosing solutions from the current population and then applying genetic operators – such as mutation and crossover – to create a new population. The algorithm efficiently exploits historical information to speculate on new search areas with improved performance [3]. When applied to optimisation problems, the GA has the advantage of performing global search. The GA may be hybridised with domain-dependent heuristics for improved results. For example, Mathur et al [6] describe a hybrid of the ACO algorithm and the GA for continuous function optimisation.

Particle Swarm Optimisation (PSO) is an optimisation procedure based on the social behaviour of groups or organisations, for example the flocking of birds or the schooling of fish [4]. Individual solutions in a population are viewed as "particles" that evolve or change their positions with time. Each particle modifies its position in search space according to its own experience and also that of a neighbouring particle by remembering the best position visited by itself and its neighbours, thus combining local and global search methods [4].

There are other SOAs with names suggestive of possibly bee-inspired operations [7-10]. However, as far as the authors are aware, those algorithms do not closely follow the behaviour of bees. In particular, they do not seem to implement the techniques that bees employ when foraging for food.

3. Foraging behaviour of Bees

3.1 Bees in the Natural Environment

A colony of honey bees can be seen as a diffuse creature which can extend itself over long distances in multiple directions in order to exploit a large number of food sources at the same time [7, 8]. In principle, flower patches with plentiful amounts of nectar or pollen that can be collected with less effort should be visited by more bees, whereas patches with less nectar or pollen should receive fewer bees [9, 10].

The foraging process begins in a colony by scout bees being sent to search for promising flower patches. Scout bees search randomly from one patch to another. During the harvesting season, a colony continues its exploration, keeping a percentage of the population as scout bees [8].

When they return to the hive, those scout bees that found a patch which is rated above a certain threshold (measured as a combination of some constituents, such as sugar content) deposit their nectar or pollen and go to the "dance floor" to perform a dance known as the "waggle dance" [7].

This dance is essential for colony communication, and contains three vital pieces of information regarding a flower patch: the direction in which it will be found, its distance from the hive and its quality rating [7, 10]. This information helps the bees to find the flower patches precisely, without using guides or maps. Each individual's knowledge of the outside environment is gleaned solely from the waggle dance. This dance enables the colony to evaluate the relative merit of different patches according to both the quality of the food they provide and the amount of energy needed to harvest it [10]. After waggle dancing on the dance floor, the dancer bee (i.e. the scout bee) goes back to the flower patch with follower bees that were waiting inside the hive. The number of follower bees assigned to a patch depends on the overall quality of the patch. This allows the colony to gather food quickly and efficiently.

3.2 Scent Marking of flowers

Empirical studies conducted on lab-based populations of the honeybees *Apis mellifera ligustica* have suggested strongly that pheromone/scent based marking of flowers may take place [11][12]. Experiments were conducted under control conditions where bees would visit sites normally and a test conditions where an extraction system was used to remove the pheromone scent. Under the control conditions bees would avoid sites that they had recently visited whereas under test conditions they visited those sites more frequently. The results presented were consistent with the existence of a repellent marking substance being left by bees after having inspected a flower.

It is believed that the reason for this is that foraging bees would benefit from not visiting a flower until nectar levels had been replenished [11] and during this time the scent marking diminishes due to evaporation. With results suggesting strongly that there is a period of time after which the scent becomes too weak and is ignored. According to the results presented in [11] it was estimated by the authors that with a visit rate of 8 flowers per minute the scent endured for an estimated 45s. In [12] this period is referred to as the half-life of the scent and is estimated to be under 1 minute.

4. The Bees Algorithm

Figure 1 shows a flow chart for the Bees Algorithm in its most simple (and initial) form[1]. The algorithm requires a number of parameters to be set, namely: number of scout bees (n), number of sites selected out of n visited sites (m), number of elite sites out of m selected sites (e), number of bees recruited for the best e sites (nep), number of bees recruited for the other (m-e) selected sites (nsp), initial size of patches (ngh) which includes site and its neighbourhood and stopping criterion. The algorithm starts with the n scout bees being placed randomly in the search space. The fitnesses of the sites visited by the scout bees are evaluated in step 2. The valuation of fitness would depend on the optimisation problem, but in general 'fitness' is taken as the value of the objective function being optimised.

In step 4, bees that have the highest fitnesses are designated as "selected bees" and sites visited by them are chosen for neighbourhood search. Then, in steps 5 and 6, the algorithm conducts searches in the neighbourhood of the selected sites, assigning more bees to search near to the best e sites. The bees can be chosen directly according to the fitnesses associated with the sites they are visiting. Alternatively, the fitness values are used to determine the probability of the bees being selected. Searches in the neighbourhood of the best e sites which represent more promising solutions are made more detailed by recruiting more bees to follow selected bees than the other selected bees. Together with scouting, this differential recruitment is a key operation of the Bees Algorithm.

However, in step 6, for each patch only the bee with the highest fitness will be selected to form the next bee population. In nature, there is no such a restriction. This constraint is introduced here to reduce the number of points to be explored.

In step 7, the remaining bees in the population are assigned randomly around the search space scouting for new potential solutions. These steps are repeated until a stopping criterion is met. At the end of each iteration, the colony will have two parts to its new population – representatives from each selected patch and other scout bees assigned to conduct random searches.

---

1. Initialise population with random solutions.
2. Evaluate fitness of the population.
3. While (stopping criterion not met)
   //Forming new population.
4. Select sites for neighbourhood search.
5. Recruit bees for selected sites (more bees for the best e sites) and evaluate fitnesses.
6. Select the fittest bee from each patch.
7. Assign remaining bees to search randomly and evaluate their fitnesses.
8. End While.

Fig. 1. Basic Bees Algorithm

5. Implementation of memory

3.1 Features of scent-based memory

As described in Section 3 the scent based marking of flowers is based on the bee scent marking individual flowers with a pheromone whose strength degrades over time, allowing nectar to be replenished.

In order to implement the degradation consideration needs to be given whether to use a time or iteration based approach. A time based approach would mimic most closely the degradation of the pheromone scent. However, with optimisation there is no uniform period of time for a function to be evaluated. Factors such as functional complexity, speed of the computer and efficiency of program code can all have an impact. Therefore degrading of memory would be better implemented through an iteration based approach.

A potential method for implementing a form of iterative degradation is to use a First-In First-Out (FIFO) or Shift Register based approach. All 'flowers' visited during an iteration are marked and therefore added to a queue of a defined size (the half-life). This queue acts as the memory for the algorithm. Visited flowers within sites are added and once the queue is full the first group of visited sites leaves the queue (FIFO), these flowers are then 'forgotten'. This is illustrated in Fig. 2 where sites are added until the queue becomes full then the 'oldest' flowers are removed from the queue in order to make room for the newest flowers.

An algorithm for implementing a FIFO queue that could used for storing the marked flowers is shown below in Fig 3.

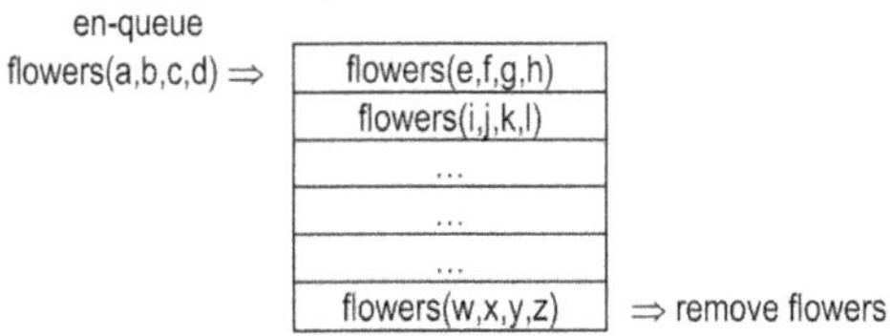

Fig.2 Queue based approach for memory

```
Mark flowers
    while (queue is full) DEQUEUE(queue)
    ENQUEUE(queue, flowers)
stop

  ENQUEUE(Q, F)
      Q[tail[Q]] ← F
      if tail[Q] = length[Q]
         then tail[Q] ← 1
      else tail[Q] ← tail[Q] + 1

  DEQUEUE(Q)
      F ← Q[head[Q]]
      if head[Q] = length[Q]
         then head[Q] ← 1
      else head[Q] ← head[Q] + 1
  return x
```

Fig.3 FIFO Queue algorithm

3.2 Revised Bees Algorithm

The revised version of the Bees Algorithm is shown in Figure 4. The algorithm is largely the same as that given earlier. The most significant differences are: Firstly sites consist of 'flowers' that are visited by a bee and evaluated; Secondly steps have been included for adding flowers to the memory and checking of flowers to ensure that they do not belong to the memory.

When the population of n scout bees is initialised they and the specific locations they evaluate (the flowers) are marked as having been visited (added to the memory / placed in the queue). No checking for membership in the set of marked flowers is required at this stage as the sites, which contain the flowers, have only just been initialised.

During the stage when bees are recruited for the best sites a check is performed to see whether any of the flowers they will visit in a site have been marked (and therefore belong to the memory). If they are marked then those flowers are ignored and new locations within the sites are selected (this is done until only new flowers get visited). These flowers are then marked.

As part of the Random Search phase membership of the group of marked flowers is checked once again to ensure flowers are not being re-evaluated. If a flower has already been visited then a new destination flower is chosen again at random (this is cross checked for membership in the group of marked sites). These selections are once again marked and added to the 'memory'.

This process continues until the stopping criteria is met. When the group of marked flowers becomes full the FIFO algorithm will de-queue flowers to make room, these de-queued flowers will then no longer be marked. This is analogous to the pheromone scent marking having evaporated.

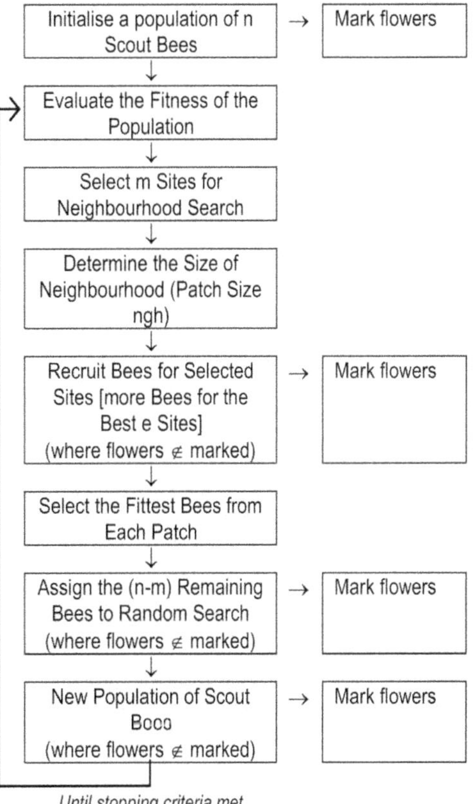

Fig.4 Bees Algorithm with memory

If further evaluation is necessary the population of scout bees are initialised and checked against the marked sites, with new sites being selected to replace already marked sites.

Once the queue that contains the list of marked sites and forms the basis of the memory becomes full the sites will be removed on a FIFO basis. The size of the queue is a user-defined parameter.

6. Discussion

It could be argued that the implementation of memory within the Bees Algorithm may benefit from no degradation. Particularly from the perspective of being able to permanently prevent repeat visits.

However, for an optimisation problem that requires a very large population of bees and/or results in a very high number of iterations the availability of RAM could become an issue. In a recent application of the Bees Algorithm [13] potentially some 20 million records of locations (flowers) visited would need to be stored if memory was permanent. This could theoretically result in a page file or swap space being used by the operating system, which may increase evaluation time. It is also plausible that it situations where a large population of bees is used there may be little benefit.

Therefore experimental testing would need to examine the size of the half life, from low to very large and the impact this has on the number of iterations required and the time taken for the algorithm to reach the optimal solution.

7. Conclusions

This paper has presented a potential solution to the issue of implementing memory within the bees algorithm.

Acknowledgements

This work was funded under the European Regional Development Fund (ERDF) Objective 1 Project SUPERMAN and by the I*PROMS European Union, Framework Programme 6 Network of Excellence.

References

[1] D.T. Pham , A. Ghanbarzadeh , E. Koc , S. Otri , S. Rahim , M. Zaidi. "The Bees Algorithm – A Novel Tool for Complex Optimisation Problems" Proceedings of 2nd Virtual Conference on Intelligent Production Machines and Systems (IPROMS 2006), July 2006.

[2] Dorigo, M. and T. Stützle, Ant Colony Optimization. 2004, Cambridge: MIT Press.

[3] Goldberg, D.E., Genetic Algorithms in Search, Optimization and Machine Learning. 1989, Reading: Addison-Wesley Longman.

[4] Eberhart, R., Y. Shi, and J. Kennedy, *Swarm* Intelligence. 2001, San Francisco: Morgan Kaufmann.

[5] Bilchev, G. and I.C. Parmee, *The Ant Colony Metaphor for Searching Continuous Design Spaces*, in Selected Papers from AISB Workshop on Evolutionary Computing. 1995, Springer-Verlag: London. 25-39.

[6] Mathur, M., S.B. Karale, S. Priye, V.K. Jayaraman, and B.D. Kulkarni, *Ant Colony Approach to Continuous Function Optimization*. Ind. Eng. Chem. Res., 2000. 39(10): 3814-3822.

[7] Frisch, K.V., Bees: *Their Vision, Chemical Senses and Language*. Revised Edition ed. 1976, Ithaca, N.Y.: Cornell University Press.

[8] Seeley, T.D., *The Wisdom of the Hive: The Social Physiology of Honey Bee Colonies*. 1996, Cambridge, Massachusetts: Harvard University Press.

[9] Bonabeau, E., M. Dorigo, and G. Theraulaz, *Swarm Intelligence: from Natural to Artificial Systems*. 1999, New York: Oxford University Press.

[10] Camazine, S., J.-L. Deneubourg, N.R. Franks, J. Sneyd, G. Theraula, and E. Bonabeau, *Self-Organization in Biological Systems*. 2003, Princeton: Princeton University Press.

[11] Martin Giurfa and Josué A. Núñez, Honeybees mark with scent and reject recently visited flowers, Oecologia 1992, Volume 89, Number 1, 113-117, DOI: 10.1007/BF00319022

[12] M. Giurfa, The repellent scent-mark of the honeybee Apis mellifera ligustica and its role as communication cue during foraging, Insectes Sociaux Volume 40, Number 1, 59-67, DOI: 10.1007/BF01338832

[13] Ang, M.C.; Pham, D.T.; Soroka, A.J.; Ng, K.W.; , "PCB assembly optimisation using the Bees Algorithm enhanced with TRIZ operators," IECON 2010 - 36th Annual Conference on IEEE Industrial Electronics Society , vol., no., pp.2708-2713, 7-10 Nov. 2010 DOI: 10.1109/IECON.2010.5675114

Innovative Production Machines and Systems
D.T. Pham, E.E. Eldukhri and A.J. Soroka (eds)

Using PSO and BA to design Optimal Speed Parameters of Wind Turbine Generators

A. A. Fahmy
Manufacturing Engineering Centre, Cardiff School of Engineering, Cardiff University
Email: fahmyaa@cf.ac.uk

***Abstract*— This paper presents an application of the Particle Swarm Optimization (PSO) and the Bees Algorithm (BA) in specifying the optimal operating speed parameters of wind turbine generators while maximizing the utilization of the wind power available at the wind farm location. The method of optimization is based on applying the PSO or the BA to search the optimum rated speed, cut-in speed, and cut-off (furling) speed of the wind turbine so that maximum generated energy can be extracted from the wind generator while reducing the cost of the units by optimizing the capacity factor in accordance to the wind speed specification available at the farm site. The study presents an improvement to the site performance by identifying the best wind turbine parameters that should be erected at the selected site to maximize the power generation using modern artificial optimization techniques to locate the three speeds. The new method depends on optimizing both the generated energy and capacity factor simultaneously so that the obtained turbine generator parameters maximize the generated energy and the capacity factor which means lower capital cost of the units. Both PSO and BA are applied for identifying the optimum wind turbine generator parameters for the wind farm promising sites in Egypt and for comparison purposes.**

***IndexTerms*— wind turbine generator, normalized power curve, capacity factor, turbine selection index, PSO, BA.**

I. INTRODUCTION

Utilization of wind turbines in large-scale electric power generation farms is a new trend in green power generation on large scale. Wind generated electric power possesses many benefits that mark it as promising source of electric power for both utility-scale generation and stand-alone distributed generation. The main benefits of wind generated electric power are mainly due to being a clean, free, and inexhaustible fuel source of power supply. It also helps in achieving local economic growth, cleaner environment, modular and scalable technology, and reduced reliance on natural resources of fuel.

However, wind generated electric power have to compete with many other energy sources, especially in the capital cost in maximum benefit from the available wind resources. So that it is important to have cost effective equipment design and should be able to adapt to load variations.

Production of wind-based electric power at a specific location depends on different factors. These factors include the mean wind speed of the site, erecting height of the turbine, and more significantly, the speed characteristics of the wind turbine itself namely, cut-in (V_c), rated (V_r), and furling (V_f) wind speeds at the hub height. There are many different models of wind turbine generators with the same kW ratings. Each of these wind turbine generators has its own specifications and speed parameters. These speed parameters affect the capacity factor at a given specific site, and subsequently affect the choice of the optimum wind turbine generator for that site[1-3].

This study presents an improvement to the site performance by identifying the best turbine and generator speed parameters that should be erected at the selected site to maximize both of the power generation and capacity factor through the use of modern optimization techniques such as PSO and BA. Using the normalized power and capacity factor curves [4-7], the best values for cut-in, rated, and cut-off speeds at a certain site will be located to achieve maximum power and capacity factor values simultaneously.

In this paper, section II introduces the problem of selecting the wind turbine parameter and the factors that affect the selection strategy. Section III introduces the Particle Swarm Optimization (PSO) and Bees Algorithm (BA). Section IV presents case studies results in the east coast of the Red Sea and in the coastal site along the Mediterranean Sea in Egypt. In this section, the wind characteristics were analyzed based on recorded information of monthly mean-wind speed of eleven meteorological stations to determine the wind characteristics. Wind Atlas Analysis Application Program (WASP) is used also to predict the complete annual mean frequency distribution for each site. Finally, section V concludes the paper.

II. MATHEMATICAL MODEL

Site matching techniques is mainly related to identifying the best turbine and generator at a particular erection site based on selecting optimum speed parameters that produce high energy at high capacity factor at that particular site. Wind speed hourly frequency distribution curve over the year for a certain site is the most common representation for wind speed yearly variation over a particular site. Statistical methods and probability density functions are usually used to describe the wind speed variation over the year at any particular site based on the hourly variations. There are several density functions which can be used to describe the wind speed frequency distribution curve. Among the most common probability density functions used to describe the wind speed values for different sites is the Weibull probability density function [8].

The Weibull probability distribution gives a very good match when used with experimental data recorded for the wind speed over the year. This probability distribution function is specified by two parameters, namely the shape parameter k (dimensionless), and scale parameter c (m/s). Four different methods for the estimation of k and c parameters were reported in literature [8].

In this paper, the most common model for electrical power extracted from wind energy P_e is used [8-9]. This model can be used in discussing wind electrical power system represented by the probability distribution function. P_e is assumed to vary as v^k between cut-in and rated wind speeds where v is the wind speed. This model can be used to obtain the normalized power and capacity factor curves as follows.

$$P_e = 0 \qquad (v < V_c) \qquad (1)$$

$$P_e = a + bv^k \qquad (V_c \le v \le V_r) \qquad (2)$$

$$P_e = P_{er} \qquad (V_r < v \le V_f) \qquad (3)$$

$$P_e = 0 \qquad (v > V_f) \qquad (4)$$

where $a = \dfrac{P_{er} V_c^k}{(V_c^k - V_r^k)}$ (5)

$$b = \frac{P_{er}}{(V_r^k - V_c^k)} \qquad (6)$$

$$P_{e,av} = \int_0^{\infty} P_e f(v) dv \qquad (7)$$

where $f(v) = \dfrac{k}{c}\left(\dfrac{v}{c}\right)^{k-1} * e^{-\left(\frac{v}{c}\right)^k}$ is the Weibull probability density function of wind speed. If the mean $(\bar{v})$ and variance (σ) of the wind speed frequency distribution are known, we can solve for the speed parameter c and the shape parameter k directly, where $k = \left(\dfrac{\sigma}{\bar{v}}\right)^{-1.086}$, $c = \dfrac{\bar{v}}{\Gamma\left(1 + \frac{1}{k}\right)}$, and Γ is the gamma function. Using the model for P_e,

$$P_{e,av} = \int_{V_c}^{V_r} (a + bv^k) f(v) dv + \int_{V_r}^{V_f} P_{er} f(v) dv \qquad (8)$$

This integral equation is mainly about evaluating two integrals, $\int f(v)dv$, and $\int v^k f(v)dv$, which can be best solved by making a change of variables by replacing $x = \left(\dfrac{v}{c}\right)^k$ and hence $dx = k\left(\dfrac{v}{c}\right)^{k-1} d\left(\dfrac{v}{c}\right)$ and $f(v) = \left(\dfrac{k}{c}\right)\left(\dfrac{v}{c}\right)^{-1} xe^{-x}$, then the two integrals can be re-written as:

$$\int f(v)dv = \int e^{-x} dx = -e^{-x},$$

$$\text{and } \int v^k f(v) dv = \int c^k \left(\frac{v^k}{c^k}\right) f(v) dv$$

$$= c^k \int xe^{-x} dx = -c^k (x+1) e^{-x} \qquad (9)$$

By substituting these changes into equation (8) and reducing the minimum number of terms, the result is as follows:

$$P_{e,av} = P_{er}\left\{\frac{e^{-\left(\frac{V_c}{c}\right)^k} - e^{-\left(\frac{V_r}{c}\right)^k}}{\left(\frac{V_r}{c}\right)^k - \left(\frac{V_c}{c}\right)^k} - e^{-\left(\frac{V_f}{c}\right)^k}\right\} \qquad (10)$$

$$P_{e,av} = 0.5\, \eta_o\, \rho A V_r^3 \left\{\frac{e^{-\left(\frac{V_c}{c}\right)^k} - e^{-\left(\frac{V_r}{c}\right)^k}}{\left(\frac{V_r}{c}\right)^k - \left(\frac{V_c}{c}\right)^k} - e^{-\left(\frac{V_f}{c}\right)^k}\right\} \qquad (11)$$

Normalizing the $P_{e,av}$ equation we will get: -

$$P_N = \frac{P_{e,av}}{0.5\eta_o \rho A c^3} = \left(\frac{V_r}{c}\right)^3 * CF \qquad (12)$$

$$CF = \left\{\frac{e^{-\left(\frac{V_c}{c}\right)^k} - e^{-\left(\frac{V_r}{c}\right)^k}}{\left(\frac{V_r}{c}\right)^k - \left(\frac{V_c}{c}\right)^k} - e^{-\left(\frac{V_f}{c}\right)^k}\right\} \qquad (13)$$

Where $P_{e,av}$ is the average power, CF is the capacity factor [10]. Figure (1) shows the normalized power and capacity factor variation curves at certain values for cut-in and cut-off speeds.

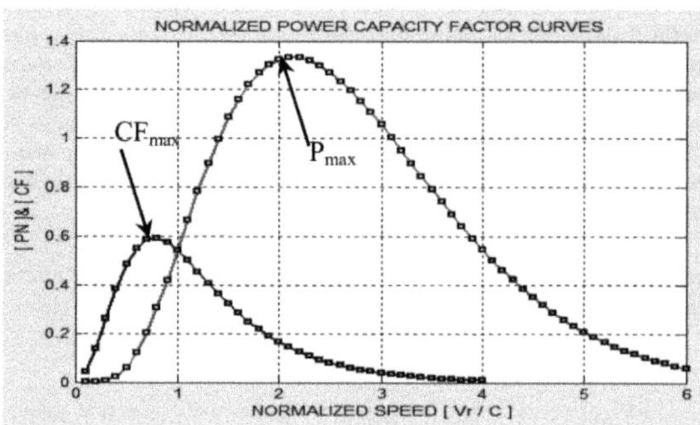

Fig.1. Normalized power and capacity factor curve at certain V_c and V_f.

Once turbine parameters are determined to maximize the average normalized power, the rated power for a turbine with a given blades area and rated overall efficiency located at an elevation related to a known average air density can calculated [8]. Therefore, the yearly energy production of such a turbine generator will be as follow:

$$\text{Energy} = P_{e,av}(\text{time}) = (CF)\, P_{er}\, (8760) \qquad (14)$$

By studying figure (1), at certain values for cut-in and cut-off speeds, it can be seen that to extract the maximum output power from the site, the rated speed of the generator should be at the normalized value where Pmax is achieved [1-7]. This selection method had the disadvantage of selecting large value of Vr for the wind turbine and generator and consequently the rated power Per, while a significant decrease in the capacity factor CF is obtained. The increased value of Per for the wind turbine raises the costs of the necessary generator, transformer, switches, circuit breakers and distribution lines, while the decrease in CF expresses less utilization time of this equipment with its rated values over the year. If the rated speed is chosen at the point where the maximum capacity factor CFmax is achieved, this choice will definitely reduce the equipment costs and eliminated the problem of minimum time usage of equipment with its rated values occurring in the previous selection method. This method is considered the ideal operating point from the view point of cost, while this will occur at a very low rated speed which leads to low rated power reducing the actual benefit from the wind power available at the site over the year. Consequently, another solution for rated speed selection should exist. So, there are values for the operating normalized speeds at which the product of the two variables, normalized power and capacity factor with respect to normalized speed, will attain a maximum at which the individual values of PN & CF being far higher than their minimum value and as close as possible to their maximum values. This value of normalized speeds which corresponds to the product (PN*CF)max falls between the two values of the normalized speeds at which PNmax and CFmax occurs [4]. Usually cut-in and cut-off speeds are chosen to be a percentage of the rated speed which was usually taken as $Vc = \delta Vr$ and $Vf = \gamma Vr$, where $\delta<1$ and $\gamma>1$. Normally the search was only performed for the rated speed, whiles the cut-in and cut-off speeds are assumed a percentage value from this speed. In this work, we extend the search space to include the cut-in speed and cut-off speed using PSO and Bees algorithms. The aim of this work is to find the optimum operating speeds V_c, V_r, and V_f that maximize the PN*CF product, given an arbitrary site described by the Weibull parameters c and k. This task entails the solution of the function optimization problem described by equation (15).

$$P_N * CF = \left(\frac{V_r}{c}\right)^3 * (CF)^2 = \left(\frac{V_r}{c}\right)^3 * \left\{ \frac{e^{-\left(\frac{V_c}{c}\right)^k} - e^{-\left(\frac{V_r}{c}\right)^k}}{\left(\frac{V_r}{c}\right)^k - \left(\frac{V_c}{c}\right)^k} - e^{-\left(\frac{V_f}{c}\right)^k} \right\}^2 \quad (15)$$

III. Particle Swam Optimization and Bees Algorithm

Swarm intelligence (SI) [11] is a fast-growing multi-disciplinary field which encompasses several iterative optimization procedures based on the collective search capabilities of large ensembles of simple interacting agents. Thanks to their global population-based optimization approach, SI algorithms are particularly suitable to search complex and multimodal surfaces such as the one generated by equation (14). This study investigates the application of two successful SI algorithms to the wind turbine generator optimization problem. In both cases, the candidate solutions are encoded via the three-dimensional vector $x=\{x_1,x_2,x_3\}$ representing the three operating speeds V_c, V_r, and V_f. The two algorithms are run for 50 times in every case study, while a stopping criteria of

A Particle Swarm Optimization

Particle Swarm Optimization (PSO) [12] is one of the first and best known SI algorithms. In this paper, the standard procedure formulated by Kennedy and Eberhart [12] was used.

The position x (i.e. the encoding) of each agent is updated as follows:

$$\Delta x_i(t) = v_i(t+1) \quad (16)$$

where t is the tth PSO cycle and $v=\{v_1,v_2,v_3\}$ is the velocity of the agent. The velocity is updated according to the following formula:

$$\begin{aligned} p_i(t) &= random_1 \cdot [pbest(t)_i - x(t)_i] \\ s_i(t) &= random_2 \cdot [gbest(t)_i - x(t)_i] \\ v_i(t+1) &= w(t) \cdot v(t)_i + c_1 \cdot p_i(t) + c_2 \cdot s_i(t) \\ i &= 1,2,3 \end{aligned} \quad (17)$$

where c_1 and c_2 are system parameters, and $random_1$ and $random_2$ are random numbers drawn with uniform probability in the interval [0,1]. $pbest(t)$ is the best position (maximum fitness) visited so far by the particle and represent the memory of the agent of its past actions. $gbest(t)$ is the best position visited so far by the neighbours of the particle, and represents the social interaction amongst agents. The third component $v(t)_i$ represents the momentum of the particle, that is, the persistence of each agent in following a direction. The weight $w(t)$ is decayed according to the following formula:

$$w(t) = w_{max} - \frac{w_{max} - w_{min}}{T} \cdot t \quad (18)$$

where w_{max} and w_{min} are system parameters, and T is the duration of the PSO search.

Every component of the velocity vector was bounded within the interval $[-v^{max}_i, v^{max}_i]$, where

$$v_i^{max} = u \cdot \frac{max_i - min_i}{2} \quad (19)$$

u is a system parameter, and max_i and min_i are the constraints on the variable x_i.

B The Bees Algorithm

The Bees Algorithm [13] is a recently developed optimization method that mimics the foraging behavior of honey bees. At the beginning, n solutions are randomly scattered on the search space. This phase simulates the scouting of forager bees in the area around the hive. The algorithm enters then the main loop.

The fitness of the solutions is evaluated and the population is ranked according to the evaluation result. A local search is performed in the neighborhood of the best m solutions. That is, nep new solutions are generated in the neighborhood of the elite $e<m$ locations, and $nsp< nep$ solutions are generated in the neighborhood of the remaining $m-e$ bees. The size of the neighborhood is a system parameter. This local search procedure simulates the recruitment of forager bees by those scouts that found rich food sources. For each neighborhood, the best solution is kept. The remaining *n-m* solutions are randomly generated. This process simulates the scouting of bees for new food sources. The main cycle is repeated until the stopping condition is met. The flowchart of the Bees Algorithm is shown in figure 2.

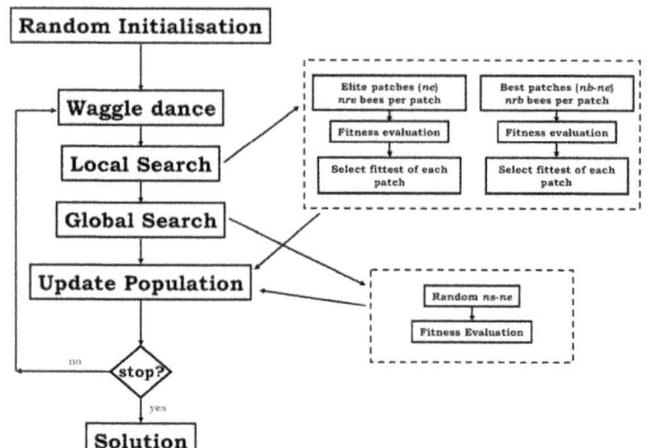

Fig.2. Flowchart of Bees Algorithm.

IV. Case Study Of Wind Energy In Egypt

Since the late 1970's, Egypt started considering the use of renewable energy where several bilateral and multilateral agreements were signed and implemented to explore potential horizons of renewable energy usage. Egypt has two coastal areas that show significant promises for wind energy exploitation, the north cost on the Mediterranean Sea and the east coast on the Red Sea. Data have been collected for 11 locations along the east coast of the Red Sea and in the coastal site along the Mediterranean Sea in Egypt. The measurements of monthly wind speed, air temperature, and air pressure were taken in an open area at a height of 10m above ground level. The Egyptian Meteorological Authority provided the data for a period of more than 10 years [14-15]. Wind Atlas Analysis Application Program (WASP) is used to predict the complete annual mean frequency distribution over the year for the sites to identify the Weibull parameter as listed in table (1).

Site	parameter c	parameter k
Sallum	4.88	1.46
Sidi Barrani	4.27	1.47
Dekhaila	4.53	1.4
Alexandria	4.36	1.49
Balteam	3.62	2.46
Damiett	3.12	2.49
Port Said	4.77	1.71
El Arish	4.56	2.39
Zafarana	8.23	2.7
Abu Darag	8.23	3.0
Hurghada	6.6	2.03

Table 1: Weibull parameters for the selected sites.

Both PSO and BA are applied to maximize equation (15), while satisfying the condition that Vc < Vr < Vf, to locate the optimum speed parameters that will achieve this aim. The domain search for each variable to maximize equation (15) is selected as follows:

$$0.2\,c \leq V_c < c \quad (20)$$

$$0.8\,c \leq V_r \leq 3.0\,c \quad (21)$$

$$2.5\,c < V_f < 5.0\,c \quad (22)$$

The setting of the PSO and BA search engines are listed in table 2 and table 3 respectivlely.

Parameter	Value
No of scout bees (n)	20
No of selected bees (m)	5
No of elite bees (e)	1
Size of neighborhood (ngh)	6.0
No of sites around selected bees (nsp)	10
No of sites around eleite bees (nep)	30
Stopping tolerance value (ε)	0.01

Table 2: Parameters of BA search engine.

Parameter	Value
No of Evaluations (ne)	200000
No of particles (m)	20
Max velocity (u)	0.01
Social neighbourhood of an agent (k)	10
Stopping tolerance value (ε)	0.01

Table 3: Parameters of PSO search engine.

For comparison purposes, the results obtained by the PSO and BA are compared to the result reported in [16]. In [16] Turbine Selection Index method (TSI) was used to determine the rated wind generator speed while the cut-in and cut-off speeds are taken as a percentage of this speed as mentioned before. The PSO and BA is now applied to search the domain for the cut-in, rated, and cut-off speeds to maximuize the energy production from the respected site using its weibull parameters listed in table (1). Figure (3) illustrates the reults for the maximum value obtained for the objective function. Figure (4) illustrates the results obtained for the cut-in speed at each location. Figure (5) illustrates the results obtained for the rated speed at each location. Figure (6) illustrates the results obtained for the cut-off speed at each location. Figure (7) illustrates the results obtained for the capacity factor at each location. Figure (8) illustrates the results obtained for the normalized power at each location. Figure (9) illustrates the results obtained for the rated electrical power at each location. Figure (10) illustrates the results obtained for the average electrical power at each location, while Figure (11) illustrates the results obtained for the total electrical energy generated over one year at each location. From the results, it is clear that the application of PSO and Bees algorithm identified better operating conditions for most sites.

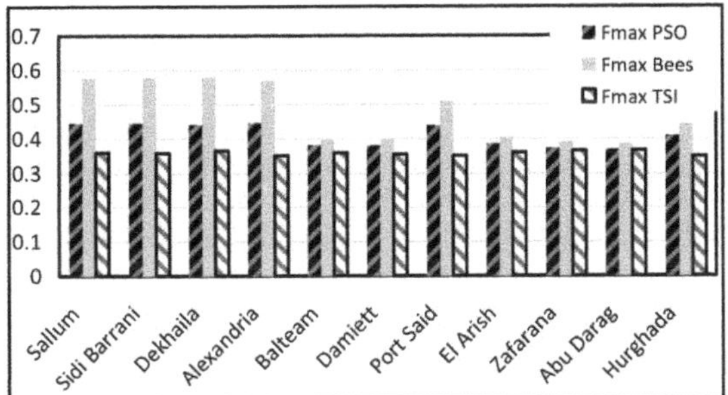

Fig.3. Results for the objective function.

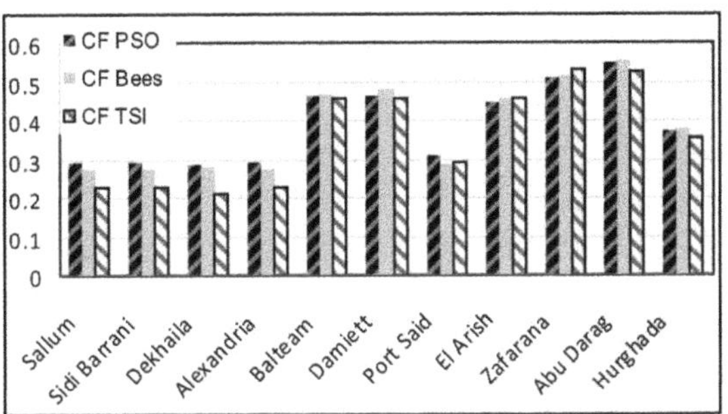

Fig.7. Results for the capacity factor.

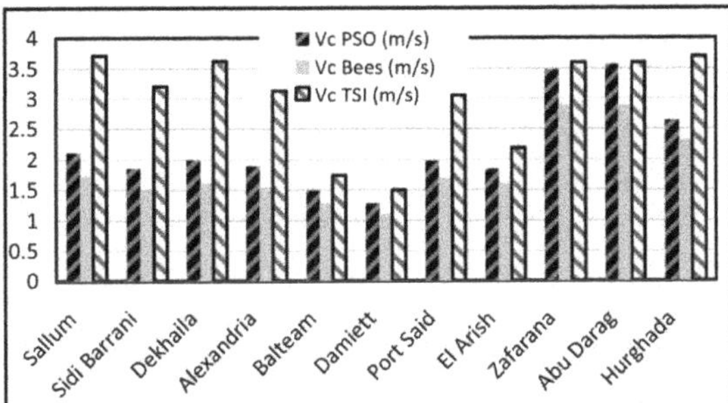

Fig.4. Results for the cut-in speed.

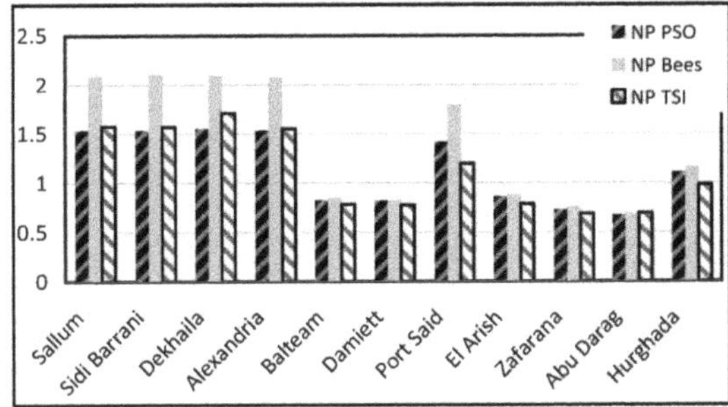

Fig.8. Results for the normalized power.

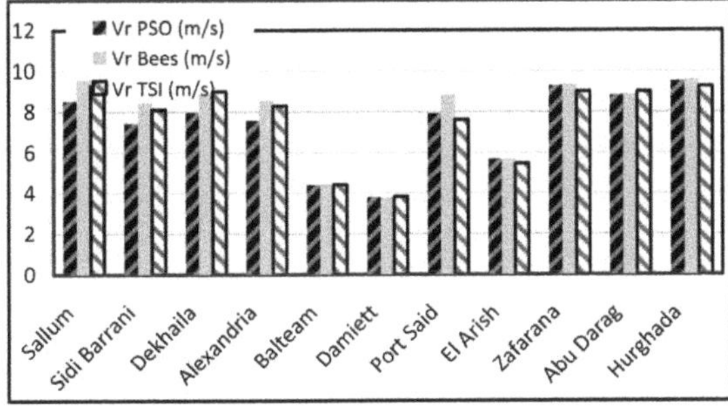

Fig.5. Results for the rated speed.

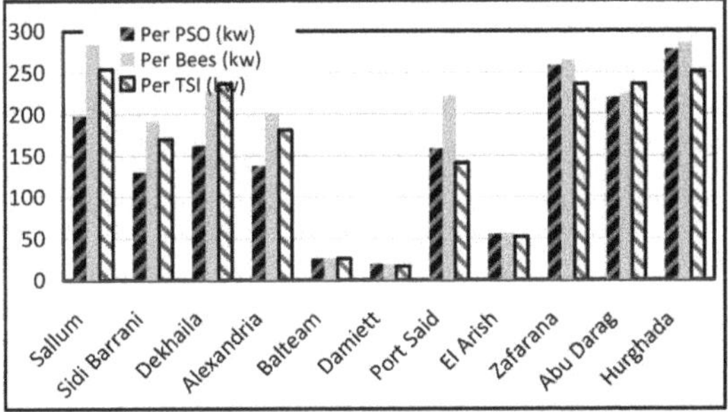

Fig.9. Results for the rated electrical power.

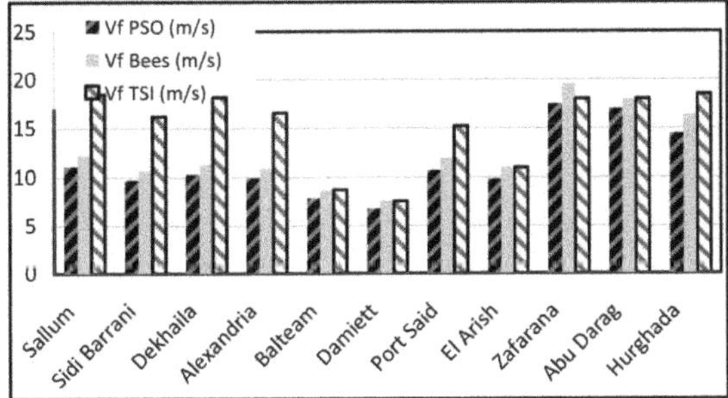

Fig.6. Results for the cut-off speed.

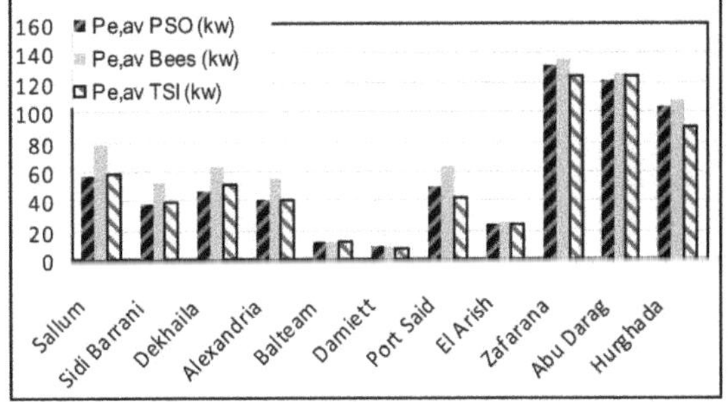

Fig.10. Results for the average electrical power.

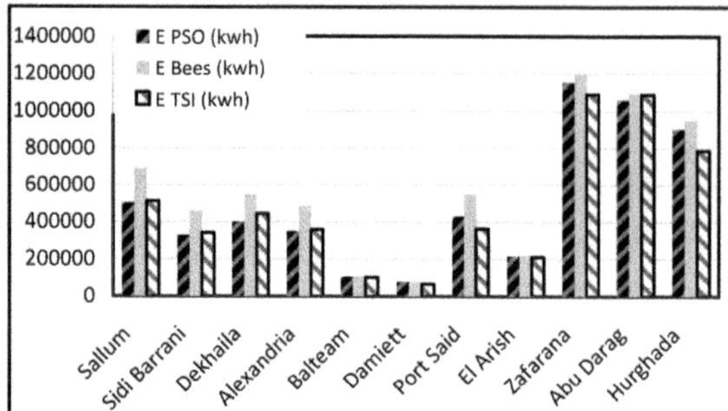

Fig.11. Results for the total energy generated in one year.

V. Conclusion

Matching site data with wind turbine generator speed parameters is an important problem in wind farm design stage. If the generator rated speed is chosen to be low, the site loses too much of the energy in the higher wind speed intervals. If the generator rated speed is too high, the turbine will seldom operate at low capacity and the capital cost will be high. The evaluation of cut-in speed and cut-off speed is normally left without specific criteria for selection. These values could also affect the total energy generated over the year from a specific site as well as the operation strategy of the wind turbine. So this paper introduces the application of artificial optimization techniques to search the domain of the speed operating parameters to match the best wind turbine generators to a specific site data using normalized powers and capacity factor curves. The new matching technique identifies optimum turbine generator speed parameters, such as rated speed, cut-in speed, cut-out speed, and rated power to maximize the energy production per capital cost. The new technique was applied on different sites in Egypt to evaluate the validity of the method. The result for the PSO and BA were compared to the result obtained using TSI method reported on [16] for comparison purposes. It can be seen from the result that be slight adjustment of the cut-in speed, rated speed, and cut-off speed, the overall generated energy from the wind turbine generator can be enhanced to a better value. The PSO and BA search the domain of the speed variables and obtained results that are better that that obtained before using TSI method. The obtained results supported the validity of the new method.

VI. References

[1] Huang, S. J.; Wan, H. H.; "Enhancement of matching turbine generators with regime using capacity factor curves strategy", IEEE Transactions on Energy Conversion, Vol.24, No.2, June 2009.

[2] Huang, S. J.; Wan, H. H.; "A study on genrator capacity for wind turbines under various tower heights and rated wind speeds using weibull distribution", IEEE Transactions on Energy Conversion, Vol.23, No.2, June 2008.

[3] Raju, A. B.; Fernandes, B. G.; Chatterjee, K.; "Estimation of optimum wind turbine generator speed parameters", IEEE Power Engineering Socity General Meeting, Vol.2, June 2004.

[4] Jansamshetti, S. H.; Rau, V. G.; "Normalized power curves as a tool for identification of optimum wind turbine generator parameters", IEEE Transactions on Energy Conservation, Vol.16, No.3, September 2001.

[5] Jansamshetti, S. H.; Rau, V. G.; "Optimum siting of wind turbine generators", IEEE Transactions on Energy Conservation, Vol.16, No.1, March 2001.

[6] Jansamshetti, S. H.; Rau, V. G.; "Site matching of wind turbine genrators: A case study", IEEE Transactions on Energy Conservation, Vol.14, No.4, March 1999.

[7] Jansamshetti, S. H.; Rau, V. G.; "Hieght extrapolation of capacity factors for wind turbine generators", IEEE Power Engineering Review, Vol.19, Issue 6, June 1999.

[8] Gary, L. J.; " Wind energy systems ", Manhattan, KS, Electronic Edition, 2006.

[9] Powell, W. R.; "An analytical expression for the average output power of a wind machine", Solar Energy, Vo.26, 1981.

[10] Albadi, M. H.; El-Saadany, E. F.; "Wind turbines capacity factor modeling- A novel approach", IEEE Transactions on Power Systems, Vol.24, No.3, June 2009.

[11] Bonabeau, E., Dorigo, M., Theraulaz, G., 1999, Swarm Intelligence: from Natural to Artificial Systems. New York: Oxford University Press.

[12] Kennedy, J., Eberhart, R., 1995, Particle swarm optimization. Proc. 1995 IEEE International Conference on Neural Networks, Perth, Australia, vol.4, pp. 1942-1948.

[13] Pham, D.T., Castellani, M., 2009, The Bees Algorithm – Modelling Foraging Behaviour to Solve Continuous Optimisation Problems. Proc. ImechE, Part C, vol. 223, no. 12, pp. 2919-2938.

[14] Shata, A. S. A.; Hanitsch, R.; "Evaluation of wind energy potential and electricity generation on the coast of Mediterranean Sea in Egypt", Renewable Energy, Vo.31, 2006.

[15] Shata, A. S. A.; Hanitsch, R.; "The potential of electricity generation on the east coast of Red Sea in Egypt", Renewable Energy, Vo.31, 2006.

[16] Abdel-Hamid, R.H.; Adma, M.A.A.; Fahmy, A.A.; Samed, S.F.A.; "Optimization of wind farm power generation using new unit matching technique", 7th IEEE International Conference on Industrial Informatics, Cardiff, UK, June 24-26, INDIN 2009.

Innovative Production Machines and Systems
D.T. Pham, E.E. Eldukhri and A.J. Soroka (eds)

Discovering facial biotype patterns using multivariate cephalometric data for treatment decision making

P. Araya-Díaz[a], G.A. Ruz[b], H.M. Palomino[a]

[a] *Departamento del niño y adolescente, Area de Ortodoncia, Facultad de Odontología, Universidad Andres Bello, Santiago, Chile*
[b] *Facultad de Ingeniería y Ciencias, Universidad Adolfo Ibáñez, Av. Diagonal las Torres 2640, Peñalolén, Santiago, Chile*

Abstract

This paper presents a methodology to find facial biotype patterns in a multivariate cephalometric database from a Chilean population, using a machine learning approach, as well as to verify if the patterns discovered agree with Ricketts biotypes classification. The cephalometric data was generated from a sample of 100 teleradiographs collected from Chilean patients, recording 30 variables which were used in a clustering analysis with the k-means algorithm to identify patterns of craniofacial morphology. Four clusters were found that had significant differences in 24 of the 30 variables, suggesting that they represent different patterns of craniofacial form. The distinct clusters appear to be caused by differences in the compensation process of the facial structure responding to a genetically determined cranial and mandible form. The proposed method allows the discovering of patterns in an unsupervised manner, which differs from previously described ones. With these patterns better treatment plans can be designed taking into account their facial biotype characteristics.

Keywords: Facial patterns, Data clustering, Craniofacial morphology, Machine learning, Vert index

1. Introduction

A child's face is not a miniature version of an adult's face, in fact, it does not have the same proportions between the different parts and regions that constitute a face. The progressive facial growth is a differential growth process, meaning that some parts or regions grow in size earlier or later than others, and they do it in different ranges, directions, and magnitudes. Therefore, the growing process of the face involves a sequence of maturing steps which finally establishes the dimensions, angles, and proportions of an adult face [1].

In orthodontics, although the percentage of adult patients have increased in recent years, the majority of the treatments are carried out in preadolescents and adolescents patients, which are experimenting significant changes due to the growth process in their occlusion (bite), facial skeleton, and profile [2]. These changes should be taken into account when deciding a treatment plan, since they can act positively in the correction of a malocclusion (bad bite), helping with the mechanics of the treatment. On the other hand, if these changes are not considered, there is a risk that the treatment will have no effect or even worse, increase the patients' anomaly [3].

With this in mind, the possibility of predicting the growth sequence, magnitude, and direction, would constitute a significant advantage which could help to establish a diagnosis and facilitate the design of better treatment plans [2][4]. Unfortunately, prediction in this domain is a difficult task due to the individual behavior of craniofacial growth [5]. Each person has a unique pattern, influenced by genetic factors (internal or biological), as well as, external, such as presence of sickness, habits, etc. [2].

One approach to address this problem is to first identify facial biotype patterns which could be used as class labels in a second stage for prediction. Therefore, clusters with patients which present a similar facial biotype should be discovered by means of an unsupervised learning approach.

The application of clustering techniques to identify patterns inherent in a data set, have been used successfully in biological information [6]. Whereas in the dental research area, these techniques have been used to study the tooth size patterns in the population [9], to find task-specific EMG patterns during functional and non-functional oral behaviours [10]; to determine mandibular form patterns and its association with facial types [11], and in the identification of craniofacial patterns [12].

The objective of this work is to discover facial biotype patterns using multivariate cephalometric data. The patterns are found using a clustering approach which can be used in a second stage for prediction, thus, helping in the decision making of the treatment plan.

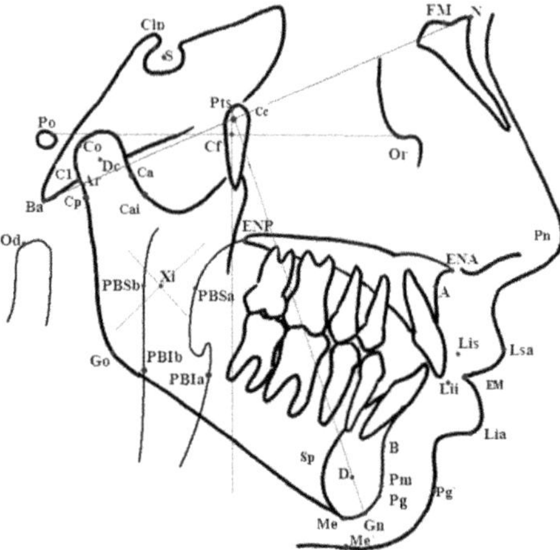

Fig. 1. Landmarks used for the cephalometric analysis

Input: $D = \{\mathbf{x}^1, \ldots, \mathbf{x}^N\}$ dataset, k = number of clusters
Output: Partition of D into k clusters
1 Randomly create k centroids $c_1, \ldots, c_k$;
2 Assign each data point $\mathbf{x}^i$, for $i = 1, \ldots, N$, to the cluster $j = \arg\min_{r=1,\ldots,k} dist(c_r, \mathbf{x}^i)$;
3 **while** *data points in some clusters change* **do**
4 Compute centroids $c_1, \ldots, c_k$;
5 Assign each data point $\mathbf{x}^i$, for $i = 1, \ldots, N$, to the cluster $j = \arg\min_{r=1,\ldots,k} dist(c_r, \mathbf{x}^i)$;
6 **end**

Fig. 2. The k-means algorithm

2. Methods

2.1. Cephalometirc data

A sample of 100 teleradiographs was collected from Chilean patients, 47 males (mean age: 14.6 ± 1.87 years old) and 53 females (mean age: 14.3 ± 2.42 years old). These patients were randomly selected from the files of two orthodontic´s clinics in Santiago, Chile. Patients that were treated with orthopedic therapies like protraction facemask, functional appliances and extra oral forces were excluded from the sample because those therapeutic appliances could modify the craniofacial morphology. Then, reference points were marked (Fig. 1) and cephalometric traces were drawn manually by a calibrated examiner (ICC: 0.87). The measures (35 variables) were taken from commonly used cephalometric´s analysis (Table 1) [15], and data like age and sex were registered as well. Nevertheless, for the clustering analysis these variables (age and sex) as well as Rickett's variables, which will be used to determine the biotype pattern, were not considered, therefore only 30 variables were used in the clustering process.

2.2. Clustering

The clustering algorithm used was the k-means [21], which is a method for finding the natural groupings of a data set. The k-means algorithm is shown in Fig. 2, where *dist* is the squared Euclidean distance.

One of the limitations of most clustering techniques, including the k-means, is that the user must specify the number of clusters (k) before the algorithm is applied. Therefore, human knowledge must be included in the clustering process, conditioning the results. To reduce this problem, several methods for selecting automatically the most plausible number of

Table 1. Cephalometric measurements (variables)

Landmarks	Variable
PBSa-PBSb	Width of upper airway
PBIa-PBIb	Width of lower airway
Nap-A	Nasion perpendicular to A
Nap-B	Nasion perpendiculat to B
Co-Gn	Mandible length
Co-A	Maxilar length
ENA-Me	Lower anterior face heigth
SNA°	SNA Angle
SNB°	SNB angle
ANB°	ANB angle
SND	AND angle
N-S-Ar°	Sella angle
S-Ar-Go°	Articular angle
Ar-Go-N	Superior gonial angle
N-Go-Gn	Inferior gonial angle
S-N	Anterior cranial base length
S-Ar	Posterior cranial base length
Ar-Go	Mandible ramus heigth
Go-Gn	Mandible base length
S-Go	Posterior facial heigth
N-Me	Anterior facial heigth
S-Ar/Ar-Go	Quotient of S-Ar and Ar-Go
Lis-Ls	Superior lip thickness
Lii-Li	Inferior lip thickness
Pg-Pg`	Pogonion thickness
Me-Me`	Menton thickness
Pts-Cp	Mandibular area
M-Pts	Maxillary area
M-Clp-Cp°	Condyle angle
C-Clp-Od°	Odontoid angle
Ena-Xi-Pm°	Lower facial heigth
Po-Or/N-Pg°	Facial depth
Ba-Na/Pt-Gn°	Facial axe
Go-Me/Po-Or°	Plane angle
Dc-Xi-Pm°	Mandible arch

clusters have been developed. In this paper, we selected k by computing a cluster validity index proposed in [22]. To compute this measure, clusters were formed using k=1,...,10. This procedure was repeated
30 times, and then the simulation that obtained the highest cluster validity index value was selected.

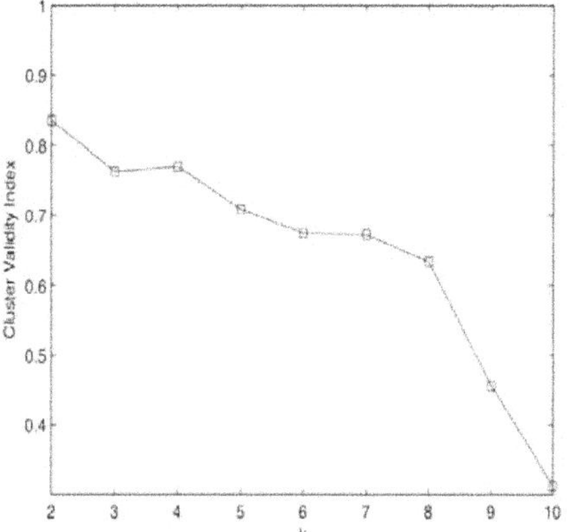

Fig. 3. Cluster validity index value (the higher the better) for different values of k

To visualize the distribution of the clusters formed, principal component analysis will be used to project the 30-dimensional data onto a 2D plane. Also, the ANOVA test is carried out to analyze the statistical differences between the clusters.

2.3. *Identifying facial biotypes*

The facial biotypes found from each cluster will be compared with the facial biotypes introduced by Ricketts [16], by means of the vert index, which uses the five variables which were left out from the clustering process. This index determines the amount of vertical growth and classifies the individuals in: brachyfacial (vert > +0.5), mesofacial (vert $\geq$ -0.5 and $\leq$ +0.5) and dolicofacial (vert < -0.5) types [16][23].

3. Results

The results of the best cluster validity index for k=1,...,10, is shown in Fig. 3. The higher the value of the index, the more plausible is k. The most plausible value is k=2, but this result limits our analysis to two biotypes, which are general and not very specific. So for this study, the next most plausible value was used, k=4.

The four clusters are shown in Fig. 4 using a PCA projection. When k=2 is considered, clusters 2, 3, and 4 merge to form one cluster.

Cluster 1 is formed by 23 individuals (mean age of 15.13±1.83), cluster 2 by 27 individuals (mean age of 14.14±1.67), cluster 3 by 29 individuals (mean age of 14.2±2.18), and cluster 4 by 21 individuals (mean age of 14.47±2.18). There was no significant difference in the age between these groups (this variable was analyzed after the clustering procedure, since it was not included for the cluster formation process). The clusters had significant differences (P<0.001) in 24 of the 30 variables (P<0.001). The values of the important variables and the mean vert index for each cluster are shown in Table 2.

Table 2. Values of important variables, age and vert index for each cluster

Variable	C1	S.D.	C2	S.D.	C3	S.D.	C4	S.D.	*P* value
Age	15.1	1.8	14.1	1.6	14.2	2.7	14.4	2.1	0.372
a. Co-Gn	131.4	6.2	115.1	5.5	122.9	4.4	119.8	5.3	0.000***
b. N-Me	133.3	4.3	123.5	7.4	128.2	6.2	119.5	5.7	0.000***
c. S-Ar-Go°	147.0	7.3	151.0	6.6	147.4	7.1	139.6	5.7	0.000***
d. ENA-Me	79.5	3.1	73.1	5.1	75.2	4.8	69.3	4.7	0.000***
e. SNA°	84.9	3.1	81.4	3.7	79.6	2.3	84.3	2.7	0.000***
f. Pts-M	56.9	2.5	52.5	3.2	58.2	3.2	51.7	3.4	0.000***
g. SNB°	81.1	3.8	76.6	3.5	76.1	2.3	80.4	2.4	0.000***
Vert	0.89	0.91	0.25	0.94	0.31	1.06	0.32	1.03	0.093

S.D.: Standard Deviation
***$P \leq 0.001$

Fig. 4. Visualization of the clusters using principal component analysis

4. Discussion

The four clusters formed, are statistically significantly different in 24 of the 30 (P<0.001) variables between them, suggesting that the groups formed represent actually different patterns of craniofacial form.

Age of the subjects (not included into the cluster analysis) was similar between the groups formed, discarding that the differences on the volume of the structures, due to the possible variation of the growth stage of the patients studied, could be a factor that affects the clustering results.

In this data set, thickness of soft tissues in menton and pogonion zone was similar between the subjects. This could suggest that, even when soft tissue can actually mask a skeletal disharmony, like its has been reported in [24], it would be unusual in this sample data; On the other hand, even when the height of the superior and inferior airways could affect the respiratory function, and therefore, the form of the structures involved in this process [26], the results show that they do not have a fundamental role in the variability of the craniofacial morphology of the data samples.

ANB° angle, that is extensively used to classify patients into skeletal class, is not significantly different between the four clusters found. Nevertheless, it is important to clarify that, in this data set, there were only two patients with negative ANB°, given that the sample was taken randomly. This point affects the representativeness of the analysis, and is a weakness of the present study that will be corrected in the future. However, the patterns identified in this work, would be valid for patients class I and II (ANB° 0-4°).

Analyzing the variables in each cluster, we can describe the four clusters formed searching for the singularity of each one:
Cluster 1, is the most different from the rest of the clusters, as we can observe in the visualization of the patients distribution (Fig. 4). They have the largest mandible and the largest and most advanced maxilla. The mean vert index, in this group, was +0.89 and they are, therefore, brachyfacial according to Ricketts biotypes, even when they have a greater anterior facial

height. This result could be explained, because these patients also have the greatest posterior facial height, giving an anterior-posterior facial height proportion of 65.6%. The mean ANB° angle was 4.1°, therefore, this group is mild class II.

Patients in cluster 2, instead, have the shorter, less height and most retruded mandible. The mean ANB° angle and vert were 4.9° and +0.25 respectively, being therefore class II mesofacial individuals.

In cluster 3, the individuals can present a larger or a shorter total mandible length (Co-Gn). In this case, we can observe that offsets also occur. Patients with smaller mandible, show smaller maxilla or a bigger mandible body length. There are also individuals with a bigger mandible and a retruded maxilla, but they also have the biggest anterior facial height of all the clusters. So, the patients maintain a class I sagital relationship (ANB° = 3.6°) with a vert index of +0.32 (mesofacial).

In cluster 4, a tendency to obtain a sagital harmony was observed, so that, a minor inferior anterior height (ENA-Me) can offset a bigger sella angle and an advanced maxilla can offset a larger mandible. On the other hand, a greater sella angle is offsetting with a minor articular angle, and a minor mandible with a more protruded position of it. Also, individuals with a bigger mandible and minor anterior facial height (N-Me) are in this cluster with the biggest SNA° of all the clusters. So, in this group mesofacial (vert = +0.3) and class II patient (ANB° = 4.4°) were included.

The groups obtained from the clustering technique, are independent of ANB° and vert variables even when cluster 1 has a strong brachyfacial tendency that could mean that this group has very singular features and a homogeneous morphology.

5. Conclusions

The clustering analysis is effective in identifying different craniofacial patterns based on a multivariate database. These patterns were found using 30 variables collected from a sample of Chilean patients, which were not completely concordant with those described by Ricketts. The distinct clusters appear to be caused by differences in the compensation process of the facial structure responding to a genetically determined cranial and mandible form.

The proposed method can be applied to several databases, creating specific classifications of craniofacial morphology adapted to different populations. Also, this method allows the discovering of patterns in an unsupervised manner, which differs from previously described ones. This is an advantage, considering that this is carried out without referencing to any standard values.

Future research will consider an automatic classification stage which will classify each patient into one of the four facial biotype patterns found in this work. Also, each cluster should have associated a treatment plan which should take into account the characteristics of their biotype patterns, thus, obtaining better results at the end of the treatment period.

References

[1] McCarthy J. Plastic Surgery Vol.4: Cleft lip and palate and craniofacial anomalies. W.B. Saunders Company, Philadelphia, 1990.

[2] Bishara S. Facial and dental changes in adolescents and their clinical implications. Angle Orthodontics. 70 (2000) 471-483.

[3] Gregoret J. Ortodoncia y cirugía ortognática. Diagnóstico y planificación. ESPAXS publicaciones médicas, Barcelona España, 1997.

[4] Chvatal B, Behrents R, Ceen R and Buschang P. Development and testing of multilevel models for longuitudinal craniofacial growth prediction. American Journal of Orthodontics and Dentofacial Orthopedics. 128 (2005) 45-56.

[5] Alhaija A and Richardson A growth prediction in class III patients using cluster and discriminant function analysis. European Journal of Orthodontics. 25(2003) 599-608.

[6] Boutros P and Okey A. Unsupervised pattern recognition: an introduction to the whys and wherefores of clustering microarray data. Brief Bioinforms. 6 (2005) 331-343.

[7] Whitfield CW, Cziko AM, and Robinson GE. Gene expression profiles in the brain predict behavior in individual honey bees. Science. 302 (2003) 296-299.

[8] Golub TR, Slonim DK, Tamayo P, Huard C, Gaasenbeek M, Mesirov JP, Coller H, Loh ML, Downing JR, Caligiuri MA, Bloomfield CD and Lander ES. Molecular classification of cancer: Class discovery and class prediction by gene expression monitoring. Science. 286 (1999) 531-537.

[9] Lee SJ, Lee S, Lim J, Ahn SJ and Kim TW. Cluster analysis of tooth size in subjects with normal occlusion. Am J Orthod Dentofacial Orthop. 132 (2007) 796-800.

[10] Farella M, Palumbo A, Milani S, Avecone S, Gallo LM and Michelotti A. Synergist coactivation and

substitution pattern of the human masseter and temporalis muscles during sustained static contraction. Clin Neurophysiol. 120 (2009) 190-197.

[11] Ogawa T, Kawasaki H, Takahashi O, Aboshi H and Kasai K. Application of a Fourier series to analysis of the relationship between mandibular form and facial morphology. J Oral Sci. 42 (2000) 93-100.

[12] Suzuki A and Takahama Y. Parental data used to predict growth of craniofacial form. Am J Orthod Dentofac Orthop. 99 (1991) 107-121.

[13] Abu Alhaija ES and Richardson A. Growth prediction in Class III patients using cluster and discriminant function analysis. Eur J Orthod. 25 (2003) 599-608.

[14] Bhatia SN, Wright GW and Leighton BC. A proposed multivariate model for prediction of facial growth. Am J Orthod. 75 (1979) 264-281.

[15] Proffit W. Contemporary Orthodontics. 3th Spanish ed, Harcourt: Spain; 2001.

[16] Ricketts RM, Bench R, Gugino C, Hilgers J and Schulhof R. Bioprogressive Therapy. 2th Spanish ed, Editorial panamericana: México D.F;1999.

[17] Jarabak JR and Fizzel JA. Technique and treatment with light wire edgewise appliances. 2nd Spanish ed, Mosby: St. Louis;1972.

[18] Mcnamara JA Jr and Brudon WL. Orthodontics and Dentofacial Orthopedics in the mixed dentition. 3th Spanish ed, Needham Press: USA; 1995.

[19] Arnett GW, Jelic JS, Kim J, Cummings DR, Beress A, Worley CM Jr, Chung B and Bergman R. Soft tissue cephalometric analysis: diagnosis and treatment planning of dentofacial deformity. Am J Orthod Dentofacial Orthop. 116 (1999) 239-253.

[20] Mercier J. Delaire's craniofacial architectural analysis. A reminder of the changes introduced by its designer in 1994. Rev Stomatol Chir Maxillofac. 101 (2000) 12-16.

[21] MacQueen, J. Some methods for classification and analysis of multivariate observations. In Proc. Fifth Berkeley Symposium on Mathematical Statistics and Probability. Berkeley: University of California Press 1967.

[22] Bezdek JC and Pal NR. Some new indexes of cluster validity. IEEE Trans. Syst. Man. Cybern. B28 (1998) 301-315.

[23] Ricketts R, Roth R, Chasconas SJ, Schulhof RJ and Engel GA. Orthodontic diagnosis and planning. 1st ed, Rocky mountain/orthodontics: USA 1982.

[24] Arnett GW and Bergmann RT. Facial keys to orthodontic diagnosis and treatment planning. Part I. Am J Orthod Dentofacial Orthop. 103 (1993) 299-312.

[25] Arnett GW and Bergmann RT. Facial keys to orthodontic diagnosis and treatment planning. Part II. Am J Orthod Dentofacial Orthop. 103 (1993) 395-411.

[26] Ung N, Koenig J, Shapiro PA, Shapiro G and Trask G. A quantitative assessment of respiratory patterns and their effects on dentofacial development. Am J Orthod Dentofacial Orthop. 98 (1990) 523-32.

[27] Mattar SE, Anselmo-Lima WT, Valera FC and Matsumoto MA. Skeletal and occlusal characteristics in mouth-breathing pre-school children. J Clin Pediatr Dent. 28 (2004) 315-318.

[28] Tourne LP. The long face syndrome and impairment of the nasopharyngeal airway. Angle Orthod. 60 (1990) 167-176.

Innovative Production Machines and Systems
D.T. Pham, E.E. Eldukhri and A.J. Soroka (eds)

Dynamic Scheduling in Job-Shop Production

S. Thiel[a], H. Tutsch[b]

[a] *Fraunhofer IAO, Nobelstrasse 12, 70563 Stuttgart, Germany*
[b] *flexis AG, Schockenriedstrasse 46, 70565 Stuttgart, Germany*

Abstract

Producers of high-level complexity products, e.g. automobiles, have a high interest on improving the process of maintenance and reconditioning. For products in these processes time-savings have direct influence on each day's turnover, as well on customer satisfaction. Since for the dynamic and complex processes in maintenance and reconditioning existing long-term planning systems are not performing appropriate, a dynamic scheduling system for job-shop production is required. This article introduces an agent-based approach for scheduling and monitoring of these jobs at runtime.

Keywords: Planning platform, flexibility, agent-based optimisation

1. Introduction

In production, reconditioning and maintenance jobs share a high factor of uncertainty about the steps required for accomplishing a job. For example, several different reasons may lead to an automobile service-job called "motor not starting". In a job-shop it typically is one out of the range: "Battery is not connected", "Battery defect", "fuel empty", "fuel is empty, because of leakage in fuel tank". Obviously, depending on the reason, quite different operations are required. These come with a correlating selection of resources and tools required to accomplish the job and the fixation process will consume a certain amount of time. Following the example, it get's clear that depending on the analysis different kind of sub-processes need to get triggered accordingly. Thus, long-term planning, known from the classical job-shop scheduling problem is not appropriate.

1.1 Issues in job-shop production

Companies are continuously investing in technology, resulting in high flexibility of their resources in comparison to international competition. However, the flexibility of the production resources only compensates for the problems arising from short-term planning. Capacities and restrictions are used as fixed constraints and the planner attempts to find an optimal solution with regard to all the different constraints.

Therefore it is a necessity to use a tool that enables the incorporation of the different flexibilities of the manifold resources employed, into short-term production programme planning. Thus, optimised production results combined with a high delivery reliability will be achieved. The efficient use of resources in the maintenance and reconditioning departments will assist in safeguarding existing jobs by tapping into available but not utilised flexibilities in the processes and in the orders itself.

The situation in job-shop production requires optimisation where both, resources and orders or even tasks within the orders, are negotiating the short-term processing of the programme with each other. Because in maintenance and reconditioning the orders and their tasks are changing e.g. during the disassembling process of a product, the priorities of the orders and the tasks within the orders are changing perpetually. To face this instability, a planning platform is required to react fast and efficient to allow short-term re-planning.

1.2 About an agent-based scheduling approach

To provide a dynamic and flexible solution for optimized scheduling flexis and Fraunhofer IAO have developed an agent-based approach. The following paragraph will give some general overview on it, followed by a more detailed view in the next section.

Having discussed the shortcomings of long-term planning systems and the specific requirements of job-shop production for maintenance and reconditioning jobs, it's clear that a new approach needs to interact directly with the processes at runtime. For this the system needs to monitor the actual progress of each job and the status of all resources taking part in the process. Based on this information a job can be replanned continuously, e.g. depending on results gained from an analysis step or on the current available resources.

This innovative approach provides major achievements in three categories:

Robustness
In the case a resource is temporally unavailable the agents of the system will instantaneously react and try to find alternative resources to accomplish their job.

Scalability/Flexibility
The system adapts automatically to the current number of jobs by creating the required amount of correlating agents. So the system can easily been tested within a restricted scenario. When adding further supported processes, resources and jobs, no particular change to the system is required.

Process-optimisation
Continuous interaction and negotiations between the agents provide process optimisation at runtime that is also taking into account complex processes including conditions and alternative branches.

Before detailing the approach, a short view on the related history of job-shop scheduling is given.

1.3 *Job-shop & Agent-based scheduling*

Since quite a while it's well-known [1] that the generic job-shop scheduling problem is NP-hard, much harder than the travelling salesman problem. Therefore several heuristics exist to calculate optimized plans solving it within a reasonable time. Most of these heuristics aim to find a general static solution on the problem, however as argued in the previous paragraph a dynamic solution is required.

Multi-Agent systems (MAS) [5] for planning and scheduling in production have been a research target for many years [2,3]. Particularly in the resort job-shop scheduling, many prominent examples can be found. E.g. ISIS by Fox [7,8]. More recent publications focus on the integration of MAS into planning and scheduling [4]. But, a closer look to the existing approaches reveals that agent-based scheduling systems in production are mainly applied to achieve long-term planning. Newer approaches also include runtime monitoring process optimisation, but are restricted to simulated job execution improving the static long-term planning.

So far, existing systems are unable to re-act on dynamically changing resources and general conditions at runtime. However, in some scenarios, especially in the case of job-shop production, dynamic planning and scheduling is required.

3. Getting the job done

3.1 The environment/resources

The system is categorized into the following items:

Process:
A process describes the steps and resources required to transform a product from an initial status into an end status (preferably a product-was-completed status). For

this, the process includes a description of interdependency of steps, as well as their pre- and post-conditions. Some steps need to be accomplished in a certain sequence, while other might be worked on in no specific order or even in parallel. Depending on the outcome of a step, there may be a decision, which step will follow. – So also decision trees can be part of a process. Typically a process references to a class of jobs that occurs in the production context.

Job:
A job always refers to a real-world application of a process. Therefore, each job correlates with the product that is worked on during the process. To identify its current status, a job contains information about the current step of the process as well as context related parameters helping to determine the proceeding steps (e.g. result of the previous operation), or parameters required for executing certain operations (e.g. the colour the product should be painted with).

Resource:
Each step of a process can require one resource for execution. Therefore resources can refer to a machine, a working station, a person, having a certain qualification, etc. On its way through the process, a job will typically pass several resources. Each resource is able to execute one or many operations. Before a job can utilise a resource, it has to make sure, the resource is not occupied by another job. Therefore each resource comes with a queue that allows the jobs to register for a resource.

Operation:
An operation refers to a real-world activity that is executed on the product. An operation is the only way to change a product's properties. However, there are also "passive" operations like a test or a decision made by an expert. Beside the product, each operation may require several input parameters and may return several result values after being accomplished.

Boundary Conditions:
In a typical production scenario, for a successful execution of a process, several boundary conditions have to be taken for granted. This is generally the case for infrastructural parameters like electricity or water. Other parameters like inventory stock of a resource depend on the long-term planning and its availability can change and partially depends on the actual operations performed. However, it's often not easy to access this information from the dynamic system. A more sophisticated model may include this information to further improve dynamic scheduling.

3.2 The system

Traditionally jobs for job-shop production are collected by a job-control system that is controlled by, or part of the ERP used by the enterprise. For job-shop

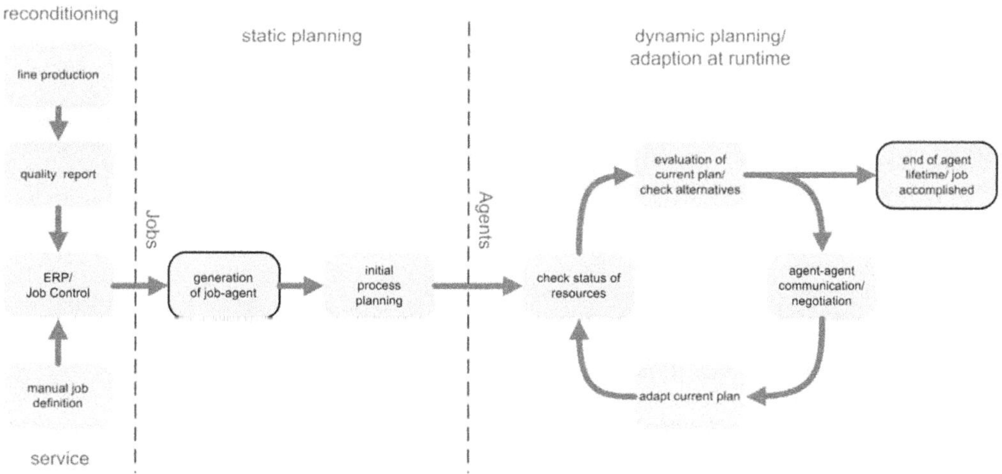

Figure 1: Agent's life-cycle

production, there are generally two potential sources filling the queue of jobs: Reconditioning and Maintenance

Products that show quality issues when manufactured at the assembly line, are moved to job-shop processes for further analysis and repair. These products are queued in for reconditioning and typically come with a formal description of the quality issue (quality report). Secondly, products returned from the customer are added from the service department. They've been analysed manually and are added with additional information about symptoms or defects identified. [Fig. 1]

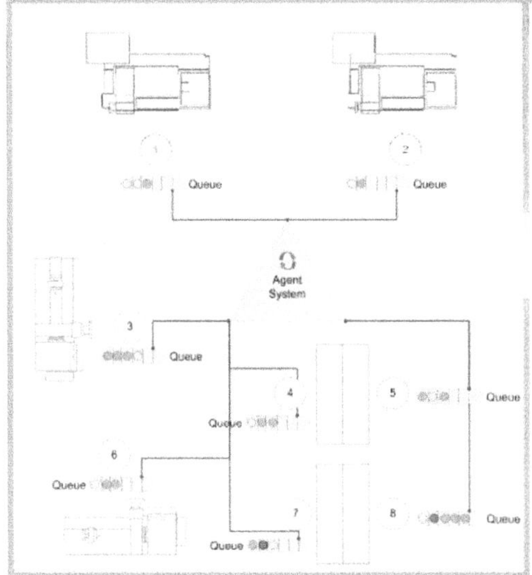

Figure 2: Job-Shop Queues at Runtime

Main approach of the system is to understand each job as an agent that proactively walks through the process. Each agent has the task to get its product completed. To minimize the process time of a single product, but also to achieve an optimized general performance, the agents receive credits for fast progressing through the process, but also for some cooperative behaviour. These credits can be collected and invested in agent-to-agent negotiations, e.g. about positions in the queue of a resource. [Figure 2]

An agent's live-time is divided into two phases; after being generated out of the job-control system the agent performs a static planning of the related process. This takes into account the current status of the resources as well as the existing queues etc. After the static planning, the agent enters the second phase, where the dynamic planning and execution takes place.

Mainly the second phase follows a loop of execution, continuous analysis, negotiation and replanning. Within this loop, the agent monitors the status of the relevant resources and accordingly re-evaluates the current (as well as alternative) plans for process execution. This implies the need of estimating the time required for executing all steps until the end of the process. Since also the waiting time for each resource needs to be considered, this is not a trivial task. Additionally, some processes include branches which execution depends on the outcome of some previous operation. So, it's clear that estimations about the remaining runtime of a process will often be inaccurate. To keep the estimation algorithm efficient, the agent uses a hybrid heuristic based on case-base reasoning and Bayes-networks. [6] While doing the planning, negotiations with other agents are ongoing to improve the agent's position in a resource waiting-queue. Once the best plan was optimized, the agent follows the new plan and starts over with the first step. When the process of an agent came to an end (step), the agent's life-time ends as well.

The continuous replanning and monitoring of the production progress lets the system react dynamically on changes in the job process, changes with the resources available or the amount of jobs executed in parallel. This allows the system to perform flexible and scalable, while doing an optimization at runtime.

5. Outlook

For evaluation purpose, a test-application is planned to be implemented at MTU Maintenance Berlin-Brandenburg. Since the production and maintenance processes at MTU are of particular high complexity and contain a high degree of uncertainty, MTU provides an optimal environment to approve performance and scalability of the system.The development and testing will be funded by the German Bundesministerium für Bildung und Forschung (BMBF) within the funding program "KMU-innovativ".

6. Summary

The paper introduced a system that enables the incorporation of the different flexibilities of the manifold resources employed, into short-term production programme planning. This helps to achieve optimised production results combined with high delivery reliability. An efficient use of resources in the maintenance as well as in reconditioning departments taps into available, but not utilised flexibilities in the processes and in the orders themselves.

References

[1] Conventional genetic algorithm for job shop problems; Nakano, R. and Yamada, T.; Proceedings of the Fourth International Conference on Genetic Algorithms, 1991

[2] Benchmarking and robust multi-agent-based production planning and control; Frey, D., Nimis, J., Wörn, H. and Lockemann, P.; Engineering Applications of Artificial Intelligence, Elsevier 2003

[3] Agent-Based Enhanced Workflow in Manufacturing Information Systems: the MAKE-IT Approach; Montaldo, E. and Sacile, R. and Coccoli, M. and Paolucci, M. and Boccalatte, A.; Journal of Computing and Information Technology, 2004

[4] Multi-agent based integration of process planning and scheduling; Li, X. and Li, W. and Gao, L. and Zhang, C. and Shao, X.; 2009

[5] Artificial Intelligence; Russell, S. and Norvig, P.; Pearson, 2004

[6] Finn V. Jensen: Bayesian Networks and Decision Graphs. Springer-Verlag, New York 2001, ISBN 0-3879-5259-4.

[7] Artificial intelligence in the factory of the future; Fox, M.S.; Proceedings of the ACM 12th annual computer science conference on SIGCSE symposium, 1984

[8] ISIS: a retrospective; Fox, M.; Intelligent scheduling, 1994

Innovative Production Machines and Systems
D.T. Pham, E.E. Eldukhri and A.J. Soroka (eds)

Lamarckian and Baldwinian learning in differential evolution for production scheduling problems

Nikos Giannopoulos and Andreas C. Nearchou

Department of Business Administration, University of Patras, 26500 Patras, Greece

Abstract

Consideration is given to Lamarckian and Baldwinian learning in the context of a hybrid population heuristic for the solution of the total weighted tardiness scheduling problem. The solution approach combines a differential evolution algorithm for population evolution and a variable neighborhood search to improve the population. Computational experiments over known benchmarks taken from the open literature demonstrate the effectiveness of the proposed approach.

Keywords: weighted tardiness problem, differential evolution, variable neighborhood search

1. Introduction

Total weighted tardiness (TWT) scheduling problem on one machine is a well-known *NP*-hard combinatorial optimization problem (COP) [1], formally defined as follows: a set of n independent jobs simultaneously available at time zero must be sequenced on a single machine. Each job i (i=1,...,n) has a positive processing time p_i, a due date d_i, and a positive weight w_i. The machine is available for continuous processing of the jobs, while preemption of a job is not allowed. For a given sequence of jobs, the tardiness of job i is defined as $T_i = \max(0, C_i - d_i)$, with C_i the completion time of job i. The objective is to minimize the summation of the weighted tardiness of all the jobs, i.e., to minimize $\sum w_i T_i$.

Instances with more than 50 jobs cannot be solved to optimality with implicit enumerative techniques such as dynamic programming, and branch and bound algorithms [2,3]. Therefore, large size instances of the problem must be faced through the use of heuristics. In the last three decades several heuristics have been proposed for solving TWT classified as either constructive, or improvement. Constructive heuristics [4-6] generate a schedule from scratch by fixing a job in a position at each step according to some simple or complex rule. Improvement heuristics start from a solution generated by some sequence generating technique, and using suitable schemes try to improve this solution with the hope of approaching the global optimum. Modern improvement methods are based on meta-heuristic approaches such as simulated annealing [6], genetic algorithms [7], tabu search [7], etc. Hybridizing meta-heuristics with local search is a promising area of research constituting a major trend in the field of combinatorial optimization.

This paper presents a new effective hybrid meta-heuristic for TWT which combines differential evolution (DE) with variable neighborhood search (VNS), referred to herein as DE/VNS. Two forms of hybridization are introduced in DE borrowed from the field of genetic local search: Lamarckian evolution and Baldwin effect [8,9]. Both of them use the metaphor that an individual learns during its lifetime.

The rest of the paper is organized as follows: Section 2 gives a brief overview of DE and VNS algorithms. Section 3 presents DE/VNS for the solution of TWT. Experimental results on benchmarks test problems are described in section 4; while section 5 concludes the contribution of the paper.

2. Brief description of DE and VNS algorithms

2.1. The DE algorithm

DE introduced by Storn and Price [10] for global optimization over continuous. DE utilizes Ns, D-dimensional parameter vectors $x_{i,k}$, $i \in \{1 \ldots Ns\}$ (k denotes the iteration number of the algorithm.) as a population to search Ω. The initial population,

$$S = \{x_{1,0}, x_{2,0}, \ldots, x_{Ns,0}\}, \quad (1)$$

is taken to be uniformly distributed in Ω. At each iteration, all vectors in S are targeted for replacement. Therefore, Ns competitions are held to determine the members of S for the next iterations. This is achieved by using mutation, crossover and acceptance operators. In the mutation phase, for each target vector $x_{i,k}$, $i \in \{1,2,\ldots,Ns\}$, a mutant vector $\hat{x}_{i,k}$ is obtained by

$$\hat{x}_{i,k} = x_{\alpha,k} + F(x_{\beta,k} - x_{\gamma,k}), \quad (2)$$

where α, β, γ, $i \in \{1,2,\ldots,Ns\}$ are mutually distinct random indices and are also different from the current target index i. $x_{\alpha,k}$ is known as the base vector and $F > 0$ is a scaling parameter. The crossover operator is then applied to obtain the trial vector $\psi_{i,k}$ from $\hat{x}_{i,k}$ and $x_{i,k}$. The crossover is defined by

$$\psi^j_{i,k} = \begin{cases} \hat{x}^j_{i,k} & \text{if } rnd \le CR \text{ or } j = I_i \\ x^j_{i,k} & \text{if } rnd > CR \text{ and } j \ne I_i \end{cases}, \quad (3)$$

where I_i is a randomly chosen integer in the set I, i.e., $I_i \in I = \{1,2,\ldots,D\}$; the superscript j represents the j-th component of respective vectors; $rnd \in (0,1)$ is drawn randomly for each j. The ultimate aim of the crossover rule is to obtain the trial vector $\psi_{i,k}$ with components coming from the components of the target vector $x_{i,k}$ and the mutated vector $\hat{x}_{i,k}$. This is ensured by introducing CR and the set I. Notice that for CR=1 the trial vector $\psi_{i,k}$ is the replica of the mutated vector $\hat{x}_{i,k}$. The targeting process (mutation and crossover) continues until all members of S are considered. After all Ns $\psi_{i,k}$ have been generated, acceptance is applied. In the acceptance phase, the function value at the trial vector, $f(\psi_{i,k})$, is compared to $f(x_{i,k})$, the value at the target vector and the target vector is updated using

$$x_{i,k+1} = \begin{cases} \psi_{i,k} & \text{if } f(\psi_{i,k}) < f(x_{i,k}) \\ x_{i,k} & \text{otherwise} \end{cases} \quad (4)$$

Reproduction (mutation and crossover) and acceptance continues until some stopping conditions are met.

2.2. The VNS algorithm

VNS introduced by Mladenovic and Hansen [11] as a general meta-heuristic for solving COPs based on two or more neighborhoods and a systematic change of the neighborhood during the search procedure. Its general mechanism is as follows.

Step 1:

Select the set of neighborhood structures N_k, for k=1,…,k_{max}, that will be used in the search. Determine an initial solution x to start with, and set a stopping condition.

Step 2: Repeat the following until the stopping condition is met:

1) Set k =1;
2) Repeat the following steps until $k = k_{max}$:
   a. Shaking. Generate a point x' at random from the kth neighborhood of x ($x' \in N_k(x)$);
   b. Local search. Apply some local search method on x' to generate a new local optimum solution x'';
   c. Move or not. If x'' is better than the incumbent solution x, move there (x=x''), and continue the search with k =1; otherwise, set k = k+1;

Shaking is the procedure of generating a new solution in the neighbourhood of the current solution. In particular, a random point x' is selected from the N_k neighbourhood of the current point x. The solution x' generated during *shaking step* is then being improved using a suitable local search heuristic. Finally, the new solution x'' created by local search heuristic is compared to that of the original solution x and accepted when its quality is higher. VNS is terminated when an appropriate user-defined stopping condition is satisfied. This condition may be a maximum permitted number of iterations, or a maximum processing time allowed, or a maximum number of iterations between two improvements, etc.

3. The proposed DE/VNS for TWT

Algorithm DE/VNS for TWT

Step 1. Generate a random initial population S of Ns solutions. Set iteration counter k=0;

Step 2. Generate the solutions of population S of the next iteration (k+1) as in the following:

for i = 1 **to** Ns **do**

Mutation phase: Generate a *mutant* vector $\hat{x}_{i,k}$ using Eq. (2).

Crossover phase: Generate a *trial* vector $\psi_{i,k}$ using Eq. (3).

Solution Interpretation phase: Build TWT schedules $\pi^{x_{i,k}}$ and $\pi^{\Psi_{i,k}}$ corresponding to $x_{i,k}$ and $\psi_{i,k}$, respectively.

Acceptance phase: **if** COST($\pi^{\Psi_{i,k}}$) < COST($\pi^{x_{i,k}}$) **then** $x_{i,k+1} = \psi_{i,k}$ **else** $x_{i,k+1} = x_{i,k}$

endfor

Step 3. Alter each solution in S by VNS heuristic with probability p_{VNS}.

Step 4. Determine the population best solution π^{best}; keep track for the *best-so-far* solution π^{bsf}

Step 5. **if** *termination condition* is not satisfied then go to ***Step 2*** else **Return** (π^{bsf}) **and** stop

---

Remarks:

1. DE/VNS combines DE to perform global exploration among a population of individual solutions, and the local improvements abilities of VNS to perform local exploitation around selected DE solutions.
2. The appropriate settings for CR and F control parameters are dynamically estimated during the evolution of the algorithm using a self-adapted scheme previously proposed by Brest *et al.* [12].
3. Mutant vectors are generated using the DE scheme DE/best/2/bin proposed in [13].
4. *Solution interpretation* phase is responsible to map every genotype vector to its associated phenotype, i.e., to an actual TWT solution. This mapping is performed using an effective encoding scheme called *Sub-range-keys* introduced in a previous work of one of the authors of this paper [14].
5. π^z denotes the schedule corresponding to the vector solution z ($z \in S$); while COST(π^z) denotes the cost of this schedule (i.e., its total weighted tardiness).
6. The VNS procedure (shown below) developed in this work is equivalent to the Reduced VNS heuristic [11]. An individual vector solution z is chosen with probability p_{VNS} from the entire population S to pass the local search improvement. This probability is not static. It is different for every individual vector solution and changes during the run of the algorithm. In particular, p_{VNS} is a component (a gene) of the genotype and adjusted by means of evolution.

Procedure *RVNS* (z, π^z, COST(π^z))
Input: a real-valued vector z (a genotype)
Output: a modified version of z, π^z, and its cost
begin
  outloop=0; *kmax*=3;
  repeat
    kcount=1; y=z;
    repeat
      if *kcount* =1 **then** shift(y);
      if *kcount* =2 **then** exchange(y);
      if *kcount*=3 **then**
        shift(y); exchange(y); shift(y); exchange(y);
      endif
      π^y = **Encoding_mechanism**(y);
      if COST(π^y)<COST(π^z) **then** *kcount*=1; z=y;
      else *kcount* = *kcount* +1;
    until (*kcount* > *kmax*) **or** (COST(π^z) $\leq$ *twt**) ;
    outloop = *outloop* + 1;
  until (*outloop* > *n/2*) **or** (COST(π^z) $\leq$ *twt**) ;
end;

As one can see from the above procedure, three different neighbourhoods are explored based on counter *kcount*. The operations of *shift* and *exchange* search models work as in the following: the former shifts a randomly selected gene to the left or right a (randomly chosen) number of positions. In latter model two positions are randomly selected along the vector and the genes contained in these positions are exchanged.

7. Altering an individual population solution (*Step* 3 of the algorithm) is performed either in both genotypic and phenotypic level (Lamarckian learning case), or only in phenotypic level (Baldwinian learning case). The case of combining the two learning forms in a mixed search strategy (i.e. partial Lamarckian and partial Baldwinian learning) is also examined and discussed in the experimental section below. The following three mixed learning strategies were examined: 95% Lamarckian and 5% Baldwinian learning, 5% Lamarckian and 95% Baldwinian learning, and 50% Lamarckian and 50% Baldwinian learning.

4. Experimental Results

DE/VNS was tested on known benchmarks problems available by the public **ORLIB** electronic library. In the experiments we include the data set instances with 40 and 50 jobs. Each data set contains 125 different instances. The algorithm was coded and compiled in Borland Delphi 5.0, and run on a PC with 2.11 GHz CPU. Four criteria were used to quantify the performance of the algorithm:

- avg.dev=average relative deviation from the global optimum in percentage,
- *max.dev*=maximal relative deviation from the global optimum in percentage,
- *#opt*:=number of test instances for which the exact optimum solution was achieved
- *avg.cpu*:=average computation time measured in seconds.

Relative deviations were computed by$(UB-twt^*)/twt^*$; where, UB is the total weighted tardiness of the best solution found by DE/VNS, and twt^* the existing optimal total weighted tardiness for a specific instance. 10 runs of the algorithm were carried out starting each time from a different random seed, and the best result obtained after these runs was reported. A time limit of $20n$ iterations for the global optimizer DE was decided to be used; with n denoting the number of the jobs to be scheduled. The size of DE population was equal to $3n$. The following five versions of the proposed DE/VNS were examined:

- with pure Lamarckian learning, denoted by DE/VNS l
- with pure Baldwinian learning, denoted by DE/VNS b
- with 5% of Lamarckian steps and 95% of Baldwinian steps, denoted by DE/VNS lb 5/95
- with 50% of Lamarckian steps and 50% of Baldwinian steps, denoted by DE/VNS lb 50/50

Table 1
Experimental results on 40-jobs and 50-jobs TWT benchmarks data set.

40-jobs TWT

algorithm	%avg. dev	max.dev	#opt	avg.cpu
DE/VNS l	0.00	0.00	**125**	3.63
DE/VNS b	1141.37	-	22	22.34
DE/VNS lb 5/95	0.11	5.41	75	48.57
DE-VNS lb 50/50	0.001	0.10	124	5.48
DE-VNS lb 95/5	0.00	0.00	**125**	5.44

50-jobs TWT

algorithm	%avg. dev	max.dev	#opt	avg.cpu
DE-VNS l	0.00	0.00	**125**	8.20
DE-VNS b	1801.77	-	21	92.75
DE-VNS lb 5/95	0.10	4.72	65	74.68
DE-VNS lb 50/50	0.00	0.00	**125**	10.67
DE-VNS lb 95/5	0.00	0.00	**125**	8.40

- with 95% of Lamarckian steps and 5% of Baldwinian steps, denoted by DE/VNS lb 95/5.

Table 1 displays the results obtained over the examined benchmarks data sets. As one can see from this Table, DE/VNS l, DE/VNS lb 50/50, and DE/VNS lb 95/5 showed the best performance provided quite similar high quality results. In particular, they managed to determine the exact optimum solution to all the instances included in the data sets. On the contrary, DE/VNS b and DE/VNS lb 5/95 were not able to perform well to the majority of the test instances.

DE/VNS with pure Baldwinian learning is clearly unsuitable for TWT. In regard to the speed of convergence, DE/VNS l found to be the fastest, spending in average 5.9 sec of CPU time until the convergence.

5. Conclusions

A hybrid meta-heuristic was developed for the solution of the total weighted tardiness problem on one machine. This meta-heuristic (called DE/VNS) combines a differential evolution (DE) algorithm together with a variable neighborhood search (VNS) heuristic. Two basic forms of hybridization were examined namely, Lamarckian learning and Bandwinian effect. Moreover, the case of a mixed hybridization strategy with partial Lamarckian and partial Baldwinian learning was examined. Experiments over public available benchmarks data sets showed a quite high performance of DE/VNS in regard to both solutions quality and speed of convergence. The main observation of the experiments is that Lamarckian DE/VNS is superior and much faster than Baldwinian DE/VNS. Future work will be focused on the development of a multi-objective version of DE/VNS to address real-life more complex production scheduling problems.

References

[1] French S. (1990), Sequencing and scheduling, an introduction to the mathematics of the job-shop, Ellis Horwood Publication.

[2] Abdul-Razaq T.S., Potts C.N., and Van Wassenhove L.N. (1990), A survey of algorithms for the single machine total weight tardiness scheduling problem. Discrete Applied Mathematics, 26, 235-253.

[3] Du, J., Leung, J.Y.T. (1990), Minimizing total tardiness on one machine is NP-hard. Mathematics of Operations Research 15, 483–495.

[4] Smith W.E. (1956), Various optimizers for single-stage production. Naval Research Logistics Quartely, **3**, 59-66.

[5] Bauer A., Bullnheimer B., Hartl R.F., and Strauss C. (1999), An ant colony optimization approach for the single machine total tardiness problem. Proc. of CEC'99, IEEE Press, Piscataway, 1445-1450,.

[6] Potts C.N. and Van Wassenhove L.N. (1991), Single machine tardiness sequencing heuristics. IEE Transcations, 23, 346-354.

[7] Crauwels H.A.J., Potts C.N., and Van Wassenhove L.N. (1998), Local search heuristics for the single machine total weighted tardiness scheduling problem. INFORMS Journal on Computing, 10, 341-350.

[8] Whitely D., Gordan V., Mathias aK. (1994), Lamarckian evolution, the Baldwin effect and function optimization, in Davidor *et al.* (eds): Parallel Problem Solving from Nature, PPSN III, Springer-Verlag, Berlin, 6-15.

[9] Lobo F., Goldberg D.E. (1997), Decision making in a hybrid genetic algorithm, in: Proceedings of the 1997 IEEE Int. Conference on Evolutionary Computation, 121–125.

[10] Storn R. and Price K. (1997), Differential Evolution – A simple and efficient heuristic for global optimization over continuous spaces, Journal Global Optimization;**11**:341-354.

[11] Mladenovir N. and Hansen E, (1997), Variable neighborhood search, Computers & Operations Research. 24 (1 I), 1097-1100.

[12] Brest J., Greiner S., Boskovic B., Mernik M., and Zumer V. (2006), Self-Adapting Control Parameters in Differential Evolution: A Comparative Study on Numerical Benchmark Problems, IEEE Trans. on Evolutionary Computation, 10/6, 646-657.

[13] Price, K. (1996), Differential Evolution: A Fast and Simple Numerical Optimizer, NAFIPS'96, pp. 524-527.

[14] Nearchou A. and Omirou S. (2006), Differential evolution for sequencing and scheduling optimization, Journal of Heuristics 12/6, 395-411.

Innovative Production Machines and Systems
D.T. Pham, E.E. Eldukhri and A.J. Soroka (eds)

Computer based generation of control programs for a high speed mobile harbor crane in a novel maritime container handling system

H.S. Park [a], D.B.H. Anh [a], L.C. Hieu [b]

[a] *School of Mechanical and Automotive Engineering, University of Ulsan, Ulsan 680-749, Korea*
[b] *School of Engineering, University of Greenwich, Kent ME4 4TB, UK*

Abstract

There is an ongoing trend to make increased use of automated cranes and driverless vehicles for handling and transportation of the maritime containers. The concept of a mobile harbor system was therefore founded and aimed at developing a novel maritime container handling solution that allows going out to a ship that anchors in the deep water to load and unload containers on sea and then transport them to their destination ports regardless of the water depth. The proposed mobile harbor system has three sub-systems: (i) a high speed mobile harbor crane (HSMHC), (ii) a mobile harbor traffic management (MHTM), and (iii) an intelligent container yard management (ICYM) system. The generation of control programs for controlling the HSMHC system plays a key role in the development of the novel maritime container handling system. In this paper, a method for computer based generation of the PLC programs for controlling the HSMHC system is presented. The 3D virtual simulation is used to describe the control sequence which is mathematically modeled by using the sequence matrix and Petri net technique. Based on the control sequence, the PLC programs are automatically generated and then verified based on the simulation tool. The experiments conducted on the actual prototype of the HSMHC system showed the stable results.

Keywords: Mobile harbor crane, maritime container handling, PLC, Petri net.

1. Introduction

The globalisation of trade leads to the rapid growth of the global shipping networks and containerisation. In addition, the increase in the trade about products and services and the movement of capital, has contributed to the economic globalisation. The global containership fleet increased by 17 million dead weight tons or 15.5% Year-On-Year, and over 60% of the world's deep-sea general cargo is transported in containers, whereas some routes, especially between economically strong and stable countries, are containerized up to 100% [1, 2]. The time in port for ships combined with low rates for loading and discharging is the key factor for the competitiveness of a container seaport.

Traditionally, the container activities can be categorized into three types: import, export, and transshipment activities [3]. In export activities, the containers are retrieved from the stored location and transported by vehicles to the quayside. Then, the fixed quay cranes are commonly used to remove the containers from the vehicles and load them onto the vessels. The import activities are similar to the export one, but in the reverse order. In transshipment activities, after unloading from the vessel, the containers are stored in the storage yard, and finally they will be loaded onto other vessels for the next shipment.

Currently, there is an ongoing trend to make increased use of automated cranes and driverless vehicles for handling and transportation of the maritime containers. With the recent tendency toward containerisation, there has been an increasing interest in development of the mobile harbor cranes that are (i) capable of handling containers as well as other general cargoes in different styles, (ii) self-traveling, and (iii) versatile [4]. Moreover, since the worldwide container and cargo handling capacity is more and more increasing. This leads to the need for enhanced port capacity and capability. However, the port capacity enhancement is not the best solution, because it brings in a number of other problems that are related to the environmental or security concerns as well as big investment. Besides, the container handling solutions especially for ports with insufficient infrastructure are also important.

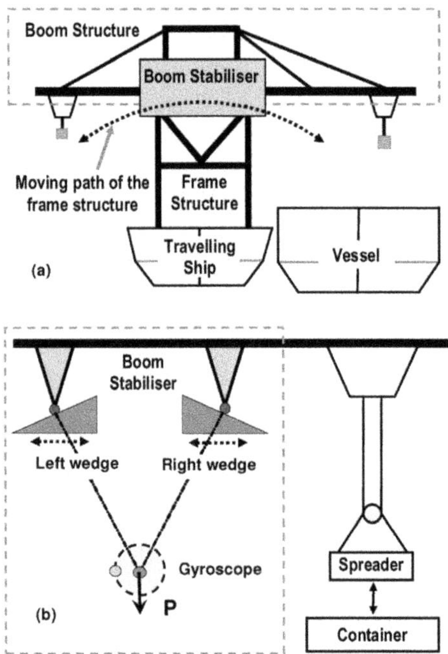

Fig. 1. (a): A high speed mobile harbor crane system with a crane is mounted on the traveling ship. (b): A ball-joint based boom stabilisation structure

The concept of a mobile harbor system was therefore founded under the KAIST Mobile Harbor project which was funded by Ministry of Knowledge Economy, Korea. The main objective of this project is to develop a novel maritime container handling solution that allows going out to a ship that anchors in the deep water to load and unload containers on sea and transport them to their destination ports regardless of the water depth, and the a deep-port for the landing of a vessel is therefore not required.

The proposed mobile harbor system has three sub-systems: (i) a high speed mobile harbor crane (HSMHC), (ii) a mobile harbor traffic management (MHTM), and (iii) an intelligent container yard management (ICYM) system.

The HSMHC system is used to load and unload maritime containers and transport them to their destination ports or vessels. The newly developed HSMHC system which includes a crane mounted on the traveling ship is shown in Fig. 1 (a). In order to avoid and miminise the effects of the ocean waves and the unstable nature of the traveling ship when loading and unloading the containers, the boom and frame structure which are connected via the ball-joint based boom stabilisation unit (boom stabiliser) are used for the crane. The frame structure is firmly mounted on the traveling ship, while the boom structure is not. Figure 1(a) presents the moving path of the frame structure under the vibration and unstable movement of the traveling ship; and the whole HSMHC system is well balanced with the use of the boom stabiliser. With the use of the ball-joint based boom stabiliser, the boom structure motion is independent to the frame structure and supports stabilization of the cargos and containers as shown in Fig. 1(b). The boom stabiliser has two independent driven wedges which are driven by the hydraulic motors and the dual-axis angular rate *sensor* (gyroscope) is used for measurement of orientation angles. The whole HSMHC system is controlled by the Programmable Logic Controller (PLC), including motion, stabilisation and crane configuration control as well as the control of the loading and unloading processes of the cargos and containers. The vibration and unstable movement of the traveling ship is detected by the gyroscope which then reports to the PLC controller to activate the movement of the wedges for compensation of the frame structure rotation. In this way, the stability of the boom structure as well as the containers is obtained.

In order to prepare the PLC programs, traditionally the programmers have to base on their knowledge and experiences. Since there is a huge amount of information that needs to be processed for such a complex HSMHC system, there is an overload in information management, and it should lead to the errors if the manual method is used. In order to solve this problem for the proposed HSMHC system, a computer based model was developed to support the process of preparation of the PLC programs for the HSMHC system; it aims at (i) minimising the programming time, and (ii) reducing the programming errors. In this paper, a novel approach that facilitates generation of the PLC programs for controlling the highly automated HSMHC system will be mainly focused on, especially the use of a PLC simulation method to describe a control sequence of the PLC programs. The sequence matrix and Petri net techniques are used to model the control sequence and automatically generate the PLC programs. Finally, the generated PLC programs are verified based on the simulation tool.

2. Materials and methods

The systematic procedure for automatic generation of a control program for the HSMHC system is shown in Fig.2, including 6 steps which are presented in the following Sections 2.1 – 2.6.

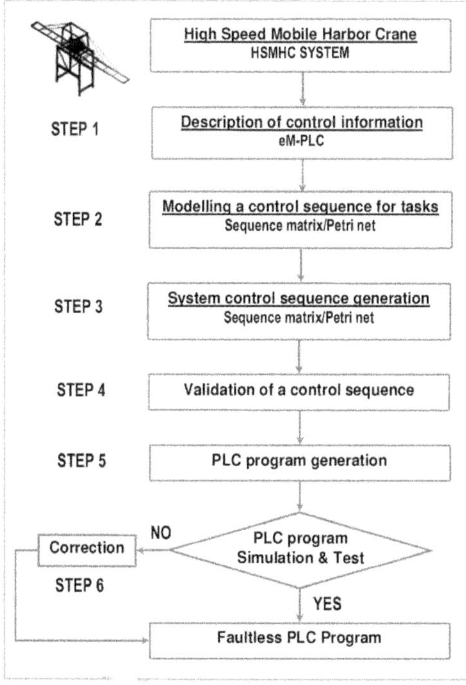

Fig. 2. The 6 steps for automatic generation of a PLC control program for the HSMHC system.

2.1. Description of the control information

A control problem is analysed based on the PLC simulation tool, eM-PLC (Tecnomatix), to get the feasible control information of the HSMHC system. The eM-PLC tool enables off-line verification and generation of PLC programs; it allows design manufacturing cells, digital special machines and automation processes in a 3D virtual environment and then introduces control information into that virtual environment.

For the proposed HSMHC system, the control information is determined by a set of *Control Tasks* for performing the operations of the HSMHC system. A *Control Task* is described by several *Control Activities* in order to increase the reusability and flexibility of the PLC programs. Figure 3 presents the eM-PLC tool for off-line simulation of the HSMHC system for obtaining the control information for *Control Tasks* of the HSMHC system. Firstly, a CAD data of the HSMHC system are imported into the simulation environment of eM-PLC. Secondly, the CAD components and their kinematics relationships are defined. Finally, based on the analysis and results from the virtual simulation in eM-PLC, the control information for *Control Tasks* of the HSMHC system is determined.

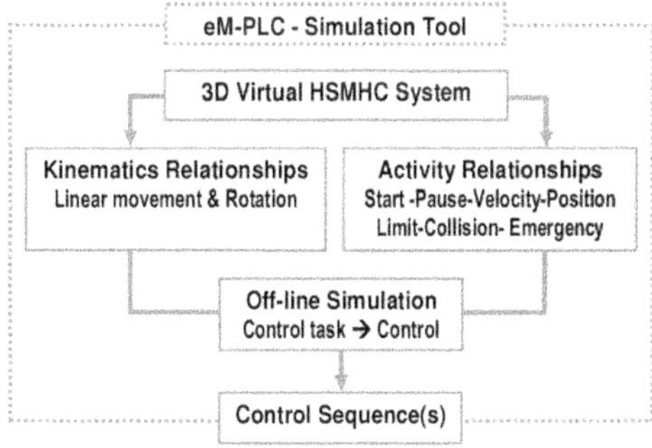

Fig. 3. The eM-PLC tool for off-line simulation of the HSMHC system

2.2. A control sequence for the HSMCH system

A control sequence for a certain system may be defined as set of *operations* to complete control tasks. A control sequence is considered as an optimal if it is minimising resources such as the time and money invested in the system, and maximising the quality of responses. In order to have an optimal control sequence for the HSMHC system, the operating simulation of the HSMHC system via eM-PLC is used; in this way, an optimal control sequence for the HSMHC system is selected based on the experiences of the PLC programmer(s). Figure 4 presents the description of the control sequence for the HSMHC system and its constituent elements: *Control Components*, *Control Tasks* and *Control Activities*. As mentioned in *Section 2.1*, a control sequence for the HSMHC system is described by *Control Components*. A *Control Component* is defined by *Control Tasks*. A *Control Task* is finally described by *Control Activities* and *Control Conditions*. A *Control Condition* is a set of rules which decide if the *Control Activity* is ended or not; it controls the status of the *Control Activity*.

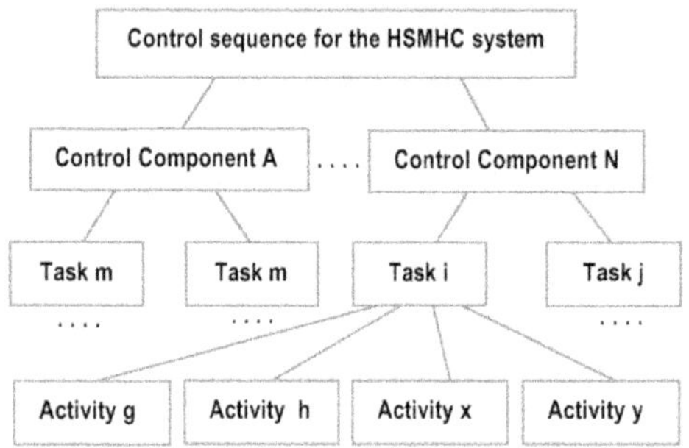

Fig. 4. Definition of the control sequence for the HSMHC system and its constituent elements: *Control Components*, *Control Tasks* and *Control Activities*.

The *Control Activities* and their sequence for a *Control Task* are derived from the analysis of controlling behavior of the actual prototype of the HSMHC system which were developed and installed at University of Ulsan in Korea (UoU) (Fig. 7). In order to present a control sequence for a *Control Task* mathematically, the sequence matrix for a *Control Task* (SMT) is used [5]. The sequence matrix lists all constituent operations and the corresponding information exchange. In this study, the sequence matrix A_{gh} is defined as follows.

$$A_{gh} = \begin{cases} 1 & Activity\ g\ before\ h \\ -1 & Activity\ g\ OR\ with\ h \\ 0 & Otherwire \end{cases} \quad (1)$$

g, h = 1,.., k (*k*: number of *Control Activities*)

In order to present the concurrence as well as to facilitate the validating of the control sequence for a *Control Task*, it is necessary to develop a model describing the control sequences interpretably. The Petri met (P/N) technique is therefore selected.

The P/N is a technique that is useful for modeling a dynamic concurrent behavior of the network-based systems. It is a directed bipartite graph, in which the nodes represent *Transitions* and *Places*. The *Directed Arcs* describe which *Places* are pre- and/or post-conditions for which *Transitions*. A *Place* may have zero or more tokens.

The application of P/N shows a lot of advantages, especially with regard to the aspects of the synchronization and the non-determination of the execution. It is also useful for detecting the errors in the control sequence via evaluation of the P/N model properties. In this study, a P/N is defined as follows:

$$\begin{aligned} &P/N = (P, T, A, W, Mo) \quad (2) \\ &P = \{Pj\}, j = 1, k; T = \{tj\}, j = 1, k; \\ &A \subseteq \{P\times T\} \cup \{T\times P\}; W: A \rightarrow \{1, 2\}; \\ &Mo: P \rightarrow \{0, 2 ...\} \end{aligned}$$

P, T and A are respectively sets of *Control Activities*, *Control Conditions*, and *Directed Arcs*. W is a set of weight on *Directed Arcs*. Mo is an initial mark which represents the initial state of a system.

2.3 Generation of a system control sequence

In the HSMHC system, the control sequence for a *Control Task* is considered as a node; it is modeled based on the sequence matrix and Petri net technique. In order to generate the control sequence for the HSMHC system, a *Control Task* which is defined by a sequence of *Control Activities* is considered as an operation in a control sequence for the HSMHC system. Based on the simulation in the eM-PLC tool, a control sequence for the HSMHC system is created. The sequence matrix for the HSMHC system is defined as T_{gh} and presented as follows:

$$T_{gh} = \begin{cases} 1 & Task\ g\ before\ h \\ -1 & Task\ g\ OR\ with\ h \\ 0 & Otherwise \end{cases} \quad (3)$$

g, h = {1,2, ... k }; k: *Control Task*

2.4 Validation of a control sequence

Based on the evaluation of the P/N model properties, the errors in control sequence for the HSMHC system are detected. These errors could be manually fixed by refining the value of the sequence matrix.

2.5 Generation of a PLC program

Based on the control sequence, a PLC program is finally generated. The *Control Activities* and *Control Conditions* in the control sequence for a *Control Task* are described by the modules of PLC codes. Figure 5 presents the modules of the PLC codes to establish the control sequence for the boom stabiliser which is one of the sub-systems of the HSMHC system (Fig.1). The module "PLC – code 1" is used for reading the position of the boom stabiliser. The module "PLC – code 2" is used to establish the *Control Condition* for the *Control Activity* "Position reading".

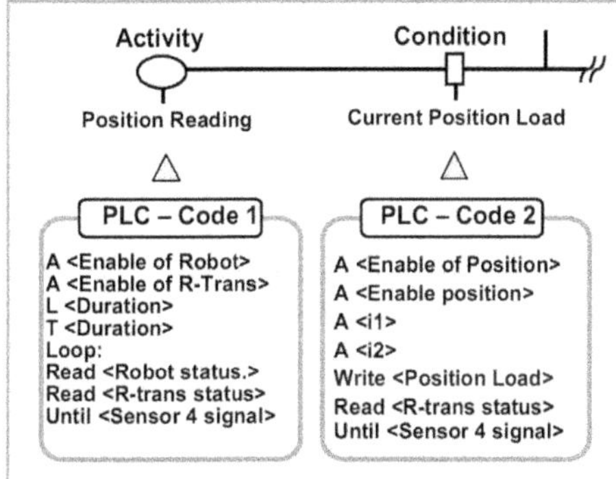

Fig. 5. The modules of PLC-codes of the *Control Activities* and *Control Conditions* for reading the position of the boom stabiliser of the HSMHC system.

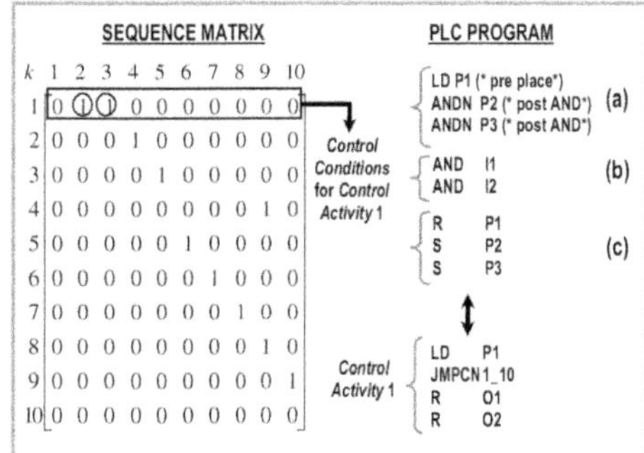

Fig. 6. A sequence matrix and establishment of *Control Conditions* for a *Control Activity*.

The modules of the PLC codes that establish the *Control Activities* for the HSMHC system are created by using the Boolean variables. The *Control Conditions* for the *Control Activity* can be established by not only the modules of the PLC-codes of the *Control Activity* but also the sequence matrix for a *Control Activity* as shown in Fig.6. In order to establish the *Control Conditions* for the *Control Activity 1*, the module of the PLC codes (b) is written. However, the additional *Control Conditions* (a) and (c) can be established by scanning of the sequence matrix for the *Control Activity*1 (Fig. 6).

2.6 Validation of the generated PLC program

The generated PLC program is finally validated and refined based on the PLC simulation and P/N model of the HSMHC system. First of all, the generated PLC program in the form of the instruction lists is imported into the eM-PLC tool. Based on the control sequences for the HSMHC system, the P/N model is created. If there is a difference when executing the P/N model and the virtual HSMHC system, the faults are detected [6].

3. Results

The software for automatic generation of the PLC program for the HSMHC system, called MHC-PLC, was developed by using Visual C++ language. The generated PLC programs were tested on the actual HSMHC prototype at UoU which was manufactured with the basic parameters shown in Table 1 and Fig.7. The PLC programs generated by MHC-PLC software were used to operate the HSMHC prototype successfully with the stable results. Figure 8 presents the menu of the developed MHC-PLC software which shows the generated P/N model for detection of the errors in the control sequence of the HSMHC system.

Table 1:
Basic parameters of the HSMHC prototype at UoU

Base length	2.0 m
Width	1.2 m
Height	3.2 m
Boom length	5.1 m
Total weight	700 kg

4. Discussion and conclusion

Although there are commercially available packages that are able to generate PLC programs automatically based on the simulation tools such as eM-PLC, there are the existing difficulties and challenges when using these packages for generation of the control programs for the complex problems, especially when the system to be controlled consists of different functional units which use different operating systems and independent control programs.

Fig. 7. The HSMHC prototype for testing the PLC programs generated by MHC-PLC.

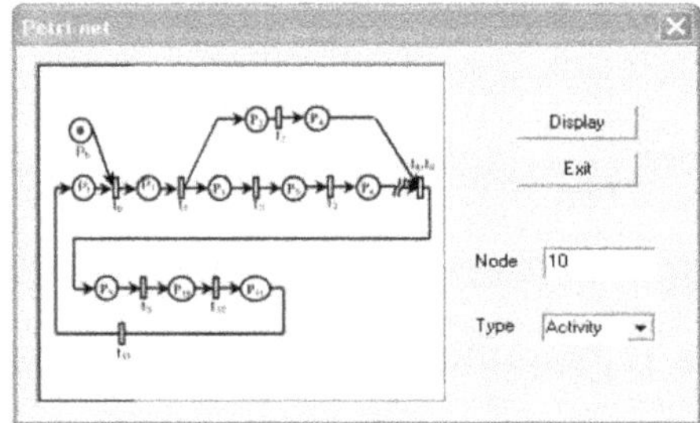

Fig. 8. The P/N menu of the developed MHC-PLC software with the generated P/N model for detection of the errors in the control sequence of the HSMHC system.

In the HSMHC system, the boom stabiliser is controlled by an embedded system in which the operating system is Windows CE that was developed by Microsoft. Meanwhile, the remaining units are directly controlled by the PLC programs. In order to control the boom stabiliser by the PLC programs, the interface between PLC and Windows CE is needed. In addition, the knowledge about the embedded controller is also necessary. These difficulties are not easy to overcome if we only base on the currently available PLC program generation packages. The proposed solutions presented in this study include 6 steps (Fig. 2) for automatic generation of the PLC program that help to solve the complex control problems and overcome the above difficulties and challenges. First of all, the control sequence for the HSMHC system is created based on the simulation in eM-PLC. In order to present a control sequence mathematically, a sequence matrix is used. The P/N technique is then applied to facilitate the process of validation of the control sequence. Finally, based on the faultless control sequence which is presented by a sequence matrix, an optimal PLC program for controlling the HSMHC system is automatically generated with the shortest ramp-up time.

The PLC programs that are generated by the use of the sequence matrix have a clear structure and they can be modified easily. Additionally, by checking the sequence matrix, the PLC programmer is able to know the structure of the PLC program. When it is necessary, the PLC program can be modified flexibly by changing the value of the *Control Activities* in the sequence matrix. Especially, the use of the sequence matrix and continuous scanning of the *Control Conditions* makes the robustness of the generated PLC programs.

Acknowledgements

This research was supported by the Ministry of Knowledge Economy, Korea, under the Industrial Source Technology Development Programmes and supervised by KEIT - Korea Evaluation Institute of Industrial Technology.

References

[1] Dirk S, Stefan V, and Robert S. Container terminal operation and operations research – a classification and literature review. Container Terminals and Automated Transport Systems: Logistics Control Issues and Quantitative Decision Support. Springer, 2004, pp 3-49.

[2] Yen-Chun JW, and Mark G. Container port efficiency in emerging and more advanced markets. Transportation Research Part E: Logistics and Transportation Review, 2010, 46 (6) pp 1030-1042.

[3] Loo HL, Ek PC, Kok CT, & Yongbin H. An optimization model for storage yard management in transshipment hubs. Container Terminals and Automated Transport Systems: Logistics Control Issues and Quantitative Decision Support. Springer, 2004, pp 106-29.

[4] Noriaki M. Development of Large-Sized Mobile Harbor Crane, Mitsubishi Heavy Industries, Ltd. Technical Review, 2003, 40 (2), pp 1-4.

[5] Kendra EM, Askiner G, and Surendra M.G. Disassembly process planning using Petri nets. In proceedings of the 1998 IEEE International Symposium on Electronics and the Environment, 1998, pp 88-93.

[6] Fredrik D, Philip M, and Patric E. Validation, off-line programming and optimisation of industrial control logic. Mechatronics, 2003, 13, pp 571–585.

Innovative Production Machines and Systems
D.T. Pham, E.E. Eldukhri and A.J. Soroka (eds)

Controller Selection for Manufacturing Systems

J. L. Victory, M. Alhamar, K.M. Ebrahimi

School of Engineering, Design and Technology,
University of Bradford, West Yorkshire, BD71DP

Abstract

The overall control strategy employed in a high-volume manufacturing systems can comprise of a multitude of different types of controllers assigned to operate different mechanisms within the line. The process of selecting a combination of dedicated controllers which enables the overall system to function in a cohesive way is the subject of this paper.

This paper attempts to demonstrate how Analytic Hierarchy Process (AHP) modelling in conjunction with simulation tools, can provide such a process, enabling validation of controller selection and also capture and utilise available controls engineering knowledge and experience.

Keywords: controller, machine tool, selection

1. Introduction

Production facilities comprising individual machine tools, inter-linking automation, washers/de-burrs, gauges, testing machines and material handling devices are required for example in order to machine components for automotive engines and to assemble and test them prior to delivery of the engines to the Vehicle Assembly plants. The machining and assembly operations are separated into dedicated lines linked together by automation or manual means. These production facilities require control systems to be selected, specified and integrated into the individual machine tools, their inter-linking automation and the other ancillary equipment that forms the component production lines. Control systems are selected specifically to meet the process requirements of the program, which can vary according to production volume requirements from a low 50,000 upa (units per annum) to a higher 500,000 upa for example. The volume of engines required by the program effectively defines the machining and assembly processes required, by virtue of the investment required in facilities and tooling. High volume capacity requires many simple machining operations and short cycle times whereas low volumes call for fewer machines but with more complex machining and longer cycle times. Within the machining operations, increasing controls complexity is reflected through the use of simple actuators and sensors, to single point-to-point axis motions, multi-axes without interpolation and multi-axes with fully interpolated CNC functions. Currently, both PLC and CNC controllers are used to control the machines and are selected on an analytical basis which considers performance, economics and customer preferences.

Communications systems between machines also becomes more complex at lower volumes. Within high volume machining lines, simple transfer means are needed to minimise both the time taken and the size of buffer stock required between operations. Handshaking between machines and automation can be simple hardwired circuits or serial data exchange, point to point between controllers.

Although the current process has been effective during past programs, it is becoming less so due to the pressures of ' lean manufacturing '. The Ford Control section's ability to document the process and analyse how the process could be improved by studying lessons learned from previous programs has been seriously affected by a drain on talent and resources. This is reducing the ability to satisfactorily complete new programs. Recently, minimal resource has been available to reflect on the performance of the Controls Selection process and to analyse the methodology used. It has also not been possible to formalise the application of all suitable Quality Tools and Methods and to fully document the results of analysis of lessons learned. There is now a high risk that opportunities may be missed to implement continuous improvements and also that the selection process would be unlikely to be able to stand up to an external audit (e.g. ISO 9000).

The control system requirements could be illustrated by consideration of the process needs of a typical multi-station transfer machine, as this normally includes all the elements of control found in simpler and smaller facilities. For high volume mass production, transfer machines with fixed operations and single part programs, are the normal process solution for major sections of the production lines. Each physical element of the transfer machine can be represented by a control object for the sake of the development of the control sequence, each object has a function controlled by logic, it has one or more inputs a set of sequence logic instructions and one or more outputs.

A description of the process of creating a simulation of a machine control system in an attempt to evaluate the effect of key attributes on overall system performance allowing precise comparison between different system features and architectures is given in next section. The theory of constructing and utilising Analytic Hierarchy Process [1,2] to assist the engineer to make a validated choice, supported by accumulated specialist knowledge and the results of lessons learned from previous iterations of the selection process is given in the following sections.

2. Method

In order to improve the current process and attempt to negate the effects of the resources issues described earlier, the intention will be to develop a controls selection tool that will assist and enable validation of the selection process. It should also enable the most effective utilisation of the controls engineering skills available and ensure that lessons learned during selection and application of the control systems are captured, allowing build-up of a knowledge base which will form the core of an expert system related specifically to the process described earlier. Table 1 illustrates the attributes that can be used to assess performance during the selection process.

A machine control system can be represented by a network of these system elements in series and parallel combinations. The element above corresponds to Open-loop control which normally forms some parts of the overall machine control system. With the addition of a feedback signal from output to input, the system can perform as a closed loop with the example of a servo-drive system, typically used for machining station feed drives.

Table 1. Controller Attributes

Element	Attribute	Parameter	Units of measure
Input	Speed data rate signal level	'off' to 'on' time serial bit transfer current (e.g.)	seconds bits/second 0-20mA
Output	Speed data rate signal level	'off' to 'on' time Serial bit transfer current (e.g.)	seconds bits/second 0-20mA
Physical Element (Examples Only, Non Inclusive)	relay PLC CNC PC Drive All	contact rating I/O capacity processor clock speed RAM size resolution cost	AC/DC duty register bit count MHz.Or seconds per instruction Mbyte mm per revolution £ capital spend
Control	logic program electrical noise	instruction set immunity	Range of instructions Micro volts/second
Change of State of Output with Input Change	'off' to 'on' speed change instruction transfer	Speed rpm data rate	seconds rpm/second baud

The establishment of a hierarchy for any particular problem involving selecting the best option or options from a set of alternatives, enables the whole problem to be viewed as a system with an overall objective, a set of criteria to meet that objective and a number of alternatives to provide for each of the individual criteria. Pairwise comparison is carried out on both the criteria and the alternatives to establish their relative weights. The determination of the relative weights is achieved by assigning preference numbers from a ratio scale to the pairwise choices to indicate the strength of one alternative over the other and then placing these comparative scores in a Matrix which is solved to produce an indication of the relative strength of all the alternatives and criteria. These are represented by the Eigenvalues and Eigenvector of the Matrix solutions. The decision hierarchy is shown in Figure 2, the example shown is quite simple with a 3 level hierarchy. Although more complex hierarchies with 4 or more layers require more calculation than this example, the analytical principle remains the same and overall priorities are established in a top-down (or bottom up) manner by applying priorities or weightings for a layer to the layer below (or above).

In this example, the alternatives are not affected by the criteria and thus a set of weights obtained from the decisions regarding the relative importance of each criteria, can be applied directly to the resulting weights obtained from the comparisons of the alternatives.

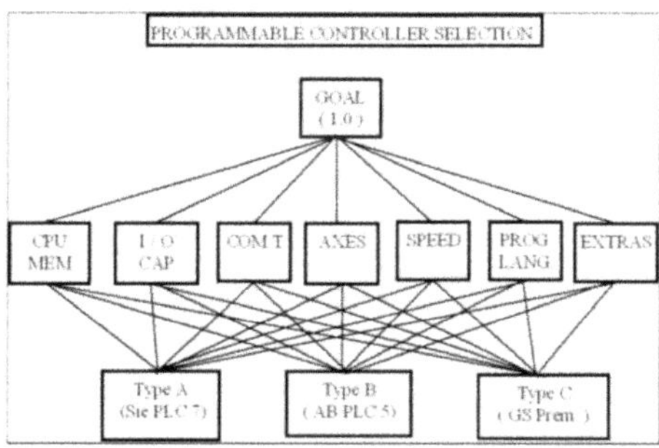

Fig. 1 The Analytic Hierarchy Principle

This can be achieved by one 7x7 Matrix for the Criteria and seven 3x3 matrices of the alternatives against each criteria as shown in figure 1.

3. The Mathematical Theory

There has been limited mathematical foundation yet offered by Saaty to the selection of the most appropriate structure of the decision hierarchy for a particular problem. The many iterations required in order to identify the most suitable hierarchy are achieved mechanically by a computer program. Although a step forward, the method still relies upon the decision makers understanding of the problem and is therefore mainly a qualitative process.

In the application of the AHP theory, there are two distinct methods of evaluating the hierarchy, i.e. Top-down or Bottom-up, in the case of the control selection procedure, we are looking at the bottom-up scenario only [3]. The justification for this is as follows:- The objective is not commutable. The only acceptable outcome is to attain the resolution of the maximum benefit from the synthesis of the hierarchy, i.e. the 'best' control system. The criteria contribute towards the objective, not the reverse. If we were to change the objective there would be an effect on the criteria, but there is no mandate for selecting other than the best system, so therefore there will be no feedback from the objective to the criteria.

The alternatives contribute towards the criteria and the criteria are in effect specifications for the alternatives. There may be minimum and maximum limits for the criteria, but there cannot be unfeasible limits on the contribution of the alternatives to the criteria, as the alternatives have to be available in the market to comply with the directives of the quality system. This system requires that all technology being applied is 'implementation ready' to prevent non-achievement of the program quality, cost and timing. This restriction effectively prevents any variation in the criteria leading to feedback from the criteria to the alternatives.

As a result of the above propositions, there is no requirement to employ the System with Feedback , [4]. Therefore the evaluation of the alternatives as far as required by the control selection problem, follows the theories laid down by Saaty [1]:-

	A	B	C
A	AA	AB	AC
B	BA	BB	BC
C	CA	CB	CC

Evaluations are carried out by comparing Row to Column as shown in the matrix e.g. Cij compares A to B, thus , If we consider the comparisons of the preferences of the alternatives Cij and their values aij they can be represented by a square matrix ;

$$A = (aij) \qquad (1)$$

For n activities, to ensure all alternatives are compared with each other we require n -1 comparisons. However, for a square reciprocal matrix , if comparing two elements A and B :-

Then:

$$Pc(A,B) = \frac{1}{Pc(A,B)} \qquad (2)$$

Where Pc is the preference for A over B or B over A for a criterion c. Thus the reciprocals can be inserted in the matrix instead of doing the comparisons. Therefore, it is only require: (n - 1) / 2 comparisons.

This is further reduced in practice as the matrix diagonal comprises self-comparisons which will always be 1. Therefore for a 3 x 3 matrix we actually only have to perform 3 comparisons.

In order to evaluate the combined results of the comparisons and the directly written entries, we need to find the characteristic values or eigenvalues of the square matrix and the eigenvector. Thus for a matrix A with eigenvalues and eigenvector u we can express the relationship as :-

$$A\,u\,z = \lambda\,z, \qquad (3)$$

(where $z \neq 0$ for non-trivial solutions of the equation)

Equation 3 can be re-written in the form :-

$$A - (I)\ u\,z = 0 \qquad (4)$$

Where I is the Identity matrix or unit matrix

4. Simulation and Results:

In order to fully evaluate the use of the Analytic Hierarchy Process in support of control system selection decision making, the software package ECPro was used [5]. A decision hierarchy was implemented as shown in figure 1. The generation a 'tree' view allows the decision maker to view the hierarchy as shown in figure 2.

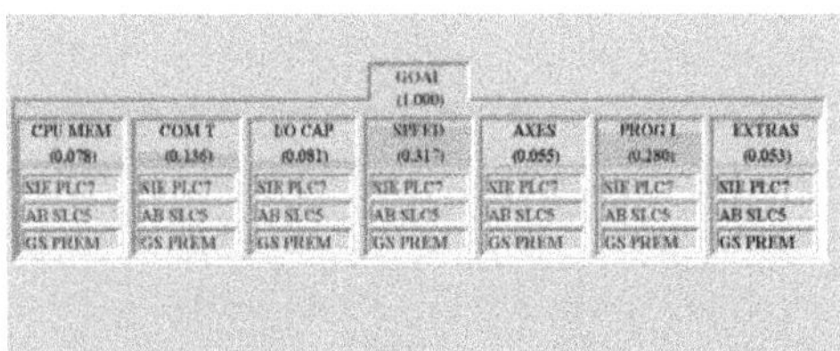

Figure 2 Decision Hierarchy for Alternatives

Once the hierarchy has been established, rating is selected and the alternatives are listed and then evaluated against the pre-established criteria and the pre-assigned attributes.

The core principle of 'pair-wise' comparison is of course central to the AHP and therefore the primary method of prioritising criteria or alternatives is within this procedure. Comparing the alternative at each level of the hierarchy is the first step in full evaluation of the final outcome of the analysis.

The next step is the synthesise the whole hierarchy in order to achieve the evaluation of the alternatives. The purpose of the synthesis is to combine the weightings of the criteria in their contribution to the alternatives so that all preferences are included in the overall calculation.

Figure 3 shows the evaluation of priorities following the synthesis of the hierarchy. The priorities are normalised so that the highest possible priority is 1.0 and all other priorities are lower but positive index. Therefore the sum of all the individual priorities is unity.

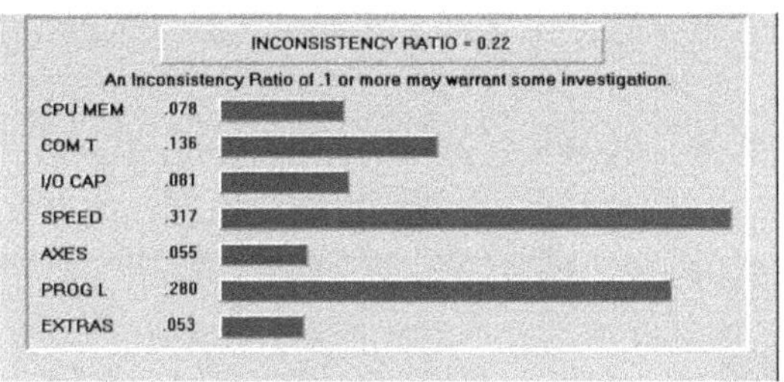

Figure 3 Priorities of Criteria

Following synthesis and development of priorities, the decision-maker (DM) can view the results of the process. The DM is provided with an index of consistency (C.I.), which illustrates how well the comparison process has matched the ideal case. As a rule of thumb, consistency should be at least 0.10, therefore, in the case above, it would be required to evaluate where the inconsistencies exist and perform new iterations of the comparison process until a good level of consistency is achieved.

The DM is also provided by other tools to assist this process. Sensitivity graphs are provided and can be manipulated to show how the change in priority of one attribute can effect any other attributes. Figure 4 shows the sensitivity graphs.

The process of decision making is normally iterative and the rework of the comparison, changes to the hierarchy and adjustments to priorities are easily carried out by the decision maker using the software package [5]. At each iteration the results are immediately presented for consideration.

The results of the comparisons between different alternatives is given in figure 5 as the priority tree achieved, resulting in hierarchy GS with highest Priority.

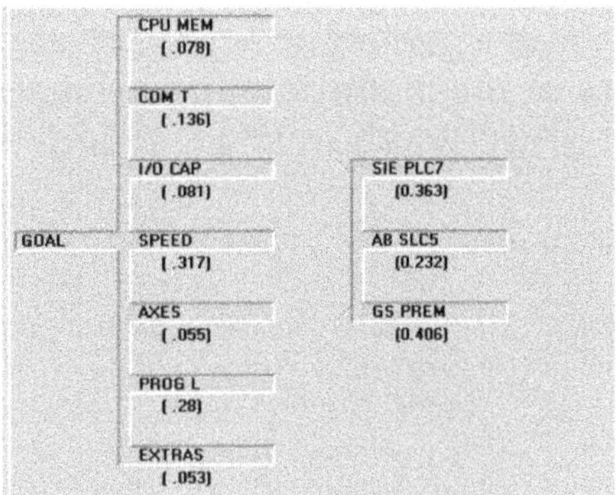

Figure 4 Case Study Results

Conclusion:

From a number of simple trials, the method has been found to be highly suitable for decision support in controls selection process for transfer lines. The hierarchy shown in figure 1 was used to emulate the decision taken in the case study described in this paper, although this was not done in the team scenario, the ranking achieved from the individual inputs was comfortingly similar.

References:

1. Saaty, T.L. , An exposition of the AHP in reply to the paper "Remarks on the Analytic Hierarchy process" , 1990 , Management Science, Vol. 36, No. 3, pp 259 – 268
2. Saaty, T.L. , Axiomatic foundation of the Analytic Hierarchy Process, 1986 Management Science, Vol. 32, No. 7, pp 841 – 855.
3. Tabucanon, M.T. et al, Decision support system for multicriteria machine selection for flexible manufacturing systems, 1994, Computers in Industry 25, pp 131 – 143.
4. Partovi, F.Y., Determining What to Benchmark: An Analytic Hiearchy Process Approach, 1994, Int. Journal of Operations & Production Management, Vol. 14 , No. 6, pp 25-39.
5. Forman, E.H., Decision Support Software, Expert Choice Pro. Version 9.0, 1997, Expert Choice Inc., PA, USA.

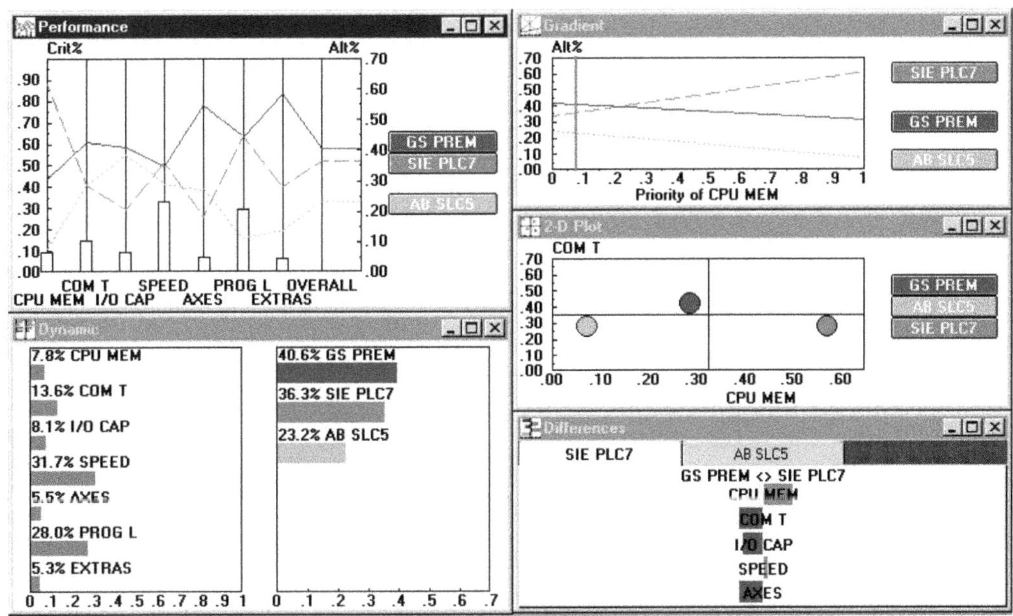

Figure 5 Sensitivity Graphs For Analysis of the Inter-relationships Between the Attributes

Innovative Production Machines and Systems
D.T. Pham, E.E. Eldukhri and A.J. Soroka (eds)

Two-stages monitoring and diagnosis of mean shifts in bivariate correlated process using an integrated MEWMA-ANN scheme

Ibrahim Masood[a], Adnan Hassan[b]

[a] *Faculty of Mechanical and Manufacturing Engineering, Universiti Tun Hussein Onn Malaysia, 86400 Parit Raja, Batu Pahat, Johor, Malaysia*
[b] *Faculty of Mechanical Engineering, Universiti Teknologi Malaysia, 81310 UTM Skudai, Johor, Malaysia*

Abstract

Various artificial neural networks (ANN)-based pattern recognition schemes have been developed for monitoring and diagnosis of mean shift variations in bivariate processes. These schemes generally perform better in classifying mean shifts and provide more diagnosis information compared to the traditional multivariate statistical process control (MSPC) charts. However, some disadvantages in terms of reference multivariate/bivariate patterns and excess false alarms may restrict attention on the scopes and development in this area. Therefore, this paper aims to investigate two-stages monitoring and diagnosis for some reference bivariate correlated patterns using an integrated multivariate exponentially weighted moving average (MEWMA)-ANN. Feature-based input representation was applied into an ANN training towards improving discrimination capability between bivariate normal and bivariate mean shift patterns. Besides comparable diagnosis performance, the proposed scheme has resulted in better monitoring performance with smaller false alarms and faster detection compared to the raw data-based ANN scheme.

Keywords: Bivariate process, Pattern recognition, Two-stages monitoring and diagnosis

1. Introduction

It is already known that variation in manufacturing processes has become a major source of poor quality. Many manufacturing processes may involve two or more correlated variables and an appropriate procedure is required to monitor these variables simultaneously. This issue is sometimes called multivariate quality control (MQC). It has led to extensive research in the field of multivariate statistical process control (MSPC) for monitoring and diagnosis of multivariate process variation in mean shifts/variances. Further discussions on this issue can be found in [1–4].

Development in soft computing technology have motivated researchers to explore the use of artificial intelligence techniques such as artificial neural networks (ANN), among others for automatically and intelligently recognizing patterns in relation to MSPC charting. Identification of these patterns coupled with engineering knowledge of the process would lead to more specific diagnosis and troubleshooting. Various ANN-based pattern recognition schemes have been proposed such as MSPC-ANN [5–8], novelty detector [9], modular-ANN [10], ensemble-ANN [11] and multi-module-structure-ANN [12]. The MSPC-ANN schemes combined the MSPC charts (i.e. for monitoring any mean/variance shifts in multivariate processes) with ANN recognizer (i.e. for diagnosing the source variable(s) that responsible for mean/variance shifts). The other schemes such as novelty detector, modular-ANN, ensemble-ANN and multi-module-structure-ANN were designed to perform continuous monitoring and diagnosis simultaneously. Further discussion on these schemes can be found in [13].

In this study, these schemes are referred as bivariate pattern recognition (BPR) since their investigations are mainly focused on two variables. These schemes generally perform better in classifying mean shifts and provide more effective diagnosis information compared to the traditional MSPC charts. Nevertheless, they have revealed some disadvantages in terms of:

(i) Reference bivariate patterns.

In monitoring and diagnosis of bivariate process mean shift patterns, the joint effect in mean shift and cross correlation between two variables should be taken into account. One approach is based on several Shewhart control charts patterns that provide useful meaning about univariate process mean shifts. However, these patterns are unable to indicate cross correlation between the two dependent variables, which often lead to a higher false alarm than assumed. The other approach is based on χ^2 control chart patterns. These patterns consider the joint effects but unable to indicate the source variable(s) of mean shifts. Generally, there are limited works reported on modeling of bivariate correlated processes and patterns that can effectively identify the joint effect.

(ii) Excess false alarms.

The existing MSPC charts / BPR schemes are mainly reported average run length, $ARL_0 \approx 200$ in monitoring bivariate stable processes. This monitoring level, however, produce rather a higher rate of false alarms ($FAR \approx 0.5\%$) compared to the univariate SPC charts such as Shewhart ($ARL_0 \approx 370$, $FAR \approx 0.27\%$), cumulative sum, CUSUM ($ARL_0 \approx 400$, $FAR \approx 0.25\%$) and exponentially weighted moving average, EWMA ($ARL_0 \approx 500$, $FAR \approx 0.2\%$). This is less economic for a practitioner to frequently conduct troubleshooting due to wrong identification of stable process as unstable.

These disadvantages may restrict attention on the scopes and development in this research area. Therefore, the purpose of this study was to develop an improved BPR scheme for enabling on-line monitoring and diagnosis of some reference bivariate correlated patterns with smaller false alarms, rapid detection and accurate recognition of the source variable(s). Details discussion is organized as follows. Section 2 presents an integrated MEWMA-ANN scheme. Subsection 2.1 describes the modeling of bivariate process and patterns. Input representation, design, training and testing of an ANN are then discussed in subsections 2.2 and 2.3. Section 3 provides performance comparison between the MEWMA-ANN against the raw data-based ANN. Section 4 finally outlines some conclusions.

2. An integrated MEWMA-ANN scheme

Process monitoring refers to the identification of process status (either in a statistically stable or unstable state), whereas process diagnosis refers to the identification of the source variable(s) of the unstable process. Process monitoring based on the traditional MSPC charts alone generally would result in an excess false alarm ($ARL_0 \approx 200$, $FAR \approx 0.5\%$) and it is unable to provide diagnosis information.

To overcome this problem, two-stages monitoring and diagnosis approach using MEWMA-ANN scheme as shown in Figure 1 was investigated. A MEWMA control chart ($r = 0.1$, $h = 8.64$) was utilized to perform first stage monitoring by triggering an out-of-control signal. Then, a feature-based ANN was applied to perform second stage monitoring and diagnosis by recognizing the bivariate patterns that contain any out-of-control signals. An ANN recognizer should be able to correctly identify patterns of bivariate process mean shifts as quickly as possible with smaller ARL_1 (small type II errors) and high recognition accuracy (RA). Besides, it should be able to leave patterns of bivariate stable processes running as long as possible with larger ARL_0 (small type I error / small false alarms).

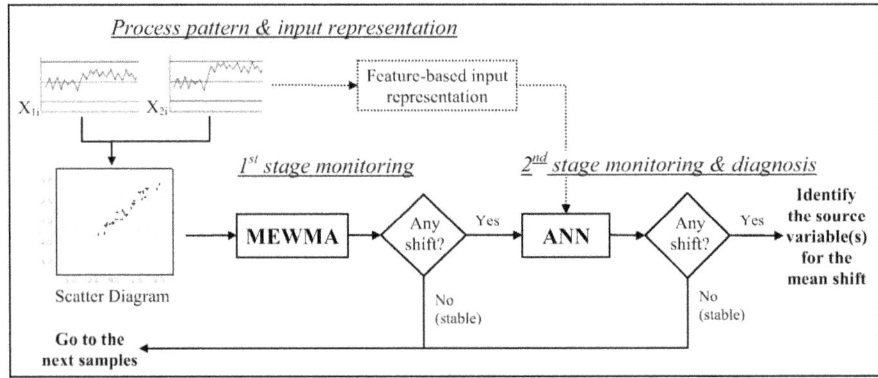

Fig. 1. Conceptual diagram of an integrated MEWMA-ANN scheme

2.1. Modeling of bivariate process and patterns

Let $X_{1i} = (X_{1\text{-}1},\ldots, X_{1\text{-}24})$ and $X_{2i} = (X_{2\text{-}1},\ldots, X_{2\text{-}24})$ represent data streams for bivariate process based on window size, 24 samples. Observation windows for both variables start with samples $i^{th} = (1,\ldots, 24)$. It is followed by $(i^{th}+1)$, $(i^{th}+2)$ and so on. In a statistically stable state, samples for both variables are identically and independently distributed with zero mean ($\mu_0 = 0$), unity standard deviation ($\sigma_0 = 1$) and zero cross correlation ($\rho = 0$). They yield random patterns when plotted separately on two Shewhart control charts and yield a circle pattern when plotted on a scatter diagram. Scatter diagram yields ellipse patterns when $\rho > 0$ as shown in Figure 2.

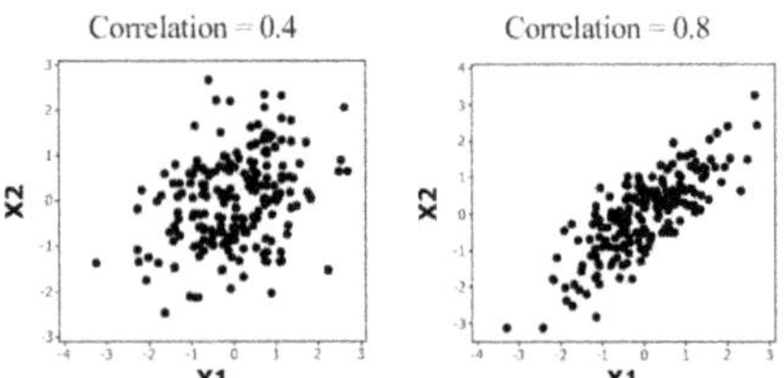

Fig. 2. Changes in cross correlation

Disturbance from assignable causes may deteriorate data streams into an unstable state. Initially, the pattern structure is in 'partially developed'. Then, it will be more obvious into 'fully developed'. This occurrence could be identified by common causable patterns such as sudden shifts, trends, cyclic, systematic or mixture. However, investigation for this study was focused on sudden shift patterns as follows:

- Normal (0,0): Both X_{1i} and X_{2i} are stable
- Up-shift (1,0): X_{1i} in up-shift, X_{2i} is stable
- Up-shift (0,1): X_{2i} in up-shift, X_{1i} is stable
- Up-shift (1,1): Both X_{1i} and X_{2i} in up-shifts
- Down-shift (1,0): X_{1i} in down-shift, X_{2i} is stable
- Down-shift (0,1): X_{2i} in down-shift, X_{1i} is stable
- Down-shift (1,1): Both X_{1i} and X_{2i} in down-shifts

Bivariate patterns that attributed to the similar assignable causes would share common structure and properties that are identifiable and recognizable. Changes in mean shift and cross correlation can be identified by center position and ellipse shape as shown in Figure 3. Discussion on bivariate patterns has also been reported in [14].

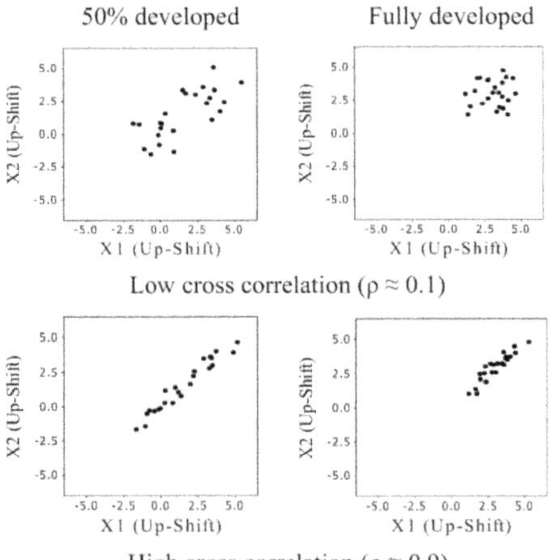

Fig. 3. Changes in mean shift and cross correlation

2.1.1 Data generator

Ideally, observation samples for training and testing the scheme should be tapped from real process environment. Since they are not economically available, synthetic data were generated based on the following steps [9–10; 15–16]:

Step 1: Generate random normal variates for process variable 1 (n_1) and process variable 2 (n_2), which are identically and independently distributed within [-3, +3]. $b = 1/3$ is a baseline noise level or random noise.

$$n_1 = b.\, r_1 \quad (1)$$

$$n_2 = b.\, r_2 \quad (2)$$

Step 2: Generate random data series (Y_1, Y_2) with means (μ_1, μ_2) and standard deviations (σ_1, σ_2).

$$Y_1 = \mu_1 + (\, n_1\, \sigma_1\,) \quad (3)$$

$$Y_2 = \mu_2 + [\, n_1\, \rho + n_2 (\sqrt{1 - \rho^2})\,]\, \sigma_2 \quad (4)$$

Step 3: Compute mean and standard deviation for (Y_1, Y_2) from the generated random data series. These values represent the statistically stable process means (μ_{01}, μ_{02}) and standard deviations (σ_{01}, σ_{02}).

Step 4: Transform random data series into normal or shift patterns to mimic real observation samples (X_1, X_2). The magnitude of mean shift (h) is expressed in standard deviation of stable process (σ_{01}, σ_{02}).

$$X_1 = h_1 (\sigma_{01} / \sigma_1) + Y_1 \quad (5)$$

$$X_2 = h_2 (\sigma_{02} / \sigma_2) + Y_2 \quad (6)$$

Step 5: Rescale pattern data series into a standardize range, approximately within [-3, 3].

$$Z_1 = (X_1 - \mu_{01}) / \sigma_{01} \quad (7)$$

$$Z_2 = (X_2 - \mu_{02}) / \sigma_{02} \quad (8)$$

A pair standardized sample (Z_1, Z_2) represents a standardized bivariate vector measured at time t (Z_t) that follows the standardized normal bivariate distribution $N(0, R)$. Zero and $R = [(1\ \rho)\ (\rho\ 1)]$ are the mean vector and a general correlation matrix for bivariate stable process with unity variances ($\sigma_1^2 = \sigma_2^2 = 1$) and covariance equal to cross correlation ($\sigma_{12} = \sigma_{21} = \rho$).

2.2. Input representation of ANN recognizer

Input representation provides a strong influence on an ANN performance. An improved set of statistical features was utilized for this study. It comprising of last value of exponentially weighted moving average ($LEWMA_\lambda$) with λ = [0.25, 0.20, 0.15, 0.10], mean (μ), multiplication of mean with standard deviation (MSD), and multiplication of mean with mean square value (MMSV). Each bivariate pattern was represented by 14 input data (i.e., $LEWMA_{0.25\_1}$, $LEWMA_{0.20\_1}$, $LEWMA_{0.15\_1}$, $LEWMA_{0.10\_1}$, μ_1, MSD_1, $MMSV_1$, $LEWMA_{0.25\_2}$, $LEWMA_{0.20\_2}$, $LEWMA_{0.15\_2}$, $LEWMA_{0.10\_2}$, μ_2, MSD_2, $MMSV_2$). This approach yields a smaller network, less computation and shorter training time compared to the raw data-based approach that requires 48 input data (i.e., 24 consecutive samples for both process variables).

2.2.1 LEWMA features

The EWMA-statistics as can be computed using Equation (9) incorporates historical data in a form of weighted average of all past and current observation samples [17]:

$$EWMA_i = \lambda\, X_i + (1 - \lambda)\, EWMA_{i-1} \quad (9)$$

X_i represents the original samples. In this study, the standardized samples (Z_i) were used instead of X_i so that Equation (9) becomes:

$$EWMA_i = \lambda\, Z_i + (1 - \lambda)\, EWMA_{i-1} \quad (10)$$

where $0 < \lambda \leq 1$ is a constant parameter and i = [1, 2,…, 24] are the number of samples. The starting value of EWMA ($EWMA_0$) was set as zero to represent the process target (μ_0). The last value of EWMA (LEWMA) with λ = [0.25, 0.20, 0.15, 0.10] were empirically selected for strengthening pattern properties towards improving discrimination capability between normal pattern and shift patterns.

2.2.2 MSD and MMSV features

The MSD and MMSV features were used to magnify the magnitude of mean shifts (μ_1, μ_1) towards improving discrimination capability between normal and shift patterns. This approach aims to reduce false alarms in monitoring bivariate stable process.

$$MSD_1 = \mu_1 \times \sigma_1 \quad (11)$$

$$MSD_2 = \mu_2 \times \sigma_2 \quad (12)$$

$$MMS_1 = \mu_1 \times (\mu_1)^2 \quad (13)$$

$$MMS_2 = \mu_2 \times (\mu_2)^2 \quad (14)$$

where (μ_1, μ_2), (σ_1, σ_2) ((μ_1^2, μ_1^2) are the means, standard deviations and mean square value respectively.

2.3. Design, training and testing of ANN recognizer

Multilayer perceptron (MLP) neural network model as shown in Figure 4 was used. Feature-based input representation yields only 14 input neurons compared to the raw data-based (48 neurons). The output layer contains seven neurons, which was determined according to the number of pattern categories. One hidden layer with 22 neurons was selected based on experiment. Such experiments revealed that initially, the training results improved in-line with the increment in the number of neurons. Once the neurons exceeded 22, excess neurons did not improve the training.

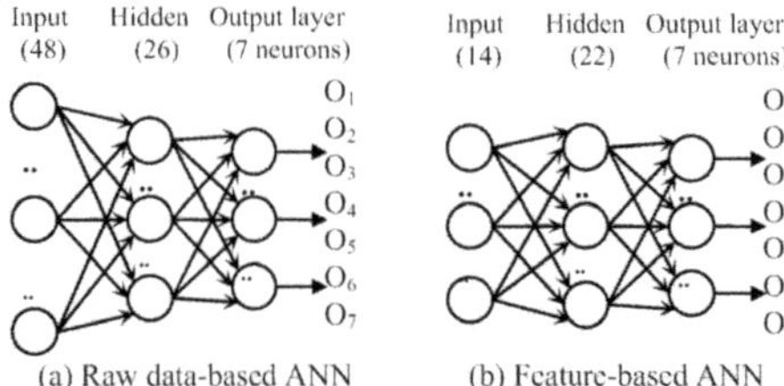

Fig. 4. Network architectures for three layer MLP models.

Partially developed patterns of bivariate process mean shifts were applied for training an ANN recognizer. Detail parameters of the training patterns are summarized in Table 1. It should be noted that for bivariate process mean shifts, the number of training pattern = [100 x (total combination of shifts) x (total combinations of cross correlation)], while for bivariate normal process, the number of training pattern = [1500 x (total combinations of cross correlation)]. On the other hand, dynamic patterns were used for testing the recognizers, which is suited for on-line process monitoring as addressed in [10].

Input representations were normalized to a compact range between [−1, 1]. The maximum and the minimum values for normalization were taken from the overall data of training patterns.

Table 1. Parameters for the training patterns

Pattern Category	Mean Shift (σ in Std Dev.)	Total Pattern (100 x σ x ρ)
N (0, 0)	X1: 0.00 X2: 0.00	1,500 x 1 x 5 = 7,500
US (1, 0)	X1: 1.00, 1.25, … , 3.00 X2: 0.00, 0.00, … , 0.00	100 x 9 x 5 = 4,500
US (0, 1)	X2: 0.00, 0.00, … , 0.00 X1: 1.00, 1.25, … , 3.00	100 x 9 x 5 = 4,500
US (1, 1)	X1: 1.00, 1.25, 1.00, 1.25, … , 3.00 X2: 1.00, 1.00, 1.25, 1.25, … , 3.00	100 x 25 x 5 = 12,500
DS (1, 0)	X1: -1.00, -1.25, … , -3.00 X2: 0.00, 0.00, … , 0.00	100 x 9 x 5 = 4,500
DS (0, 1)	X2: 0.00, 0.00, … , 0.00 X1: -1.00, -1.25, … , -3.00	100 x 9 x 5 = 4,500
DS (1, 1)	X1: -1.00, -1.25, -1.00, -1.25,.., -3.00 X2: -1.00, -1.00, -1.25, -1.25,.., -3.00	100 x 25 x 5 = 12,500

Based on back propagation (BPN) algorithm, 'gradient decent with momentum and adaptive learning rate' (traingdx) was used for training the MLP model. The other training parameters setting were learning rate (0.05), learning rate increment (1.05), maximum number of epochs (1500) and error goal (0.001), whereas the network performance was based on mean square error (MSE). Hyperbolic tangent function was used for hidden layer, while sigmoid function was used for an output layer. The training session was stopped either when the number of training epochs was met or the required MSE has been reached.

3. Results and discussion

The performance of MEWMA-ANN scheme compared to the raw data-based ANN was evaluated based on average run length (ARL) and recognition accuracy percentage (RA). Three thousand testing patterns of the specific magnitudes of mean shift and cross correlation were simulated. The ARL results were also compared with the MEWMA [18] as reported in the literatures.

The ARL_1 results as summarized in Table 2 are the average results from six shift categories. Type I error (rate of false alarms, FAR) and type II error can be computed directly from the ARL results [19]. In monitoring bivariate stable process, the proposed MEWMA-ANN scheme gave a longer ARL_0 (smaller false alarms) compared to the existing BPR/MSPC schemes. On the other hand, the shorter ARL_1 results indicated that the MEWMA-ANN scheme is more sensitive in detecting moderate and large shifts (≥ 1.5 standard deviations) while the raw data-based ANN is more sensitive in detecting smaller shifts. Generally, the proposed scheme exhibits a strong ability for on-line monitoring bivariate processes as good as the Shewhart control chart in identifying stable state and as sensitive as the MEWMA control chart in triggering unstable state. The results support the conclusion that the mean shift with larger magnitudes can be triggered more quickly.

Table 2. Average run lengths

Mean Shift (std dev)	MEWMA-ANN	Raw data-based ANN	MEWMA r = 0.10 h = 8.64
	ρ = 0.1 / 0.9	ρ = 0.1 / 0.9	ρ = 0.0
0.0	362.71 / NA	240.7 / NA	200.0
1.0	10.47 / 11.32	10.02 / 10.06	10.15
1.5	5.99 / 6.35	6.69 / 6.68	6.11
2.0	4.23 / 4.43	5.17 / 5.15	4.42
2.5	3.31 / 3.43	4.25 / 4.23	
3.0	2.76 / 2.82	3.65 / 3.62	2.93

The RA measure how accurate is a scheme could perform diagnosis for the source variable(s). The RA results as summarized in Table 3 are the average results from six shift categories. The proposed MEWMA-

ANN scheme exhibits a strong ability for diagnosing the source variable(s) particularly in dealing with moderate and large shifts ($\geq$ 1.5 standard deviations). This scheme is capable to accurately classify the source variable(s) patterns. However, the raw data-based ANN performed slightly better in dealing with high correlated processes. The results also support the conclusion that the mean shift with larger magnitudes can be more accurately diagnosed.

Table 3. Recognition accuracy percentage

Mean Shift (std dev)	MEWMA-ANN	Raw data-based ANN	MEWMA $r = 0.10$ $h = 8.64$
	ρ = 0.1 / 0.9	ρ = 0.1 / 0.9	ρ = 0.0
0.0	NA	NA	NA
1.0	91.5 / 91.7	91.5 / 93.9	NA
1.5	94.4 / 95.2	94.7 / 97.0	NA
2.0	95.0 / 96.1	95.0 / 97.4	NA
2.5	95.4 / 96.8	95.9 / 97.7	NA
3.0	96.0 / 97.2	96.0 / 97.7	NA
Average	94.5 / 95.4	94.6 / 96.7	

4. Conclusions

This paper proposes an integrated MEWMA-ANN scheme to perform two-stages monitoring and diagnosis of bivariate process mean shifts. In comparison with the raw data-based ANN scheme, this approach produced smaller false alarms, more sensitive in detecting mean shifts and maintained accurate classification for the source variable(s) patterns. For the future work, investigation will be extended for other causable patterns such as trend and cyclic.

References

[1] Lowry CA, Montgomery DC (1995). A review of multivariate control charts. IIE Transactions 27(6): 800–810.

[2] Kourti T, MacGregor JF (1996). Multivariate SPC methods for process and product monitoring. Journal of Quality Technology 28(4): 409–428.

[3] Mason RL, Tracy ND, Young JC (1997). A practical approach for interpreting multivariate T^2 control chart signals. Journal of Quality Technology 29(4): 396–406.

[4] Bersimis S, Psarakis S, Panaretos J (2007) Multivariate statistical process control charts: an overview. Quality and Reliability Engineering International 23: 517–543.

[5] Chen LH, Wang TY (2004) Artificial neural networks to classify mean shifts from multivariate χ^2 chart signals. Computers and Industrial Engineering 47: 195–205.

[6] Niaki STA, Abbasi B (2005) Fault diagnosis in multivariate control charts using artificial neural networks. Quality and Reliability Engineering International 21: 825–840.

[7] Cheng CS, Cheng HP (2008) Identifying the source of variance shifts in the multivariate process using neural networks and support vector machines. Expert Systems with Applications 35: 198–206.

[8] Yu JB, Xi LF, Zhou XJ (2009) Identifying source(s) of out-of-control signals in multivariate manufacturing processes using selective neural network ensemble. Engineering Applications of Artificial Intelligence 22: 141–152.

[9] Zorriassatine F, Tannock JDT, O'Brien C (2003) Using novelty detection to identify abnormalities caused by mean shifts in bivariate processes. Computers and Industrial Engineering 44: 385–408.

[10] Guh RS (2007) On-line identification and quantification of mean shifts in bivariate processes using a NN-based approach. Quality and Reliability Engineering International 23: 367–385.

[11] Yu JB, Xi LF (2009) A neural network ensemble-based model for on-line monitoring and diagnosis of out-of-control signals in multivariate manufacturing processes. Expert Systems with Applications 36: 909–921.

[12] El-Midany TT, El-Baz MA, Abd-Elwahed MS (2010) A proposed framework for control chart pattern recognition in multivariate process using artificial neural networks. Expert Systems with Applications 37: 1035–1042.

[13] Masood I, Hassan A (2010) Issues in development of artificial neural network-based control chart pattern recognition schemes. European Journal of Scientific Research 39(3): 336–355.

[14] Masood I, Hassan A (2009) Synergistic-ANN recognizers for monitoring and diagnosis of multivariate process shift patterns. International Conference of Soft Computing and Pattern Recognition: 266–271.

[15] A. M. Law and W. D. Kelton, Simulation modeling and Analysis, 3rd ed., Mc Graw-Hill, 2000.

[16] Lehman RS (1977) Computer simulation and modeling: an introduction. Lawrence Erlbaum, London.

[17] Lucas JM, Saccucci MS (1990) Exponentially weighted moving average control schemes: properties and enhancements. Technometrics 32(1): 1–12.

[18] Prabhu SS, Runger GC (1997) Designing a multivariate EWMA control chart. Journal of Quality Technology 29(1): 8–15.

[19] Montgomery DC (2005) Introduction to statistical quality control. John Wiley & Sons, Inc., New Jersey.

Innovative Production Machines and Systems
D.T. Pham, E.E. Eldukhri and A.J. Soroka (eds)

Fuzzy-knowledge based approach for multi-criteria supplier selection problem

A. Zouggari and L. Benyoucef

INRIA Nancy-grand Est, COSTEAM Project, Bat. A, Ile du Saulcy, 57000, Metz, France
{akram.zouggari, lyes.benyoucef}@inria.fr

Abstract

This paper presents a new decision making tool for automatic fuzzy based knowledge acquisition, which clubs supplier selection process with order allocation for dynamic supply chains to cope market variations. More specifically, the developed tool imitates the knowledge acquisition and manipulation in a manner similar to the decision makers who have gathered considerable knowledge and expertise in procurement domain. As per this concept, those decision criteria for supplier selection are considered first through four criteria (C1: Performance strategy, C2: Quality of service, C3: Innovation and C4: Risk), which are qualitatively meaningful. Thereafter, using fuzzy logic, the criteria are quantitatively evaluated. As a result, the tool generates decision-making knowledge, and thereafter, the developed combination of rules for supplier selection with order allocation can easily be interpreted, adopted and at the same time if necessary, modified by decision makers.

Keywords: Supplier selection, fuzzy TOPSIS, fuzzy AHP, multi-criteria, knowledge

1. Introduction

Nowadays, supplier selection with order allocation represents one of the most important functions to be performed by the purchasing decision makers, which determines the long-term viability of the company [1, 2]. With the increase in use of quality management and JIT concepts by a wide range of companies, the supplier selection decision has become even more crucial [3, 4]. As customer's demands are always uncertain, companies tend to manage their suppliers in different ways leading to a *supplier's-supplier development, supplier evaluation, supplier selection, supplier association, supplier coordination*, etc., [5, 8].

This paper presents a new tool for automatic fuzzy based knowledge acquisition, which clubs supplier selection process with order allocation for dynamic supply chains to cope with market variations. More specifically, the developed tool imitates the knowledge acquisition and manipulation in a manner similar to the decision makers who have gathered considerable knowledge and expertise in procurement domain. As per this concept, those decision criteria for supplier selection are considered first through four criteria (C1: Performance strategy, C2: Quality of service, C3: Innovation and C4: Risk), which are qualitatively meaningful. Thereafter, using fuzzy logic, the criteria application is quantitatively evaluated. As a result, the tool generates decision-making knowledge, and thereafter, the developed combination of rules for supplier selection with order allocation can easily be interpreted, adopted and at the same time if necessary, modified by decision makers.

The rest of the paper is organized as follows: Section 2 gives a brief overview of the existing literature related to supplier selection problem. Section 3 describes the problem under consideration and the proposed approach. Section 4 illustrates some of the experimental results. Section 5 concludes the paper with some perspectives.

2. Supplier Selection problem: state of the art

2.1. Factors impacting supplier selection decisions

A number of research work have been published in the last decades emphasized the strategic importance of the supplier selection process and evaluated the relative importance of the most used selection criteria and sub-criteria. Recently, Van der Rhee et al. [9] explored how executives' trade-off amongst various dimensions when selecting a supplier. An empirical study is presented to assess the marginal trade-off for cost, delivery performance, flexibility, and value-added service/support in a supplier selection process using a market utility-based approach (DCA).

Kannan and NoorulHaq [10] analyzed the interaction of criteria and sub-criteria that are used to select suppliers for the built-in-order supply chain environment. They presented an approach using Interpretive Structural Modeling (ISM) methodology to identify, rank and find out the interaction among the criteria and sub-criteria. They state that (i) business structure is the key criterion in selecting the supplier for the considered industry, (ii) services have strong drive power and weak dependence power, (iii) both quality system and adherence to quality tools are the key sub-criteria in the quality criteria.

2.2 Existing computational intelligent approaches

Wu et al., [11] proposed a two-stage integrated multi-objective decision-making approach using analytic network process (ANP) and mixed integer programming (MIP) to select suppliers when the bundling strategy existed. Moreover, Lee et al., [12] presented a fuzzy multiple goal-programming (FMGP) model to help downstream companies to select thin film transistor liquid crystal display (TFT-LCD) suppliers for cooperation.

Amid et al. [13] developed a fuzzy weighted additive and mixed integer multiobjective linear programming model for the supplier selection problem under price breaks that depend on the sizes of order quantities. The model aggregates weighted membership functions of objectives to construct the relevant decision functions, in which objectives have different relative importance.

To evaluate and select the most suitable suppliers for a telecommunication company in the GSM sector, Onüt et al. [14] proposed an integrated fuzzy analytic network process (FANP) and fuzzy technique for order performance by similarity to ideal solution (FTOPSIS) methodology. Sanayei et al., [15] proposed an integrated approach using multi-attribute utility theory (MAUT) and linear programming for rating and selecting the best suppliers, and defining the optimum order quantities among selected ones in order to maximize the quantity of purchase from the desired suppliers.

2.3. Why the need of a new approach?

Based on a synthesis of the literature and our expertise, gained from literature review and involvement with industrial partners such as Renault, Fiat, Hi-Tech, Profactor, VTT, Robosoft, Schneider Electric, etc. in the European projects respectively Growth-ONE and I*PROMS, the motivations for a new approach are as follows:

- The majority of the existing approaches/models are based on functional criteria like quality, price, delivery time, etc., and does not take into account the repercussions of the company strategy on the decisions by taking into account several novel criteria such as innovation, flexibility, risk, agility, responsiveness, etc. [1, 2 and 5].
- The multi-criteria aspect of the supplier selection problem is essential when we deal with a global approach [2, 5].
- The supplier selection problem is a decision or a set of decisions, which requires the intervention of different services of the company [2, 5].
- If a supplier cannot satisfy a minimal threshold compared to certain criterion, it cannot be selected in spite of its possible effectiveness with respect to other criteria.
- There are several constraints to take into account. In addition, if the company would wish several suppliers, it must select suppliers and the order allocation strategy. The current approaches cannot cover the various aspects of the supplier selection problem with order allocation [2, 5 and 6].
- The future behaviors of the suppliers and the environment are uncertain. Consequently, it is difficult to assign a fixed mark to the suppliers with respect to a criterion.

- An approach, which can take into account all the characteristics of the problem: the strategy of the company, the multi-actors aspect, the subjective and objective criteria, the constraints of suppliers and customer, etc., *is a lack in the literature* [8].

3. Problem description and proposed approach

3.1. Problem setting

Our supply chain comprises of a set of K potential suppliers, a unique distribution center (DC) and a unique market (customer). Each supplier is connected to the DC by a single transportation link. The transportation activity is taken care by the sub-criteria related to the quality of delivery. Our problem consists of two sub-problems. The first sub-problem concerns with the selection of suppliers during a fixed decision horizon called 'selection horizon', which is by assumption more than 1 year. Where, the second sub-problem deals with the order allocation strategy among the selected suppliers. A supplier that is selected in a given 'selection horizon' is not necessarily selected in another one; a new list of potential suppliers is created at the end of each 'selection horizon' for the next. The 'selection horizons' are independent.

3.2. Decision criteria

For ranking the potential suppliers the following three terminologies are used:

1. **Satisfactory:** A supplier who has sustained superior performance by enhancing his abilities to address the marketplace is considered satisfactory. It provides the best value and is the one with whom the company (i.e. DC) wants to focus for as much business as possible.
2. **Conditional:** A supplier who has an acceptable level of performance but may fall short of superior performance levels required for a preferred rating.
3. **Unsatisfactory:** A supplier whose performance is poor or commitment to the company (i.e. DC) is weak. This supplier will not be considered in the selection. The DC feels that doing business with this supplier decreases his efficiency and position in the existing challenging and dynamic market.

1. *C1: Performance Strategy*

Quality	Percentage of products which qualify the expectations of the DC in terms of quality
Pricing	Unit price communicated by each potential supplier
Age and position in the market	Age of the supplier and its position in the market in terms of number of customers or in terms of the transaction amounts
Environmental engagement	Percentage of supplier's profit dedicated to the environmental safety

2. *C2: Quality of service*

Delivery	Percentage of the products delivered in time guaranteed by the supplier
Service after sale	Response time to any service after sale action, guaranteed by supplier
Preventive actions	Necessary actions, which are defined and planned by the supplier in order to keep the quality of service highest
Corrective actions	Necessary actions, which are defined and not planned by supplier in order to keep the quality of service highest

3. *C3: Innovation*

Research and development	Percentage of the total supplier's profit dedicated to R&D activities
Service innovation	Percentage of total supplier's profit dedicated to innovation

4. *C4: Risk*

Geographical location	Supplier location, physical and social status
Political and economical stability	Political status of the supplier's country and its nature towards the business policies pay affecting the long term relationship

3.3. Proposed approach

- *Overview*

The proposed approach consists of mainly three stages I, II and III.

Stage I: Suppliers selection
Begin
Step1: Select the used criteria for selecting the best suppliers among criterion I, criterion II, criterion III and criterion IV
Step2: Call fuzzy-AHP method
Step3: Provide the lest of selected suppliers
End.

Stage II: Simulation
Begin
Step1: Initialize the data, market configuration; storage political; simulation parameters
Step2: Call the knowledge-based simulation platform
Step3: Update the database (i.e. new values of criteria; evaluation of the stock level, etc.)
End.

Stage III: Order allocation
Begin
Step1: Select the criteria for order allocation respectively cost, quality and delivery.
Step2: Initialize the data: new values of criteria; list of suppliers
Step3: Call the fuzzy-TOPSIS
Step4: Update the database (i.e. order allocation of each supplier)
End.

Stages I, II and III use the same knowledge-based platform, which consists of four modules, respectively, the network module, the knowledge-based discrete event simulator module; the fuzzy module; and finally the knowledge database module (see Figure 1). A triangular fuzzy membership function (TFN) is used for the fuzzification and defuzzification. Five linguistics variables are used to assess the importance weights: very high (VH), high (H), medium (M), poor (P) and very poor (VP).

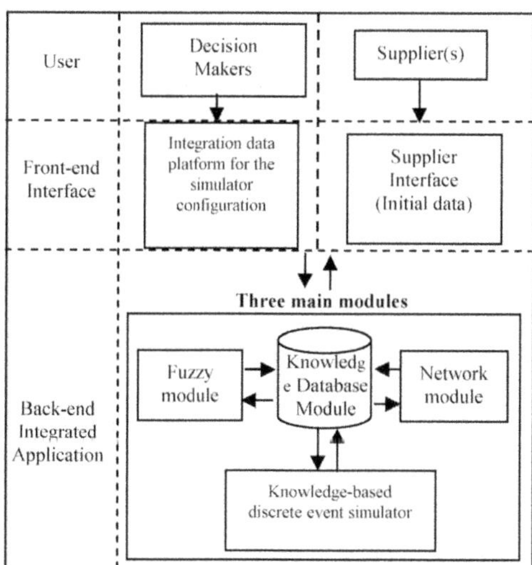

Fig. 1. Knowledge-based simulation platform architecture

4. Experimental results and Analysis

Ten potential suppliers are considered where the number of suppliers to select is fixed to three. Each, three months, new orders allocation rates are calculated.

4.1 Supplier selection (Fuzzy-AHP)

Table 1 presents the used fuzzy comparisons matrix of criteria with respect to the overall objective (global supplier selection).

Table1
Fuzzy comparison matrix of criteria

	C1	C2	C3	C 4
C1	(1, 1, 1)	(0.73, 0.87, 1)	(0.73, 0.87, 1)	(0.73, 0.87, 1)
C2	(1,1.15,1.37)	(1, 1, 1)	(0.42,0.52,0.57)	(0.55,0.68,0.78)
C3	(1.28, 1.47, 1.81)	(1.75,1.92,2.38)	(1, 1, 1)	(0.73, 0.87, 1)
C4	(1, 1.15, 1.37)	(1, 1.15, 1.36)	(1, 1.15, 1.37)	(1, 1, 1)

Using the extent analysis method [5], we have the following results where Ai corresponds to the fuzzy synthetic extent value with respect to criterion Ci (i=1, 2, 3, 4) and M the minimum degree of possibility.

A_1= (3.19, 3.61, 4) × (18.11, 16.67, 23.02)-1
= (0.14, 0.21, 0.22)
A_2= (2.97, 3.35, 3.72) × (17.89, 16.67, 22.74)-1
= (0.13, 0.20, 0.21)
A_3= (4.50, 4.95, 5.75) × (19.41, 16.67, 24.78)-1
= (0.18, 0.30, 0.29)
A_4= (4.28, 4.77, 5.55) × (19.20, 16.67, 24.58)-1
= (0.17, 0.28, 0.29)
$M(A_1 \geq A_2, A_3, A_4)$= min (1, 0.33, 0.40)= 0.33
$M(A_2 \geq A_1, A_3, A_4)$= min (0.81, 0.22, 0.280)= 0.22
$M(A_3 \geq A_1, A_2, A_4)$= min (1, 1, 1)= 1
$M(A_4 \geq A_1, A_1, A_3)$= min (1, 1, 0.91)= 0.91

The weight vector W with respect to the 4 criteria is given by (0.14, 0.9, 0.40, 0.37) where:

$$w_i = \frac{M(\widetilde{A}_i \geq \widetilde{A}_j | j = 1,..., 4; j \neq i)}{\sum_{k=1}^{4} M(\widetilde{A}_k \geq \widetilde{A}_j | j = 1,..., 4; j \neq k)}$$

Similarly, the corresponding weight vector of each sub-criterion with respect to the four criteria and the corresponding weight vector of each supplier with respect to the sub-criteria are also calculated. At the end, we compute the weight of each supplier with respect to main four criteria (see Table 2).

4.2. Simulation and knowledge management

By assumption, the simulation horizon is one year (48 weeks) and for the order allocation horizon is three months (12 weeks). The DC inventory control policy is (Q, s), where Q= 500 units and s=200 units. Moreover, each week, the DC receives a certain demand quantity with respect to the market planning demand.

Table 2
Combination of priority weights

	C1	C2	C3	C4	Supplier priority weight
W	0.13	0.09	0.40	0.37	
S1	0.07	0.06	0.07	0.71	0.07
S2	0.78	0.08	0.09	0.83	0.08
S3	0.79	0.08	0.08	0.07	0.08
S4	0.09	0.09	0.08	0.09	0.09
S5	0.09	0.1	0.11	0.1	0.1
S6	0.11	0.1	0.1	0.1	0.1
S7	0.1	0.11	0.1	0.1	0.1
S8	0.11	0.12	0.12	0.12	0.12
S9	0.12	0.12	0.13	0.12	0.13
S10	0.13	0.14	0.13	0.13	0.13

We can conclude that suppliers S8, S9 and S10 are the best three suppliers among the list of ten potential suppliers. Table 3 illustrates the evaluation of the delivery for each supplier given in linguistics values.

Table 3
Delivery linguistics evaluation values

	S8	S9	S10
Period 1	P	H	H
Period 2	P	P	P
Period 3	H	M	M
Period 4	M	P	P

Tables 4 and 5 illustrate the training examples for respectively sub-criterions quality and price. Attributes related to quality are: quality of packaging, finishing and size, where for sub-criterion price, attributes are respectively price duty-free and duties. By assumption, 'I', 'a' and '1' correspond to "good" and 'II', 'b' and '2' correspond to "medium".

The used training rules are respectively:

- **Quality**

If *Quality of packaging =I* ***then*** *Quality = H*
If *Finishing = B* ***then*** *Quality = M*
If *Weight =1* ***then*** *Quality =VH*

- **Price**

If *Price duty-free =I* ***then*** *Price = VH*
If *Price duty-free = II* ***then*** *Price = M*
If *Duties =1* ***then*** *Price=H*
If *Duties =2* ***then*** *Quality =M*

Table 4
Training example for quality

Example	Quality of packaging	Finishing	Size	
1	II	a	1	VH
2	I	a	2	H
3	I	a	2	H
4	II	a	2	H
5	II	b	2	M
6	II	a	2	M
7	II	a	2	VH

Table 5
Training example for price

Example	Price duty-free	Duties	
1	I	1	VH
2	II	2	M
3	II	1	H

For each supplier and each delivery, new linguistics values are calculated for price and quality.

4.3. Order allocation (Fuzzy-TOPSIS)

Tables 6, 7 and 8 show the comparison matrix of the three criteria respectively price, quality and delivery, with respect to the overall objective (order allocation). FPIS (Fuzzy Positive Ideal Solution) and FNIS (Fuzzy Negative Ideal Solution) are also calculated [16].

FPIS(A*) and FNIS(A-) are given by :

$$A^* = (\tilde{v}_1^*, ..., \tilde{v}_n^*) = [(1,1,1),(1,1,1),(1,1,1)]$$

$$A^- = (\tilde{v}_1^-, ..., \tilde{v}_n^-) = [(0.32, 0.32, 0.32), (0.21, 0.21, 0.21), (0.28, 0.28, 0.28)]$$

Table 6
Fuzzy comparison matrix

	S8	S9	S10
Price	(0.7, 0.93,1)	(0.4,0.6,0.8)	(0.4,0.6,08)
Quality	(0.4,0.6,0.8)	(0.3,0.4,0.5)	(0.3,0.4,0.5)
Delivery	(0.4,0.6,0.8)	(0.4,0.6,0.8)	(0.4,0.6,0.8)

Table 7
Distance between supplier Si and A*

	Pricing	Quality	Delivery
d(S8, A*)	0.26	0.33	0.92
d(S9, A*)	0.47	0.45	1.26
d(S10, A*)	0.47	0.45	1.26

Table 8
Distance between supplier Si and A-

	Pricing	Quality	Delivery
d(S8, A-)	0.54	0.20	0.33
d(S9, A-)	0.19	0.33	0.84
d(S10, A-)	0.19	0.34	0.85

Now, we can compute the closeness coefficients of the three selected suppliers as (Table 9):

$$CC_i = \frac{d(\text{supplier}_i, A^-)}{d(\text{supplier}_i, A^-) + d(\text{supplier}_i, A^*)}$$

Table 9
Closeness coefficients

Supplier	S8	S9	S10
CC_i	0.60	0.40	0.40

By normalizing CCi, we calculate the order allocation (Oi) of each supplier. Table 10 illustrates the obtained results, the order allocation of each selected supplier for the 4 periods.

Table 10
Order allocation rates for the three selected suppliers

	Period 1	Period 2	Period 3	Period 4
S8	0.40	0.28	0.36	0.37
S9	0.30	0.36	0.30	0.31
S10	0.30	0.36	0.34	0.32

5. Conclusions and perspectives

In this paper, we have presented a new tool for automatic fuzzy based knowledge acquisition, which clubs supplier selection process with order allocation for dynamic supply chains to cope market variations. More specifically, the developed tool imitates the knowledge acquisition and manipulation in a manner similar to the decision makers who have gathered considerable knowledge and expertise in procurement domain. Some experimental results are presented to illustrate the applicability of the tool.

As a perspective for the future, we are making an effort in checking the practical implications of the proposed approach in the real life industrial problems for global supplier selection. Another important line of research can be using fuzzy pattern matching approaches and also the definition of other criteria to the considered case study or more generic for several sectors of industrial activities.

References

[1] Weber, C. A., Current, J. R. and Desai, A. Vendor selection criteria and methods. *European Journal of Operational Research*, N°. 50, 1991, pp. 2-18.

[2] Mobolurin Multi-hierarchical qualitative group decision method: consensus building in supplier selection. *International conference on applied modeling, Simulation and Optimisation*, USA, 1995, pp. 149-152.

[3] Muralidharan, C., Anantharaman, S. P. and Deshmukh, S. G. Vendor rating in Purchasing Scenario: A confidence Interval Approach. *International Journal of Operations and Production Management*, 10(2), 2001, pp. 1305-1325.

[4] Muralidharan, C., Anantharaman, S. P. and Deshmukh, S. G. A Multicriteria Group Decision making Model for Supplier Rating. *The Journal of Supply Chain Management*, 38(4), 2002, pp. 22-35.

[5] Chan, F. T. S. Interactive selection Model for supplier selection process: an analytical hierarchy process approach. *International Journal of Production Research*, 41(15), 2003, pp. 3549-3579.

[6] Deng, S.J. and Elmaghraby, W. Supplier Selection via Tournaments. *Production and Operations Management*, 14(2), 2005, pp. 252-267.

[7] Jain, V., Benyoucef, L. and Deshmukh, S. G. Strategic Supplier Selection: Some Emerging Issues and Challenges. *International Journal of Logistics Systems and Management*, Vol. 5, N°1, 2009, pp. 61-88.

[8] Chan, F.T. S., Kumar, N., Tiwari, M. K., Lau, H. C. W. and Choy, K. L. Global supplier selection: a fuzzy-AHP approach. *International Journal of Production Research*, 2007, pp. 1 – 33.

[9] Van der Rhdd, b., Verma, R. and Plaschka G. Understanding trade-offs in the supplier selection process: The role of flexibility, delivery, and value added service/support. *International Journal of Production Economics (in press available online*, 2009.

[10] Kannan A. and Noorul Haq A. Analysis of interactions of criteria and sub-criteria for the selection of supplier in the built-in-order supply chain environment. *International Journal of Production Research*. 45 (17), 2007, pp. 3831-3852.

[11] Wu W-Y, Sukoco B. M., Li C-Y and Chen S. H. An integrated multiobjective decision-making process for supplier selection with bundling problem. *Experts systems with applications*, Vol. 36, 2009, pp. 2327-2337.

[12] Lee A.H.I, Kang H-Y and Chang C-T Fuzzy multiple goal programming applied to TFT-LCD supplier selection by downstream manufactures. *Experts systems with applications*, Vol. 36, 2009, pp. 7917-7927.

[13] Amid A., Ghodsypour S.H. and O'Brien C. A weighted additive fuzzy multiobjective model for the supplier selection problem under price breaks in supply chain. *International Journal of Production Research*, 2009, (in press, available online).

[14] Onut S., Kara S.S. and Isik E. Long-term supplier selection using a combined fuzzy MCDM approach: A case study for a telecommunication company. *Expert systems with applications*, Vol. 36, 2009, pp. 3887-3895.

[15] Sanayei A., Mousavi S. F., Abdi M.R. and Mohaghan A. An integrated group decision-making process for supplier selection and order allocation using multi-attribute utility theory and linear programming. *Journal of Franklin Institute*, Vol. 345, 2008, pp. 731-747.

[16] Wang, Y-M., Luo, Y., Hua, Z., On The Extent Analysis Method For Fuzzy AHP And its Application, European Journal Of Operational Research, 2007, pp.735-747.

Innovative Production Machines and Systems
D.T. Pham, E.E. Eldukhri and A.J. Soroka (eds)

Integration of Flexible RIA Based GUI in I-SOAS

Z. Ahmed[a,b], V. Popov[b,c]

[a] *University of Wuerzburg, Germany*
[b] *Vienna University of Technology, Austria*
[c] *Technical University Sofia, Bulgaria*

Abstract

It is necessary to improve the concepts of present web based graphical user interface (GUI) for the development of more flexible and intelligent user interface to provide ease and increase the level of comfort at user end like most of the desktop based applications. This research is conducted targeting the goal of implementing flexible GUI consisting of a visual component manager with different components by functionality, design and purpose. In this research paper we present a Rich Internet Application (RIA) based graphical user interface for web based product development, and going into the details we present a comparison between existing RIA technologies, adopted methodology in the GUI development and developed prototype.

General Terms: RIA, GUI, I-SOAS

Keywords: Graphical User Interface, Semantic Web, Rich Internet Application

1. Introduction

Currently, the World Wide Web is mainly based on documents written in Hypertext Markup Language (HTML), a markup convention that is used for coding a body of text interspersed with multimedia objects such as images and interactive forms. HTML has many limitations for searching and processing data from it. The main objective of this research is to provide prototype environment for web based intelligent graphical interface.

The Graphical User Interface (GUI) must consist of a visual component manager with different components by functionality, design and purpose. The GUI must be flexible and dynamic, to enable moving of objects, changing colors and fonts, choosing between components in the interface, changing user details, extract and view specific information from the database, communicate with other users in the system. Moreover the GUI can also be saved and restored after user logoff and login.

This research paper is divided into six sections. In section 2 we present RIA frameworks in details for better communication between the graphical user interface and the server. In section 3 we present some available RIA technologies and perform a comparison to choose one the best RIA technology. In section 4 we explain the association of this research work with I-SOAS. We also discuss the implemented design methodology, application flow and structure of GUI of developed prototype version of I-SOAS. In section 5 we briefly present the prototype and in section 6 we conclude the discussion.

2. RIA Development

Rich Internet Application (RIA) is web application

that is similar to desktop applications, mostly delivered by web browser plug-ins. Some of them are running independently via sandboxes or virtual machines. Present RIA technologies include Adobe Flex/AIR, JavaFX, Microsoft Silverlight etc. with the new versions of HTML and XHMTL appeared frameworks like jQuery, based on JavaScript and HTML DOM (Document Object Model) interfaces. The biggest developing companies in this sector are committed to enable RIA developers to innovate and create the next generation of Web and desktop applications. Usually an appropriate RIA framework is required to run the application, and needs to be installed using the computer's operating system before launching the application. This software is typically responsible for downloading, updating, verifying and executing the RIA.

To work with a system, users have to be able to control the system and assess the state of the system, also and to get results from it in the proper format. RIA present a new way of developing web based Graphical User Interface (GUI) including new features and more capabilities than the standard web pages. The design and the good usability of the user interface is a significant part of the research process of the project. The Intelligent User Interface (IUI) is intended to provide innovative implementations, systems and technologies and apply the new ideas about how people and computers could "intelligently" interact. Consequently, the design of effective and efficient human-computer interfaces becomes ever more critical to overall system performance. Main issues addressed by intelligent user interface research are the following:

- How can interaction be made clearer and more efficient?
- How can interfaces offer better support for their users' tasks, plans, and goals?
- How can information be presented more effectively?
- How can the design and implementation of good interfaces be made easier?

The design of a user interface affects the amount of effort the user must expend to provide input for the system and to interpret the output of the system, and how much effort it takes to learn how to do this. Usability is the degree to which the design of a particular user interface takes into account the human psychology and physiology of the users, and makes the process of using the system effective, efficient and satisfying.

Usability is mainly a characteristic of the user interface, but is also associated with the functionalities of the product and the process to design it. It describes how well a product can be used for its intended purpose by its target users with efficiency, effectiveness, and satisfaction, also taking into account the requirements from its context of use.

3. RIA Technologies

RIA is the next generation of web and desktop applications and combines the best of the user interface of desktop applications plus low-cost deployment of Web applications and rich interactive multimedia communication. It eases the design and development of Intelligent User Interface applications with new ideas and a lot of features of the IDE and for the developed application. In the beginning of the research we explored the present RIA frameworks and RIA development approaches. Because of the fact it cannot be compared with all frameworks, we chose the most wide-spread and stable ones: Flex/AIR (Adobe), Silverlight (Microsoft) and Java FX (Sun). We compared them in several trends: world-wide usage, speed, interface capabilities, IDE and development easiness and platform dependency. Also a valuable requirement is the integration with java because of that the project I-SOAS is being built on the java server platform. We summarized the results in the following table 1.

We chose Flex because of some big benefits of it ahead of the other two frameworks. First of all it is based on flash and it runs in plash player plug-ins, so it is very easy to user without additional plug-ins, because most of the computers have installed plash player in their browsers. Also both it and JavaFX have advantage that they can run on any operating system, beating Silverlight that it runs only on Windows platforms with .Net support. Flex has very good performance and have video and graphic acceleration, JavaFX don't have such. Flex builder have a lot of components for use in the developed application and is better IDE than the other two's. After the research we concluded that it is best for the communication between Flex and Java server, to use BlazeDS server plug-in, which is based of the AMF (Action Message Format). AMF is free and have server side includes for a lot of languages, including Java. It is binary format based loosely on the Simple Object Access Protocol (SOAP). It eases the management of the requests to the server-side application, using Remote Procedure Call (RPC) and the returned result is expressed as an ActionScript Object.

Table 1.Comparison RIA Technologies

Feature	Flex	Silverlight	Java FX
IDE GUI	yes	yes	no, source code only, new script language
Project UI declarations	XML based (MXML)	XML based (MXML)	declarative expressions
OS integration/cross-platform	yes (Flash or FlexAIR)	Windows only (requires .Net support)	yes (Java runtime library)
Server side integration	object based, AMF	object based, AMF	AJAX, java services
Worldwide usage	best	poor	good
Loading time / Boot	fast	good	slow
3D	good (HW support)	good (HW support)	poor (no HW support)
Components & Tools	better	good	good
Component integration with OS	good	bad (no cross platform)	good

4. I-SOAS

The project Intelligent Semantic Oriented Agent based Search (I-SOAS); is a project proposed for the development of a web based flexible graphical user interface and intelligent search in Product Data Management Systems [1, 2]. The conceptual architecture of I-SOAS is categorized into four different sequential iterative parts i.e. Intelligent User Interface (IUI), Information Processing (IP), Data Management (DM) and Data Representation (DR), as shown in Figure 1. In this research paper, we are not going into the detail of any of the four parts of I-SOAS but IUI, the main graphical user interface [3] as the development of IUI using RIA technologies is the main goal of this research (presented in this research paper).

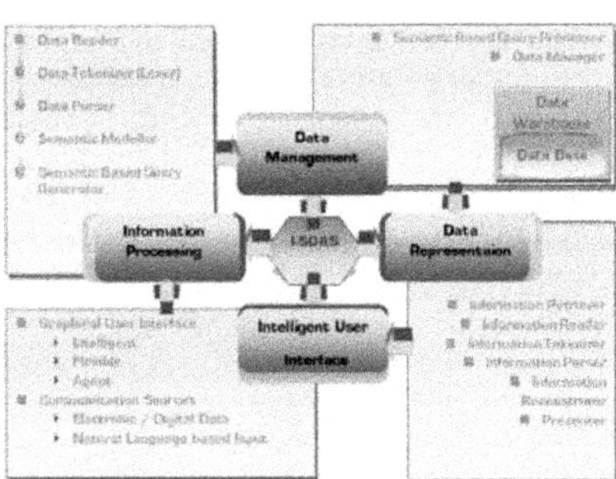

Figure.1. I-SOAS Conceptual Architecture

4.1. Design Methodology

The project structure consists of 3 basic parts: Flex files (SWF flash files), Java classes and a database. In future it is also planned to support external database, so the user may work in different areas from one place. Every part has different ways of working and developing process. These parts are integrated as one application like product line applications. The working process includes a request from the user to open the page for which the client browser receives the SWF files (flex compiled application) from the web server. Every data request is managed by the BlazeDS library (java remote include classes framework). After that the data is retrieved from the database and is sent as SWF data (compiled flex / ActionScript objects) to the interface. The communication with the database is made only with the java classes, as shown in Figure 2.

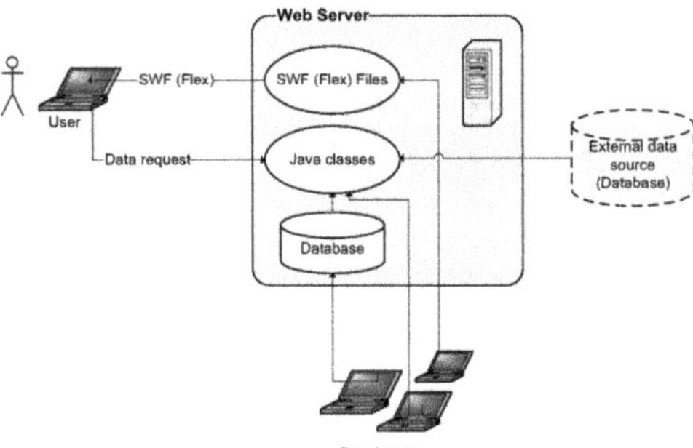

Figure.2. I-SOAS Project Design Methodology

4.2. Layer Communication

The communication between layers is present in the following Figure 3. The communication is managed by AMF (Action Message Format). Its data format is binary based using the Simple Object Access Protocol (SOAP). It is versioned with server side includes for most of the programming languages.

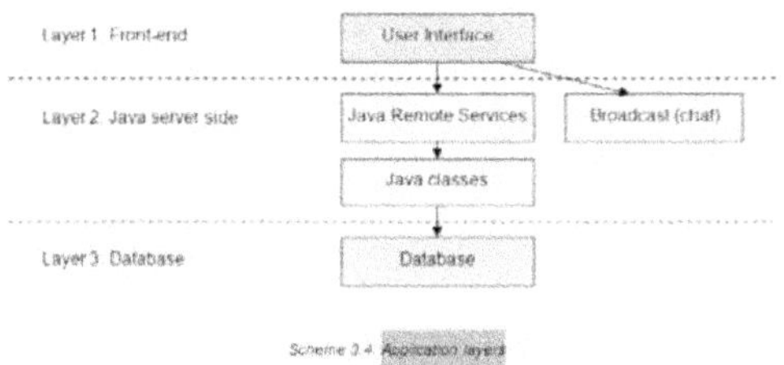

Figure.3. Application layers process

AMF eases the management of the requests to the server-side application, using Remote Procedure Call (RPC). With this technology, methods and classes at the server side (Java) can be called and be sent easily direct from the front-end application (Flex), without any additional code. There is only a configuration file mapping the class names. The result from the execution is expressed as an ActionScript Object. For now best Java based Server Side Includes (SSI) is BlazeDS developed by Adobe. The request is asynchronous, so the application does not wait for the reply from the server and result can be displayed at any time. When a data request is to be executed, e remote call is made from the user interface in the remote services' (via the server side includes) class members and the result is sent as an object of a Java class. If there have to extract data from the database (for ex. Search, login / logout ...) a permanent connection to the database is created and the query and data is processes via the Java-MySQL connector add-on. The retrieved data is processed in the remote services and returned to the user interface module as object of a specified class type. When a user is logged on, he/she is subscribed for the BlazeSD broadcasting service and can send and retrieve messages in the chat component.The remote services is used in the following cases: log-in or register user, loading and saving components' saved states, searching phrase or SQL query, loading and saving user interface settings, get actions log and chat. Some of the components do not use the remote services, because they do not require user data to work and can be accessed without login.

4.3. Application Flow

The main flow describes almost all flow of the available features in the system. It describes the begin point, how to proceed to search in I-SOAS and the work with the components. The work flow is presented in Figure 4. The interface is made like that so the user can feel free to navigate from anywhere he/she wants to anywhere else. The data and the state of the pages are preserved and shown again after going back to the page. If the user wants to search for something, he/she must login first and after that the search is available. The "Logout" button is visible only after successful login or register. When the user opens the components page, only 2 components are shown in the component manager if the user is not logged. If is logged all components are shown and after that the user interface settings are loaded and applied. Next if there is saved components, they are restored.

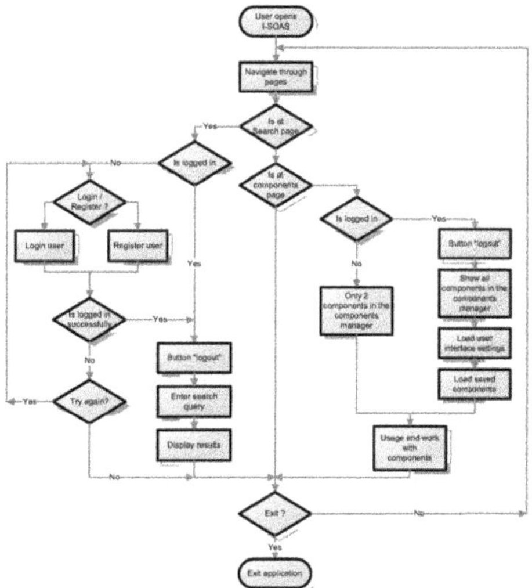

Figure.4. Application flow

4.4. GUI Structure

Front-end is completely developed with Flex. The flex application has main module for page container and by one module for every page in the project. Every component in the application is also separated as shown in Figure 5. There is a menu bar on the top of the page container, which carry out the navigation functionality. If the user is logged-in, a "Logout" button is shown in the right of the menu bar. In the scheme below is displayed the structure of the interface of I-SOAS prototype.

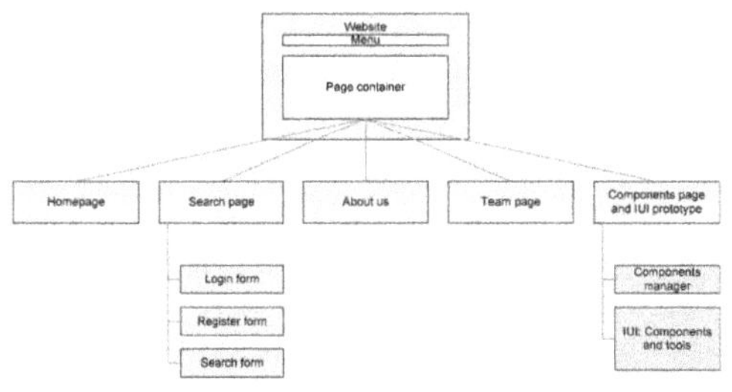

Figure.5. Application GUI structure

Every page opens in the page container and every component in the page can be moved and shown in this container. The page module is created on first use, instead of creating in load time. This speeds up a bit the loading for slow client computers and if user does not open some pages, they will not be created. Once created, the page is not destroyed and created again, but only hidden and shown again. That is how the view and components inside the page are preserved at runtime, as shown in Figure 6.

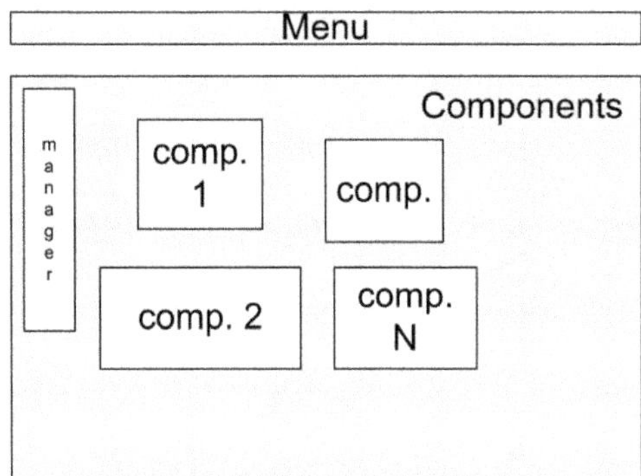

Figure.6. GUI components page structure

Homepage is the main page, Search page provides search and login options, About us page – contains contacts, Team page – contains a list of out team worked on this project and Components page – contains the prototype of the IUI with a component manager and user area. If the user is not logged in, only the components that do not require user data to work are available. If the user is logged-in, all components are available in the component manager. Also if there is saved component's state from the previous session, the saved state is restored and the user interface settings are loaded and applied. There is a counter of the displayed components in the page in the bottom of the components manager window. Components are organized in own windows. Every component window, including the component manager's window, can be moved in the user area. The component windows have animations for showing and hiding events, for showing status message in the bottom of the window and animations for the tool tips. The components page's scheme is illustrated in figure 6. All components are visual classes based on one main class, called MovablePanel, which is based on the Panel component in Flex framework. This class is responsible for the ability of the component to interact as a desktop window. It enables the movement with mouse dragging, adds close button and handles the connection between the component and the user interface. Only the clock component is based on the Label visual component in Flex framework.

4. Involved Technologies

We used following software technologies: Flex Builder 3 IDE, Eclipse IDE 3.4, Java SDK, Apache Tomcat 6 server, BlazeDS 3.2 – java remoting services plug-in, MySQL 5 Database Server and MySQL Connector 3.0. for the development of I-SOAS prototype.

5. Prototype

This is IUI of I-SOAS Web Application as shown in figure 7, capable of doing following tasks

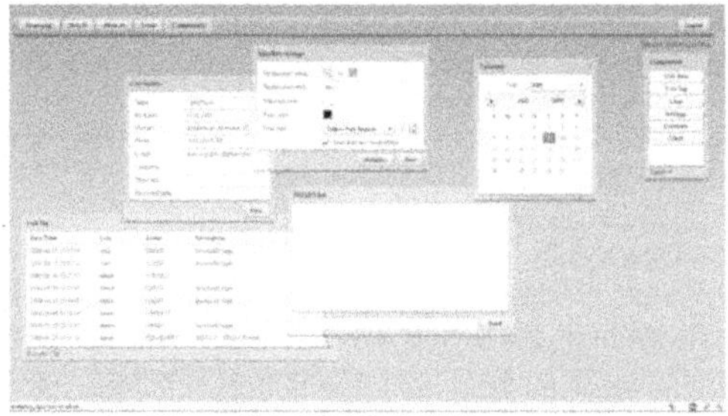

Figure.7. I-SOAS GUI Prototype

1. Providing a complete user authentication mechanism which let new user register and old user login by providing valid user name and password.
2. Providing a Visual components manager for component management.
3. Providing options to user to redesign a new graphical user interface by adding deleting and arrange components
4. Providing option to save newly build graphical user interface.
5. Providing Visuals effects in graphical user

interface development.
6. Providing options to search records by entering natural language or SQL queries

6. Conclusion

In this research paper we have discussed an approach to integrate intelligent user interface in a semantic web based application. This interface obeys the assigned tasks and criteria and also we even added additional features and effects to it. It is flexible enough to manage different type of components, different user preferences for design, enable the user to search in two different ways, change between different states and interface settings whether if the user is logged-in or not. The interface has many visual effects and flexible and has very easy changeable design. I-SOAS is Flex and Java based application and uses a relational database with MySQL database server, where stores and manages its data. The front-end (the user interface) communicates with the server and gathers data from the database thorough BlazeDS java plug-in, which is based on the AMF (Action Message Format).

In the beginning of the paper we made a research of the present RIA technologies and compared them to show the benefits and weaknesses of every particular technology framework. Also we researched some of the available PDM applications and compared them to the concept of I-SOAS. We got some ideas from those applications, so to improve the prototype of I-SOAS Intelligent User Interface.

In future we are hoping to focus on depth of architectural details, mathematical or algorithmic experiments and currently available tools and technologies for the development of the project and to improve some modules in it. As we are promoting I-SOAS as a dynamic concept of Intelligent User Interface we will also try to prove the importance of the IUI web integration in different fields in the business process and work, also and in the entertainment area.

Acknowledgements

We are thankful to Vienna University of Technology Austria, Technical University Sofia Bulgaria and University of Wuerzburg Germany for giving us the opportunity to keep working on this project i.e. I-SOAS. We are thankful to Erasmus for funding Vasil Popov for working on this project as the part of his Master Thesis research. We are thankful to Prof. Dr. Detlef Gerhard for his supervision during this research and pay our gratitude to Prof. Dr. Thomas Dandekar for his generous support.

References

[1] Zeeshan Ahmed, Detlef Gerhard: "Semantic Oriented Agent based Approach towards Engineering Data Management, Web Information Retrieval and User System Communication Problems", In the proceedings of 3rd International Conference for Internet Technology and Secured Transactions, ICITST 08, IEEE - CST, June 23–28 2008

[2] Zeeshan Ahmed, Detlef Gerhard: "Design Implementation of Semantic Oriented Agent and Knowledge based approach for Intelligent Human Machine Data Manipulation", In the proceedings of 4th Virtual International Conference on Innovative Production Machines and Systems (IPROMS 2008), Manufacturing Engineering Centre Cardiff University, organized by FP6 I*PROMS Network of Excellence by European Commission, 1st – 14th July 2008

[3] Zeeshan Ahmed: "Intelligent Human Machine Interface Design for Advanced Product Life Cycle Management Systems", In the proceedings of International Conference on Frontiers of Information Technology, ACM – FIT09, Session 5: Computational Science/Artificial Intelligence, ACM Digital Library, ISBN 978-1-60558-642-7, 16-18 December, 2009.

Supplementary Web Links

1. Rich Internet, Last reviewed 06 November 2009, <http://en.wikipedia.org/wiki/Rich_Internet_applicationY
Java, Last reviewed 06 November 2009, <http://www.java.com/en/>
2. Flex, Last reviewed 06 November 2009, <http://www.adobe.com/products/flex>
Silverlight, Last reviewed 06 November 2009, <http://en.wikipedia.org/wiki/Silverlight>, 2008
3. Apache Tomcat, Last reviewed 06 November 2009, <http://tomcat.apache.org/>
4. BlazeDS 3.2, Last reviewed 06 November 2009,<http://opensource.adobe.com/wiki/display/blazeds/Release+Notes>
5. MySQL, Last reviewed 06 November 2009, <http://www.mysql.com/>
6. Action Message Format, Last reviewed 06 November 2009, <http://en.wikipedia.org/wiki/Action_Message_Format>
7. SOAP, Last reviewed 06 November 2009, <http://en.wikipedia.org/wiki/SOAP>

Innovative Production Machines and Systems
D.T. Pham, E.E. Eldukhri and A.J. Soroka (eds)

Measurement Uncertainty in Practice

Shirley Y Coleman[a]

[a] *Industrial Statistics Research Unit, ISRU, School of Maths and Stats, Herschel Building, Newcastle University, Newcastle upon Tyne, NE1 7RU*

Abstract

Most companies need continuous quality improvement. Many have adopted the "six sigma" approach which includes statistical process control (SPC), design of experiments (DoE) and reliability studies. Each of these techniques is based on measurement and requires a reliable measurement system. Measurement systems analysis (MSA) can be used to assess precision through Gauge Repeatability and Reproducibility (Gauge R&R) studies. Training in MSA can be facilitated by a practical exercise and graphical output can be used to detect problem areas. Many companies find that their measurement systems are too poor for further problem analysis, for example measurement uncertainty may be considerable in analytical methods. Understanding the components of variation can help solve inter-departmental disputes and is a good example of statistical know-how improving communication. As well as modifications for destructive testing, there are different approaches depending upon the measurement type. Binary data may arise from continuous, ordinal or nominal information, for example go/no-go gauges in manufacturing. Attribute agreement issues arise in manufacturing and also in other areas, such as sensory panels, legal decision making and sports refereeing, for example, a 4-man panel is responsible for decisions on the football field. Improving consistency in such weighty decisions as these is of paramount importance.
Keywords: measurement, reliability, statistics

1. Introduction

Most companies have a need for continuous quality improvement. The six sigma approach to quality improvement is to step methodically through analytical procedures to detect and mend problem areas such as excessive waste and substandard products. Six sigma is equally applicable in manufacturing, healthcare, financial and other sectors. Statistical techniques are pivotal to six sigma and include statistical process control (SPC), design of experiments (DoE) and reliability studies (see, for example, [1]). Each of these techniques is based on measurement and requires a solid and reliable measurement system. Organisations are well advised to check the accuracy and precision of their measurement systems. Meticulous definition of each six sigma project takes place in the D (define) phase and includes selecting representative measurements which are accessible and meaningful. Measurement systems analysis (MSA) is part of the six sigma DMAIC toolkit and is usually introduced in the M (measure) phase. MSA can be used to assess precision through Gauge Repeatability and Reproducibility (Gauge R&R) studies and accuracy through Gauge Linearity and Bias studies. The A (analyse) six sigma phase includes root cause analysis, I (improve) includes DoE, recommending improvements, piloting and change management and C (control) focuses on establishing a control mechanism usually including SPC, full deployment and publication of results.

In practice, measurement error can be extremely poor. For example, during a factory visit, a dimension measure for TV components was 10.1 on first reading and 10.8 on second reading; the tolerance was 0.5. Thickness of electroplating was measured at 5 microns and then 8 microns depending upon where the measure was positioned; in this case, the thickness was not critical to the customer, however, the variation represents a loss of opportunity for increased profits and viability. Further investigation, however, showed that time in electroplating bath varied enormously around the standard, but, standardising the timing would have required extra personnel as the operation was fitted in by one person who also had other duties. This demonstrates the complexity of quality improvement projects and shows how a holistic approach is needed to avoid improvement in one area leading to deterioration in another. Measurement variation of slot width in turbine blades was 43% of overall slot width variation. Weld failures in radiators was traced to the first operation in the process which was cutting the shape in metal sheets. It was found that measurements at this stage were being made by ruler but the operator was considered to be a difficult character and no-one wanted to upset him by making him change. Statistical analysis can help in change management by producing clear evidence in support of change.

Some practical aspects of measurement uncertainty are given in section 2. A practical can help facilitate training and an example is given in section 3. Section 4 explores the use of graphical displays to illustrate R&R results. Section 5 looks at different approaches depending on measurement type and conclusions are given in section 6.

2. Practical aspects of measurement uncertainty

Many problems in manufacturing and business are buried under confusion of what is actually going on. Statistical thinking helps step through problems in a logical manner. Once the problem is understood, then statistical techniques can help investigate and provide solutions. The first step before introducing SPC or undertaking designed experimental trials is to check the level of measurement uncertainty. Often measurements are obscure and poorly defined, for example the ovality or internal roughness of pipes or the tilt on a component. Not only does the metric chosen to represent the issue need to be assessed to make sure it is the most appropriate but also the precision needs to be checked.

Gauge R&R involves looking at the variation in measured results and quantifying the components of variation due to the parts being measured, the people or equipment doing the measurement, and the random error of the method. It is a controlled experiment involving the people and equipment and typical parts. Preferably it has to be possible to do multiple measurements but destructive testing can also be accommodated. Any number of factors can be tested, for example the effect on the measurement of people, equipment, time of day, temperature, atmospheric pressure, humidity etc. More factors, however, make the R&R study design more complicated.

R&R testing is very important, and because it can get rather complicated, standard methodologies have been provided. One of these is the Ford R&R method described in AIAG, 1995. This recommends that 10 parts are chosen which represent the usual spread of size and type of parts. Two or three operators who usually do the measurement or 2 or 3 measurement devices, which are usually used, are then employed. The parts are all measured two or three times. Any other numbers of operators, equipment, parts or replicates can be used.

Gauge R&R studies are useful because they

- focus attention on the metric used and make sure it is the best
- encourage formal documentation of all aspects of the measurement process
- highlight measurement error
- show the effect of different measuring equipment and personnel
- show where training is needed
- show where errors are creeping in
- compare errors for different sizes of unit via a linearity study

An R&R study should be carried out on all measured parts of the process and at all measured stages. Prior to conducting R&R studies, measurement system fundamentals such as consistency (repeatability through time), stability (bias through time) and sensitivity (discrimination) of the measurement system are checked. If discrimination is poor then the measurement process will not have the ability to detect variation in the parts.

R&R studies involve quantifying the components of variation. This is the field of analysis of variance (ANOVA) but an alternative method based on comparing means and ranges is often used as it may be conceptually easier. In [2], the ANOVA method requires a two way ANOVA with parts and people as the two factors. The Xbar/R method obtains estimates of variation by converting ranges to standard deviations. The two methodologies give slightly different results as they make different assumptions about the data and summarise the variability in different ways [3].

It is quite challenging to carry out randomisation correctly in practical applications. The tendency is for one person to measure the parts one at a time and then do it again in a different random order. This leaves open the possibility of a time effect. The measurements should be done in a random order, but that requires considerable organisation and also would not reflect the usual work practice of measuring several parts in one session. Any less of a randomisation, however, brings in issues of split plot designs with possible relationships between measurements due to their proximity. Further discussion on design is given in [4].

In many quality improvement programmes, the repeatability is so poor that it makes up nearly all of the variation in the ANOVA method. In the Xbar/R method, the reproducibility can be negative and needs to be put to zero. Indeed many projects stall and never proceed because the measurement system is so poor.

The aim of R&R is to compare the measurement variability with the part to part variation to see if the discrimination of the measurement system is adequate. If the measurement system is very precise then the number of categories that the measurement system can distinguish will be large, if it is not very precise relative to the variation of parts, then the number of categories will be low. If the number of categories is only 2, for example, it means that the measurement system can only distinguish between big and small values.

Typical conclusions by companies conducting MSA studies include:

- the company must increase the ability to detect variation by increasing the discrimination of the measurement devices
- high reproducibility error has been detected and the different appraisers methods need to be standardised according to best practice
- high bias error has been detected and the instrument must be recalibrated

Replacing or improving the measurement system has a cost implication and some companies are not willing to commit the necessary funds even though better measurement would lead to opportunities for significant increases in quality.

3. A practical to facilitate training in MSA

Analysis of measurement uncertainty is a vital first step in any quality improvement project. It is well known that learning by doing is most effective. Introducing a complicated issue such as R&R through a practical exercise has the advantage of allowing teamwork in which team members can discuss the issues in a relaxed environment and therefore help each other. To demonstrate the method, our R&R practical looks at the R&R of a number of different household scales. Typically the scales include a good (precise) set, a medium set and a poor set. The variety of scales helps to bring out different features of R&R and shows the different precision. Teams of 3 to 5 people are each allocated a set of scales.

Three sets of 10 ordinary tins of about 440gms are obtained. Different sets of tins are used for each team and set of scales to make the experiment flow more easily. The tins contain foodstuff such as baked beans, rice pudding or vegetables. The tins are numbered from 1 to 10 by sticking labels on them.

The teams generate for themselves 6 random orders of 1 to 10. Three different people then weigh the tins in random order two times each. The weights are read off by one team member and recorded directly into a spreadsheet by another. The formulae required to analyse the data can be supplied with the spreadsheet.

A typical set of data obtained from the old analogue set of scales is shown in Table 1.

Table 1
Extract of raw data for R&R practical (Operator A only)

Operator	A			
Tin	1st trial	2nd Trial	Mean	Range
1	450	435	442.5	15
2	450	430	440	20
3	445	425	435	20
4	445	425	435	20
5	440	425	432.5	15
6	450	440	445	10
7	450	440	445	10
8	445	425	435	20
9	450	410	430	40
10	450	430	440	20
Mean	447.5	428.5	438	19

The first step is to plot the data and this can be done in a number of ways. It is straightforward to program the analysis and plots in a spreadsheet but if the MINITAB statistical software package is available it can do it as well. Figure 1 shows the output from MINITAB for the data in Table 1.

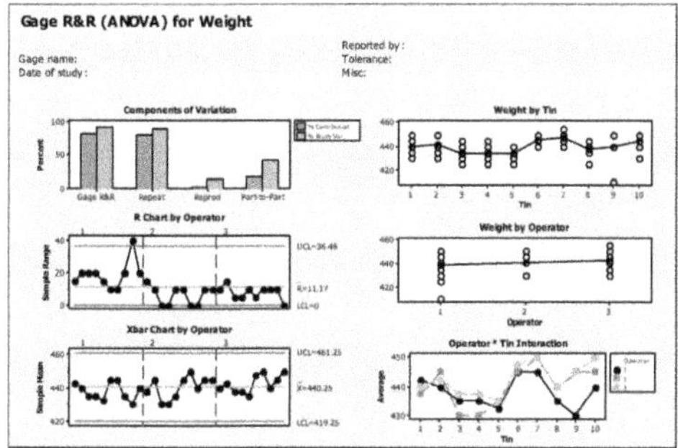

Fig. 1. Variation of tin weights (using MINITAB)

Figure 1 shows the mean measurements by each operator for each tin. It can be seen in the Operator*Tin interaction plot that the lines for each operator are separate from each other. This shows that a lot of measurement error is present.

The ANOVA for the old analogue scales is given in Table 2. The ANOVA can include the interaction of operator and part if it is thought to be significant. If it is included, the degrees of freedom for the repeatability standard deviation (sd) will be reduced which may inflate the sd if there is no significant interaction, thereby giving a falsely high value. The interaction for the data in Table 1 is insignificant and has been omitted.

Table 2
Two way ANOVA without interaction (from MINITAB)

Source	DF	SS	MS	F	P
Tin	9	1433.7	159.306	2.29285	0.0312
Operator	2	202.50	101.250	1.45727	0.2430
Repeat	48	3335	69.479		
Total	59	4971.25			

The repeatability variance component is estimated by the mean square (MS) in column 4 row 4 of Table 2. The reproducibility variance component is estimated by the operator MS in column 4 row 3 of Table 2. The R&R variance component is the sum of the repeatability variance and the reproducibility variance. The parts variance component is estimated by the MS in column 4 row 2 of Table 2. The corresponding standard deviations are the square roots of the variances.

The repeatability is defined in AIAG, 1995 as 5.15 times the repeatability sd. The multiplier 5.15 is the number of sds needed to capture 99% of the process (recall 6 sigma includes 99.7%, 5.15 includes 99%). The reproducibility is defined in AIAG, 1995 as 5.15 times the reproducibility sd. The study variation is 5.15 times the total variation. These values are shown in Table 3.

Table 3
R&R from two way ANOVA without interaction (from MINITAB)

Source	VarComp	StdDev	5.15*Sigma
Total R&R	71.068	8.43017	43.4154
Repeat	69.479	8.33542	42.9274
Reproduce	1.589	1.26037	6.4909
Operator	1.589	1.26037	6.4909
Part-part	14.971	3.86925	19.9266
Total	86.039	9.27571	47.7699

Table 4 gives the % contribution of variances. These are given as fractions of the total variation and add up to 100%. The % study variations are obtained by dividing by the total study variation and do not add up to 100%. It can be seen that the %R&R contribution is 82.60% of the variation in measurement of the tin weights. This % is very high showing that the household scales introduce a lot of measurement variation. A rule of thumb is that it should be no more than 10% of overall variation but in this practical it is obviously much too high to be useful.

Table 4
R&R from two way ANOVA showing % contributions

Source	%Contribution	%Study Var
Total Gage R&R	82.60	90.88
Repeatability	80.75	89.86
Reproducibility	1.85	13.59
Operator	1.85	13.59
Part-To-Part	17.40	41.71
Total Variation	100.00	100.00

The number of categories is calculated by dividing the standard deviation for parts by the sd for R&R and multiplying by 1.41 and rounding. It represents the number of non-overlapping confidence intervals that will span the range of product variation. The number of categories here is 1.41*3.9/8.4 which rounds to 1. This shows that the old analogue scales are not able to discriminate between the different weight tins and is only sufficient to say that they all weigh the same (i.e. all have the same single category of weight). The scales would be able to discriminate a wider range of weights, for example, 250grams vs 450grams but cannot differentiate reliably between tins more similar in weight.

The alternative Xbar/R method is preferred by some people as it avoids the use of ANOVA. In this example, Xbar/R gives slightly different results but the overall message is the same.

The experiment takes about 2 hours to perform including the planning, measuring and calculations. The groups can compare their results using the different scales. The practical gives a light-hearted but meaningful introduction to gauge R&R in a short space of time. The SMEs who worked through the practical felt confident to try R&R studies on their own. In several cases, the R&R study prevented the company from doing doomed experiments. The study alerted them to timing issues as well as training issues.

4. Graphical analysis

Graphical displays are very important and Figure 1 includes the main features of the R&R study. An alternative display is obtained using the versatile linked histograms in JMP software. Figure 2 shows the data with the results for repetition 1 highlighted. Any of the histogram bars can be highlighted and the corresponding values in the other linked histograms will also be highlighted. It can be seen that the weights are generally higher for the first repetition. This shows another feature of the measurement error.

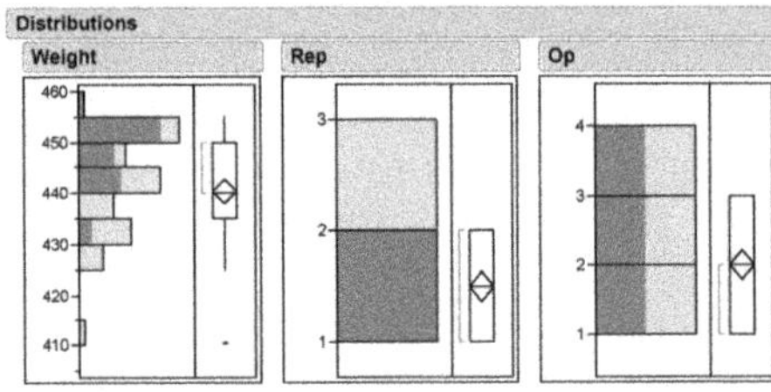

Fig. 2. Variation of tin weights (using JMP)

As well as using statistical analysis (ANOVA and Xbar/R), an interesting view can be obtained using partition analysis classification and regression trees (C&RT) in data mining. Weight is designated as the target variable and repetition and operator identity (converted to 3 indicator variables) are explanatory variables. The decision tree in Figure 3 identifies repetition as the variable most highly related to differences in weight.

The difference between repetitions is the best explanatory variable for the difference in weights. This is the same message as the high repeatability in the statistical analysis above, however, Fig. 2. shows that the weights in the first repetition are consistently higher than in the second.

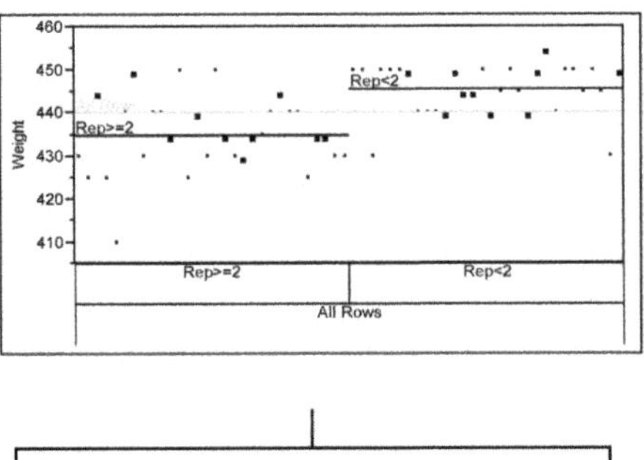

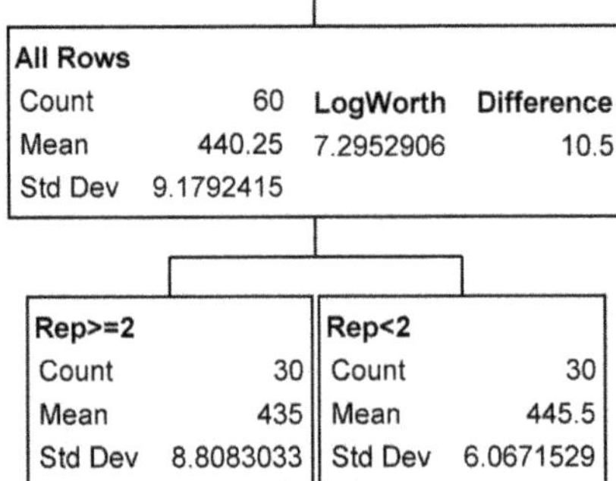

Fig. 3. Partition for Weight

5. Wider implications of measurement uncertainty

Destructive testing requires a modification to the standard MSA study. One solution is to have many similar parts and allocate them randomly to person, method and replication. This adds noise to the message. Customer opinion data may suffer from a similar problem as companies are reluctant to over burden their customers. For example, in a discrete choice study (see for example, [5]), where customers are invited to choose between two competing scenarios, each customer only gives their opinion once. Complex experimental design considerations are used to determine the combination of test scenarios but the testing is vulnerable to variation between people.

Ordinal data is very common in manufacturing, for example in subjective assessment of quality [6] as well as in measurement of opinion. Questionnaire reliability is often assessed by a Cronbach's alpha score [7]. This evaluates the internal consistency of the questionnaire. There are various assumptions and it is best applied to a homogeneous subset of the questionnaire. At the very least, however, Cronbach's alpha gives a lower bound to the reliability of the questionnaire. Sometimes ordinal data is "approximated" by normality. This is not necessarily a problem; for example, analysing noisy ordinal data using ANOVA may give similar results to using logistic regression.

Binary data may arise from continuous, ordinal or nominal information, for example go/no-go gauges in manufacturing. In [8], the authors consider the case where standard (true or golden) values are not available; they describe a method for analysing binary data and suggest sample sizes. If binary data such as go/no-go are approximately equally distributed then a sample of 50 of each gives good confidence.

Attribute agreement issues arise in manufacturing and also in other areas, such as sensory panels, legal decision making and sports refereeing. ISRU recently sent a proposal to improve consistency of Football Association (FA) panel decision-making to a number of football clubs referring to the FA's decision to add an extra game to the three-game ban in an incident at a Liverpool match. Chelsea's appeal against a three-game ban for one of their players was rejected by the FA but not thought to be "frivolous" (as in the Liverpool case) and so the FA deemed it unnecessary to add an extra game ban. Many football clubs have expressed frustration about such incidents as there appears to be a lack of consistency in FA decisions. Incidents are judged by a 4-man panel and a statistical approach could improve the consistency of the panel's decisions. The letter asked for support to take a proposal to the FA but only one club offered their support and no further action has been taken. Improving consistency in such weighty decisions is of paramount importance.

MSA leads to some deep statistical questions about sampling and the nature of variability, for example when the order of measurement is likely to have an effect, such as in taste panels where the assessor's discrimination can only be relied upon for a short time.

Measurement uncertainty may be considerable in analytical methods. Understanding the components of variation can help solve inter-departmental disputes and is a good example of statistical know-how improving communication. Analytical variation was 80% of product variation [9].

Measurement variation can be caused by uncertainty about the meaning of words. In [7], the authors propose a confusion index based on the number of meanings a word can have, the level of confidence felt by experts in using the word and the diversity of feelings expressed by users of the word. Words with high confusion index need to be defined carefully.

6. Conclusions

Analysis of measurement uncertainty is a vital first step in any quality improvement programme. It also provides an accessible way to start analysing the problem and can be a good team exercise.

Measurement uncertainty is fundamental to all data collection and analysis projects and affects every method of input. Measurement uncertainty can be high as in analytical methods, such as titration or very low as in co-ordinate measuring systems.

A practical exercise helps non experts to understand MSA and convince them of the importance of checking their measurement systems before embarking on expensive quality improvement projects.

Measurement uncertainty affects every aspect of our life and it is fortunate that extensive research is underway to ensure that the correct messages are extracted from the chaotic real world.

Acknowledgements

The author acknowledges the contributions of colleagues in ISRU to the work described in this paper.

References

[1] Henderson GR. Six sigma quality improvement with MINITAB. Wiley, 2006.

[2] Automotive Industry Action Group (AIAG) Measurement Systems Analysis, Reference Manual. Chrysler, Ford. General Motors Supplier Quality Requirements, 1995.

[3] Vardeman SB and Van Valkenburg ES. Two-way random effects analyses and gauge R&R studies. Technometrics, 1999, 44(3), pp 202-211.

[4] BS5497 Precision of methods. British Standards, 1987.

[5] Irwin DA and Pavcnik N. Airbus versus Boeing revisited: international competition in the aircraft market. Journal of International Economics, 2004, 64(2),pp223-45.

[6] De Mast J and van Wieringen WN. Modeling and Evaluating Repeatability and Reproducibility of Ordinal Classifications. Technometrics, 2010, 52(1), pp 94–106.

[7] Coleman SY and Putnik GD. Investigations to Improve Communication between Experts. In proceedings of 5th IPROMS virtual conference 6-19 July, Springer, 2009.

[8] van Wieringen, WN and De Mast J. Measurement System Analysis for Binary Data. Technometrics, 2008, 50(4), pp 468-478.

[9] Gardner JA, Coleman SY and S.G. Farrow. Why worry about your analytical methods? Analytical Proceedings of the Royal Society of Chemistry, 1993, 30, pp 183-5.

Innovative Production Machines and Systems
D.T. Pham, E.E. Eldukhri and A.J. Soroka (eds)

A fuzzy approximation to the Jacobian condition number

A. Synodinos, N.A. Aspragathos

Department of Mechanical and Aeronautics Engineering, University of Patras, Rion 26500, Greece

Abstract

This paper presents a method to approximate the Jacobian condition number based on computational intelligence methods by training a fuzzy system. To start with, a brief overview of the concept of robotic isotropy and the use of Jacobian norms to characterize the dexterity of a robot is presented. Then, the computational cost of the condition number is shown for both the l_2-norm and the Frobenius norm. A fuzzy system is then trained with an input of the robot's configuration and an output of the condition number. The calculation of the output of a fuzzy system is a simple task, with a computational cost depending to the number of rules and membership functions. The success of the method is examined and the results are presented. Lastly, topics for further research are discussed as well as examples for the use of this method.

Keywords: Manipulability, Condition Number, Fuzzy Approximation

1. Introduction

The dexterity of a robotic manipulator has been introduced to characterise the performance of the anatomy and configuration by using the kinematic, static and dynamic models. These measures express the robot's capability for motion and force transmission. As the kinematic and static models are directly associated with the Jacobian matrix, it is expected that the properties of these measures are directly associated with the properties of the Jacobian.

The velocity ellipsoid is defined as the mapping of the joint velocity unitary sphere to the operational velocity space [1]. The velocity polytop [2], on the other hand, is defined as the mapping of the joint velocity n-dimensional cube of unit edge-length to the operational velocity space. Both these mappings are directly linked to the Jacobian matrix and therefore to the manipulators configuration.

Each manipulability measure that has been introduced is associated with a particular property of the mechanical intuitive equivalent. The manipulability index represents the absolute value of the determinant of the Jacobian, and measures the magnitude of the manipulators ability to move along directions, as it is proportional to the volume of the ellipsoid [2]. The performance index measures the lowest radii of the ellipsoid [3], and thus the lowest speed of the manipulators end-effector. The condition number [4] measures the isotropy of the manipulators ellipsoid.

These indices have been proposed by applying matrix norms to the Jacobian, such as the Euclidean norm, the max norm or any p-norm. For example, the unit circle in the 1-norm is a square with its vertices on the coordinates axes, in the Euclidean norm is a circle and in the ∞-norm is a square whose edges are perpendicular to the axes. In this manner, the matrix norms are applied to the Jacobian and characterise the dexterity accordingly.

A manipulator is called isotropic if its Jacobian matrix is isotropic in at least one configuration [1][5]. A matrix is called isotropic if its singular values are all equal and nonzero. The Jacobian condition number is defined as the ratio of the greatest and the smallest singular values of the matrix. In numerical analysis, the condition number of a function with respect to an argument asymptotically measures the worst case of how much the function can change in proportion to small changes in the argument.

The use of a fuzzy system for the calculation of the condition number of a three degree of freedom mobile manipulator OMNi has been proposed by Trivino and San Martin [6]. Using expert knowledge and a plot of the mapping of the workspace for best manipulability, rules have been composed on an intuitive manner and have then been verified with Matlab.

In this paper, a method is proposed to systematically train a fuzzy system in order to calculate the condition number. This method can be applied on a manipulator of six or more degrees of freedom (DoF) to decrease the computational cost significantly. The proposed method could be used for the real time dexterous path planning of a robot. Embedded fuzzy systems can be programmed to enhance the capabilities of real time control in robotic manipulators.

2. Jacobian Matrix

The forward instantaneous kinematics problem for a serial-chain manipulator is defined as the calculation of the total velocity of the end-effector given the positions and the velocities of all joints. This linear mapping is set up by using the Jacobian matrix, as described by Eq. (1), and many methods can be used to calculate it, such as the theory of reciprocal screws [7], motor algebra [8], Plücker line coordinates [9], etc.

$$\begin{bmatrix} v \\ \omega \end{bmatrix} = J(q)\cdot\dot{q} \tag{1}$$

Where the vectors v and ω are the linear and angular velocities of the end-effector respectively, J is the Jacobian matrix and $\dot{q}$ the joint velocities vector. For a manipulator of n degrees of freedom (D.o.F.) and m generalized coordinates, the Jacobian is an m by n matrix [10].

2.1. Condition Number

The condition number of a matrix is defined as

$$\kappa(J) = \|J\|\|J^{-1}\| \tag{2}$$

The same definition is used for any consistent norm and therefore, this definition depends on the choice of norms. In robotics, the condition number of the Jacobian mostly refers to the l_2-norm and can be calculated by

$$\kappa(J) = \frac{\sigma_{\max}(J)}{\sigma_{\min}(J)} \tag{3}$$

In general, we are unable to calculate an analytic formula as a function of the robot's configuration. Other norms have also been proposed, such as the Weighted Frobenius condition number

$$\kappa_F(J) = \frac{1}{n}\sqrt{tr\left(JJ^T\right)}\sqrt{tr\left[\left(JJ^T\right)^{-1}\right]} \tag{4}$$

This uses the Frobenius norm of the weighted Jacobian matrix.

The condition number measures the Jacobian-induced distortion of the ellipsoid. The smaller the distortion, the higher the quality of the velocity transmission from the joints to the end-effector. However, it requires the calculation of the singular values of the Jacobian matrix, which is computationally more intensive than computing eigenvalues. Most of the times, the condition number is used to describe the accuracy of a robot and as a measure of its' dexterity, but it can also be used as an estimate of the closeness to a singularity [11].

The use of the condition number is critically reviewed for manipulators with translational joints. To begin with, it is not mathematically possible to define the distance to a singularity for a manipulator whose D.o.F. are a mix between translation and orientation. Also, the Jacobian of such a robot is not homogeneous in terms of units. Ma and Angeles [12] suggested to define a normalized Jacobian matrix by dividing the rotational elements by a length such as the one of the corresponding links. However, that length is arbitrary so it can vary according to the designers choice but cannot be obviated due to the definition and properties of SE(3) [13].

2.2. Computational Analysis

The computation of the SVD of a matrix can be reduced to computing the eigenvalues and a polar decomposition. This method is not stable, so the matrix is first reduced to tridiagonal form, and then that matrix is diagonalized. The first phase requires $O(mn^2)$ flops and the second phase converges after $O(n^2)$ flops. Overall, this method costs $O(mn^2+n^2)$ flops, where m and n are the dimensions of the matrix [14].

To calculate the Frobenius condition number, the SVD is not needed, however, a matrix inversion is necessary. Using the LU method, the total computational cost is $O(-\frac{5n}{6}+\frac{3n^2}{2}+\frac{4n^3}{3})$ [15].

Table 1 shows the computational cost in flops for the most common methods used by the corresponding calculations for various Jacobian matrices according to the aforementioned formulas. Although it is generally believed that the SVD is an intensive task process, and designers prefer the Frobenius condition number over the standard condition number, it is shown that it is advantageous to prefer the latter from the computational time reduction point of view.

Table 1
Computational cost in flops for Jacobian matrices of common sizes

Jacobian	SVD	Inverse $(J \cdot J^T)$
4x4	80	112,6666667
4x5	125	208,3333333
4x6	180	347
5x5	150	208,3333333
5x6	216	347
6x6	252	347
6x7	343	536,6666667

3. Fuzzy System

The proposed method is based on the property of fuzzy systems as universal approximators. Utilizing data from the analytic calculation of the condition number using eq. (3), we trained a Takagi-Sugeno-Kang (TSK) fuzzy system with the gradient method. This method, in contrast to the least or batch least squares method, can tune all the parameters of a fuzzy system with predefined quantity of rules [16][17].

For a fuzzy logic system of the form

$$f(q \mid \theta)=\frac{\sum_{i=1}^{R} g_i(q)\mu_i(q)}{\sum_{i=1}^{R} \mu_i(q)} \qquad (5)$$

Where R is the number of fuzzy sets for all inputs, the membership function μ_i is Gaussian and the output affine coefficients are g_i. The input q is the robots' joint position vector. The parameters θ that the method tunes are the input membership function centers and spreads and the output affine coefficients. The training data set, G, is a subset of the robots workspace that we use to train the system of eq. (5). The output used for the training process is the standard condition number, because of its better engineering and intuitive application on kinematic chains.

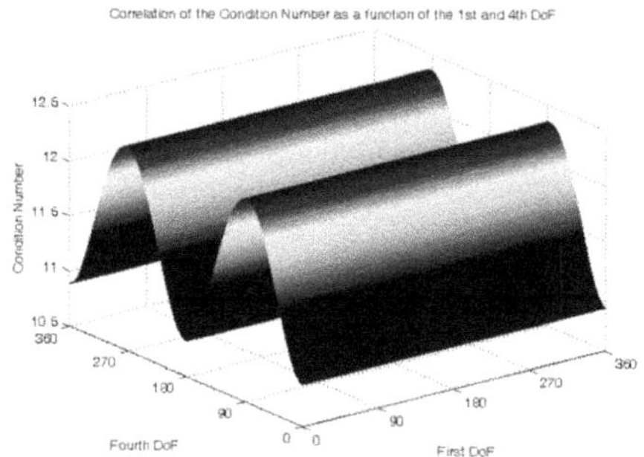

Fig. 1 The Condition Number as a function of the first and fourth DoF.

4. Joint Study

The joint study is implemented on the 6 DoF manipulator Fanuc AM120iB/10L. The manipulator is studied for each joint to analyze the effect of each DoF to the overall standard and Frobenius condition number.

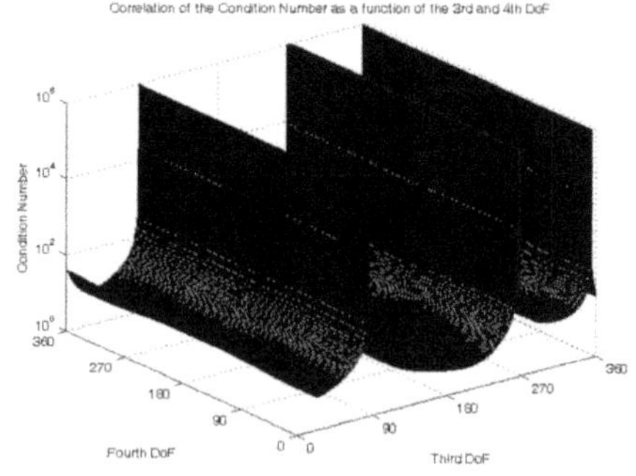

Fig. 2 The Condition Number as a function of the third and fourth DoF.

As shown in fig. 1and 3 the condition number is neither a function of the first nor sixth joint for this particular manipulator. However, for the sake of generality, this knowledge is not taken advantage of for the training of the fuzzy system. In fig. 2-4, which are on a logarithmic scale, the effect of each joint can be assessed to further understand from an intuitive perspective the configurations where the robot has its best dexterity. Although it appears in fig. 2 and 4 that the condition number is not affected largely by the corresponding DoF that is not accurate, as in fig. 1 and 5 the condition number axis is linear in contrast to the other figures.

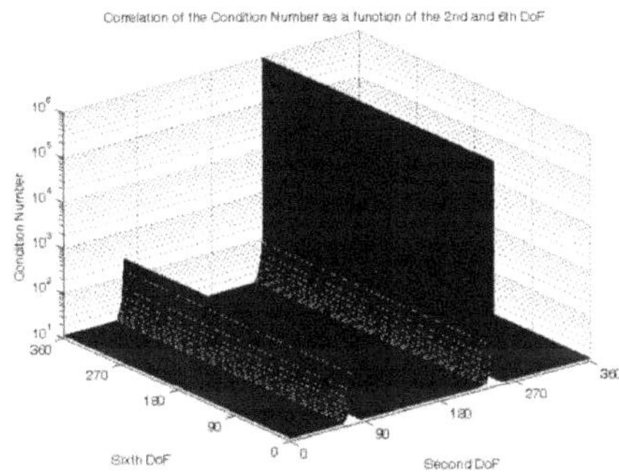

Fig. 3 The Condition Number as a function of the second and sixth DoF.

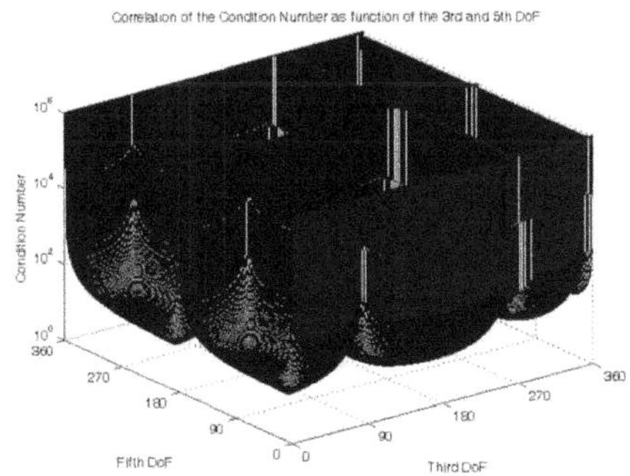

Fig. 4 The Condition Number as a function of the third and fifth DoF.

The singular configurations can easily be spotted, because the condition number approaches infinity. In fig. 5-8 the Frobenius condition number is similarly examined. Although the two condition numbers have a resemblance, there are significant variations close to singular configurations.

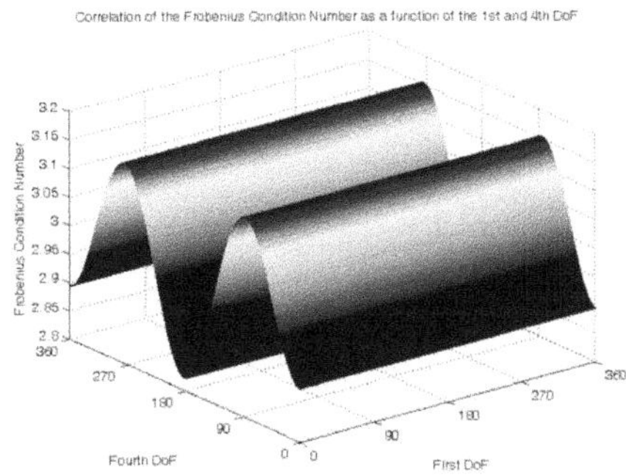

Fig. 5 The Frobenius Condition Number as a function of the first and fourth DoF.

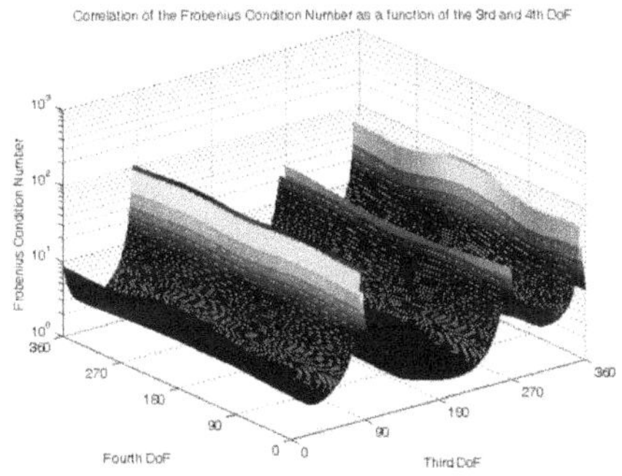

Fig. 6 The Frobenius Condition Number as a function of the third and fourth DoF.

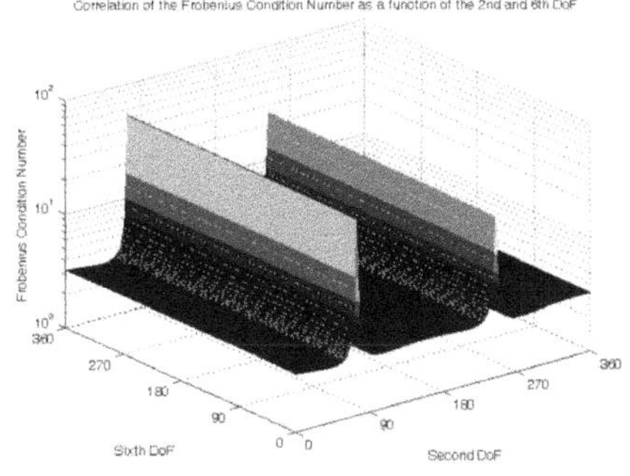

Fig. 7 The Frobenius Condition Number as a function of the second and sixth DoF.

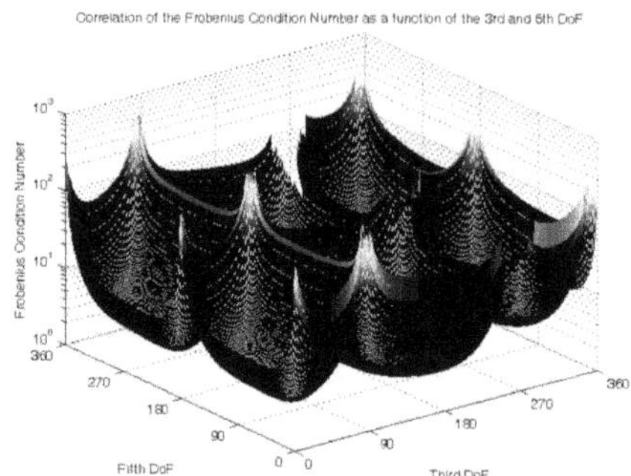

Fig. 8 The Frobenius Condition Number as a function of the third and fifth DoF.

5. Results

The training dataset is arbitrary defined in the following manner. The first five DoF range from 0° to 90° with a step of 10°, while the last one ranges from 0° to 45° with a step of 5°. All the possible different combinations for these data is inserted in eq. (3) and the condition number is calculated. The resulting dataset is sufficiently comprehensive without increasing the population immoderately.

Overall, the dataset is constituted of 10^6 elements of 6 inputs and one output. Although we could define the fuzzy system by only utilizing four inputs, due to the manipulators' condition number being unaffected by the position of the first and last joint, we are not doing so, to avoid loss of generality. The initial parameters for the fuzzy system are defined as

$$
\begin{aligned}
&R=100\\
&\text{Sweeps of G, Ngrad}=10;\\
&a_{ij}=1,\ \forall\, i\in(1,2,\ldots,R), j\in(1,2,\ldots,7)\\
&c_j^i\in\{0,\pi/4,\pi/2\}, \forall\, i\in(1,2,\ldots,R),\\
&\qquad j\in(1,2,\ldots,6)\\
&\sigma_j^i=1,\ \forall\, i\in(1,2,\ldots,R), j\in(1,2,\ldots,6)\\
&\lambda_{2,3,4}=0.1
\end{aligned}
\tag{5}
$$

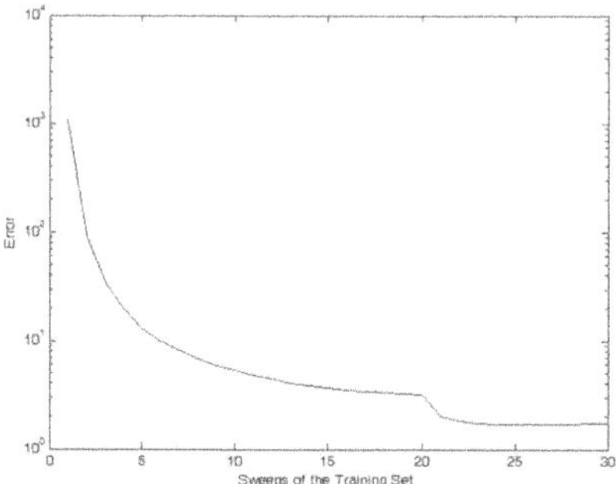

Fig. 9 Error plot in logarithmic scale as a function of the sweeps of the dataset

For the versification of the approximation error, eq. (6) is used for a random generated test set. The methods' convergence is shown in fig. 9 and is achieved after 25 sweeps of G. In fig. 10 the plot of the error of the approximation according to eq. (6) is shown after 30 sweeps.

$$e=\frac{1}{2}\left[f(q)-w\right]^2 \tag{6}$$

Table 2
Computational cost in seconds for the calculation of the condition number

	Analytic	Fuzzy	Gain
Time for the entire test set	292.8	38.9	752.6%
Time for one configuration	0.033	0.0052	634.6%

All the computations are implemented using Matlab, on a Pentium E5400 personal computer with 4GB of RAM and Windows 7 64bit. The algorithm used for the calculation of the Jacobian matrix is developed by Paul, Shimano and Mayer [18] and used in Matlab Robotics Studio. The computational cost of the proposed method is compared for a random set of 10^5 configurations as shown on Table 2.

In fig. 11 the condition number is plotted as a function of the training set for the first 10^5 data, since the first DoF does not affect it.

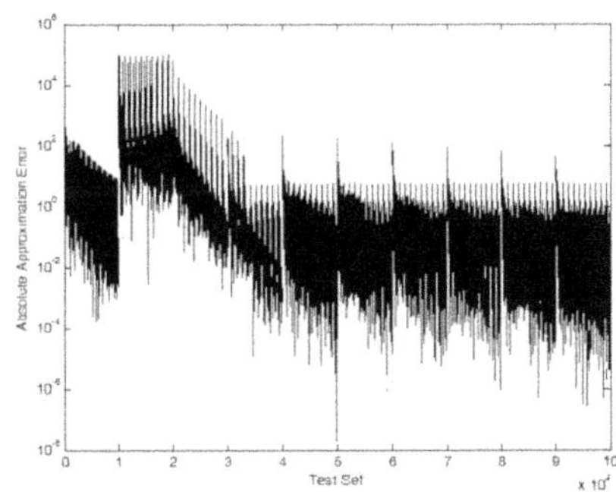

Fig. 10 The absolute approximation error in logarithmic scale of the fuzzy system after 30 sweeps of the training set

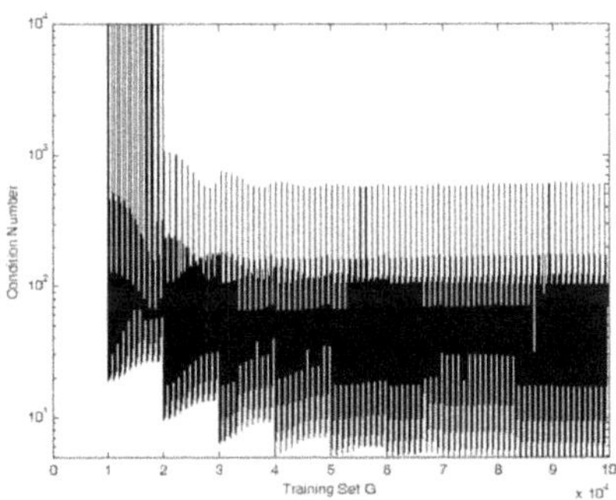

Fig. 11 The condition number as a function of the training set G

6. Conclusions and further work

In this paper, a method for the approximation of the condition number of a robotic manipulator has been proposed that manages to reduce the computational cost seven times for a 6 DoF manipulator. The relative error of the approximation at all times is kept minimal.

Further research includes the application of this method to the calculation of different manipulability measures and the analysis of their properties. An implementation of this method on an embedded fuzzy controller with the possibility of real-time dexterous manipulation is also considered for study. The dexterous motion planning of mobile manipulators on partially known environments is another topic for research.

References

[1] Salsbury, JK and Craig, JJ., "Articulated hands: force and kinematic issues." The International Journal of Robotics Research, 1982, pp. 4-17 .

[2] Yoshikawa, T., "Manipulability of robotic mechanism." The International Journal of Robotics Research, 1985, Issue Int J Robot Res, pp. 3-9.

[3] Klein, CA and Blaho, BE., "Dexterity measures for the design and control of kinematically." The International Journal of Robotics Research, 1987, pp. 73-83.

[4] Golub, GH and Van Loan, CF., *Matrix computations.* Baltimore : John Hopkins University, 1983.

[5] Angeles, J and López-Cajún, C., "The dexterity index of serial-type robotic manipulators." Kissimmee : s.n., 1988. Proceedings of ASME 20th Bienal Mechanism Conference. pp. 79-84.

[6] Trivino, G and San Martin, J., "A fuzzy logic approach to the concept of manipulability in mechanics." 2007. IEEE International Fuzzy Systems Conference. pp. 1-5.

[7] Mohamed, MG and Daffy, J., "A direct determination of the instantaneous kinematics." ASME Journal of Mechanisms, Transmissions and Automation in Design, 1985, pp. 226–229.

[8] Sugimoto, K., "Kinematic and dynamic analysis of parallel manipulators by means of motor algebra." ASME Journal of Mechanisms, Transmissions and Automation in Design, 1987, pp. 3-7.

[9] Hunt, KH, Samuel, AE and McAree, PR., "Special configurations of multi freedom." International Journal of Robotics Research, 1991, pp. 123-134.

[10] Orin, DE and Schrader, WW., "Efficient computation of the jacobian for robot manipulators." International Journal of Robotics Research, 1984, pp. 66-75.

[11] Angeles, J., "The design of isotropic manipulator architectures in the presence of redundancies." International Journal of Robotics Research, 1992, pp. 196-201.

[12] Ma, O. and Angeles, J., "Optimum architecture design of platform manipulator." Pise : s.n., 1991. ICAR. pp. 1130-1135.

[13] Park, M.K. and Kim, J.W., "Kinematic manipulability of closed chains." Portoroz-Bernadin : s.n., 1996. ARK. pp. 99-108.

[14] Trefethen, Lloyd N. and Bau III, D., *Numerical linear algebra.* Philadelphia : Society for Industrial and Applied Mathematics.

[15] Trahan, J, Kaw, A and Martin, K., *Computational Time for Finding the Inverse of a Matrix: LU Decomposition vs Naive Gaussian Elimination.* 2004.

[16] Passino, K.M. and Yurkovich, S., *Fuzzy Control.* Menlo Park : Addison Wesley Longman, 1998.

[17] Wang, L.X., *A course in fuzzy systems and control.* Upper Saddle River : Prentice Hall, 1997.

[18] Paul, R, Shimano, B and Mayer, G., "Differential kinematic control equations for simple manipulators." 1981. IEEE Transactions on Systems, Man and Cybernetics. pp. 456-460.

Innovative Production Machines and Systems
D.T. Pham, E.E. Eldukhri and A.J. Soroka (eds)

Biomechanical analysis of human body segments in human-robot interaction

D.T. Pham, M.H. Saw, F. Cao

Manufacturing Engineering Centre, Cardiff University, Cardiff CF24 3AA, UK

Abstract

During the rapid growth of practical robotic application in recent years, more advanced human-robot interactions are made possible by the development of engineering and computer technology. In order to further expand the prospect of robotic application, a research into the feasibility of human body manipulation is currently being conducted. One of the main objectives of this research is to identify the limits of physical human-robot interaction, the biomechanical characteristics of human body that decide these limits and essential end effector specifications required to theoretically carry out human-robot interactions to said limits. The current stage of the research is focusing on the geometric properties of various human body parts, defined as body segments, and the minimum end effector specification needed to manipulate these segments.

Keywords: Biomechanics, human-robot interaction, end effector

1. Introduction

This paper focuses on analysing the shapes, sizes and hardness of various segments of human bodies which will help to implement an end effector suitable for handling said body parts. The purpose of this end effector is to allow easy accommodation and firm gripping action on human body segments of different geometrical features. During direct, physical interactions with human, it is vital to avoid physical damage on human body; thus both safety and comfort issues become major concerns. It is essential to specify human body segments shapes, sizes and hardness in order to implement safe and painless handling.

With the rapid advances in technology, unique opportunities arose to allow us consider new approaches in delivering effective and sustained first aid health care service. This could potentially lead to the development of an advanced cost-effective robotic system in order to assist the provision of first aid. Furthermore, this technology could be used to assist elderly people with impaired mobility in their daily living [1,2,3]. One of the major difficulties in such implementation is a versatile end effector capable of carrying out tasks both delicate and heavy-duty, such as moving unconscious casualty into recovery position (i.e. arranging limbs and torso into desirable positions for further medical assistance).

2. Segmentation and geometric model of human body

The human body can be represented by a set of rigid

bodies of simple geometric shapes. Whitsett (1962) refined the geometrical model of human body segments developed by Simons and Gardner which comprised 14 body segments. Whitsett's model consisted of spheres, ellipsoids, cylinders, frustums of cones and rectangular parallelepipeds and allowed estimation of the mass distribution, centre of mass, mass moments of inertia and mobility of the human body. The model was primarily based on the body segment data from Dempster, 1955 and the regression equations from Barter, 1957. Hanavan (1964) proposed another geometric model similar to Whitsett's, consisting of 15 segments as the torso was modelled as two segments. In Hanavan's geometric model, the human body segments are a series of cylinders, frusta and ellipsoids [4,8].

Throughout the current stage of this research Hanavan's geometric model has been adopted as reference for determining the shape of the human body segments. Hanavan's segmentation system of the human body consists of 16 segments, namely head, two forearms, two upper arms, two palms, fingers (all 10 fingers count as one segment), upper torso, lower torso, two thighs, two shanks, and two feet. The geometric shapes of these segments were concluded from data collected during an evaluation conducted on a sample group containing 4 females and 8 males aged between 18 and 50.

Hanavan (1964) developed a geometric model of the human body, modelling segments as a series of cylinders, frusta and ellipsoids. The human body can be represented by a set of rigid bodies of simple geometric shape and uniform density. The limbs move about fixed pivot points when the body changes position. The head is a right circular ellipsoid of revolution, the upper and lower torsos are right elliptical cylinders, the hands are rectangular parallelepiped, and the remaining segments are the frusta of right circular cones.

The hand and head were defined as ellipsoids of revolution while other body segments were defined as variations of elliptical solid such as truncated circular cones (forearm, upper arm and shank), elliptical columns (upper torso and lower torso), and elliptical solids with one face being circular (foot and thigh) [5,6,7,8].

Three geometric groups commonly used in defining the body segments are cylindrical, ellipsoid and rectangular parallelepiped. The cylindrical group consists of elliptical cylinder, elliptical solid, truncated circular cone, elliptical cylinder and circular cylinder. The members of the ellipsoid group are the semi ellipsoid, sphere and ellipsoid of revolution. The rectangular parallelepiped group consists of parametric solid, prismatic and stadium solid [8,9]. These three groups and their constituent members are presented in table 1. The geometric shapes of human body segments were concluded from data collected during an evaluation conducted on a single human's body parts. The geometric shapes of various body segments and the count of these segments are presented in table 2.

Table 1
Geometric groups and group members

Geometric group	Member of geometric shapes
Cylindrical (C)	Elliptic Cylinder Elliptical Solid Truncated Circular Cone Elliptical Cylinder Circular Cylinder
Ellipsoid (E)	Semi Ellipsoid Sphere Ellipsoid of Revolution
Rectangular Parallelepiped (R)	Parametric Solid Prismatic Stadium Solid

3. Anthropometry

Anthropometry refers to the measurement of living human individuals for the purposes of understanding human physical variation. Anthropometric measurements are usually done through the use of bony landmarks to which heights, breadths, depths, distances, circumferences, and curvatures are measured. For engineers, the relationships of these dimensions to skeletal "link-joint" systems are of importance, so that the human body can be placed in various positions relative to workstations and equipment [9].

There are four different ways to measure the stature. In general, the subject standing naturally upright, but not stretched; standing freely but stretched to maximum height; leaning against a wall with the back flattened and stretched to maximum height; and lying supine. The differences between the measures when the subject stretches to maximum height either standing freely or leaning are within 2cm. Lying supine results in a taller measure, but standing slumped reduces stature by several centimetres.

The body measurements are usually defined by the two end-points of the distance measured, such as elbow-to –fingertip. The stature starts at the floor on which the subject stands and extends to the highest point on the skull. There are seven different terms which are used in anthropometry they are height, breadth, depth, distance, curvature, circumference, and reach. Brief descriptions of these terms used in anthropometry are as followed:

- Height is a straight-line, point to point vertical measurement.
- Breadth is a straight-line, point to point horizontal measurement running across the body or a segment.
- Depth is a straight-line, point to point measurement between landmarks on the body.
- Distance is a straight-line, point to point measurement between landmarks on the body.
- Curvature is a point to point measurement following a contour; this measurement is neither closed nor usually circular.
- Circumference is a closed measurement that follows a body contour; hence this measurement usually is not circular.
- Reach is a point to point measurement following the long axis of the arm and leg.

For the purpose of this study, the age of people whose body dimensions measured were limited to a range of ages between 18 years to 65 years old. Data included in this document have been taken reference from 'National Aeronautics and Space Administration' [11], research project 'DINBelg 2005', 'National Anthropometry survey of Fire-fighters, UK and United States'[12], and Advances in Human Factors/Ergonomics 4, K.H.E. Kroemer, H.J. Kroemer, and Kroemer-Elbert [10]. These documents provided comprehensive human body dimensional specifications which were used in the research. Anthropometric dimensional data from various ethnic backgrounds of both male and female are given in table 2. Dimension values are given in millimetre (mm) scale.

4. Hardness of human body segments

This project involves avoiding physical damages to human is vital to human-robot interaction. For security and safety reasons it is necessary to determine the hardness of the human body. In this study, hardness of the human body is divided into three categories namely soft, medium and hard. Hard parts are more sensitive to pressure and pain. Medium and soft parts are less sensitive to pressure and pain. Same amount of pressure is likely to cause more pain for harder body part and less pain for soft and medium parts. It is important to develop the end effector with measurement sensor (tactile/force sensor) to detect maximum allowable force to avoid any pain and damage while the gripping action take place.

The human has an estimated average weight between 60 and 100 Kg. The end effectors must be able to securely hold the load in any position. The gripping surface must have a large area in contact with the human body segments to reduce the stress caused by manipulator.

The design methodology will be treated as optimization problem to search for the optimal material arrangement which will be suitable for the purpose of handling.

The basic design requirement of the end effector is that there must be physical connections between input actuator, output end effectors and ground supports to allow the transmission of force and motion. The hardness data of the human body segment are described in table 2.

Table 2
Summary of human body segments geometric shape, size and hardness

Segment and shape description	Dimension	Min. value	Max. value	Part Count	Hardness
Head (Ellipsoid)	Head length (H_L)	169	211	1	Hard
	Head breadth (H_B)	130	168		
	Head height (H_H)	190	261		
Upper Arm (Cylindrical)	Shoulder to elbow length (L)	272	418	2	Medium
	Upper arm diameter (ØD)	69.7	129.87		
Forearm (Cylindrical)	Elbow to wrist length	215	297	2	Medium
	Forearm diameter (ØD)	37.87	104.09		
	Wrist diameter (ØD)	43.61	61.43		
Palm (Parametric solid)	Palm length (L)	87	119	2	Medium
	Palm breadth maximum (B)	69	107		
	Thickness (T)	19	36		
Finger (Cylindrical) 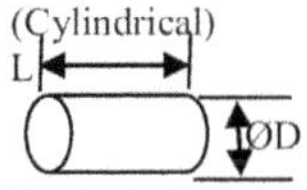	Finger length (L)	44	94	1	Hard
	Diameter (ØD)	16	25		
Upper Torso (Cylindrical)	Diameter (ØD)	218	386.75	1	Medium
	Upper torso height (H)	317	516		
Lower Torso (Cylindrical) 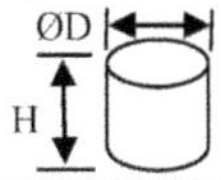	Length (waist to crotch) (H)	210	273	1	Soft
	Diameter (ØD)	165	396		
Thigh(Cylindrical) 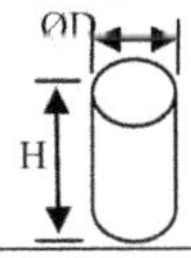	Thigh height (H)	324	397	2	Medium
	Diameter (ØD)	133.05	234.28		

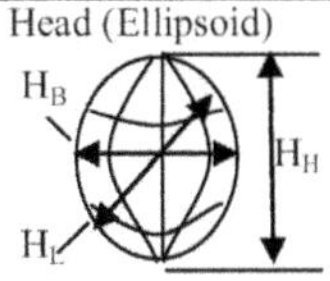

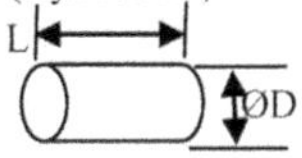

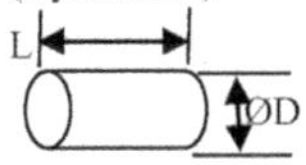

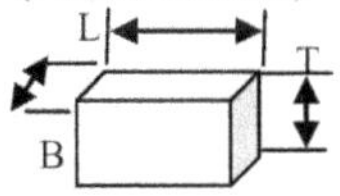

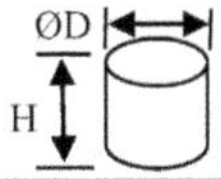

Table 2
Summary of human body segments geometric shape, size and hardness (continued)

Segment and shape description	Dimension	Min. value	Max. value	Part Count	Hardness
Shank(Cylindrical)	Shank height (H)	364	451	2	Medium
	Diameter (ØD)	96.45	136.55		
Foot (Parametric Solid)	Foot length(L)	209	293	2	Medium
	Foot breadth(B)	79	111		

From the information presented above we can conclude the distribution of different geometric shapes and hardness levels amongst human body segments, presented in table 3.

Table 3
Percentage of sizes, shapes, and hardness

Shape		Cylindrical	Ellipsoid	Rectangular Parallelepiped
Count		88%	4%	8%
Size[a] (in mm)	16 - 36	40%	-	8%
	36 - 261	40%	4%	-
	261 - 396	8%	-	-
Total		88%	4%	8%
Hardness	Hard	40%	4%	8%
	Medium	44%	-	-
	Soft	4%	-	
Total		88%	4%	8%

[a] For any body segments other than head, only diameter/breadth/thickness were considered since it is not likely that an end effector will be needed to cover the full length/height of these body segments during operation.

5. Pain threshold

Another important factor in physical human-robot interaction, especially human body manipulation, would be the force and pressure directly applied on human body. Take the 'recovery position' in first aid procedures for example; the most likely and frequent contact areas where gripping action would take place on the human body are shoulder, upper arm, forearm, hand, knee and shank. It is imperative that the gripping force does not exceed the pain threshold, which is the maximum pressure that can be applied to human body before causing painful sensation; and in a less favourable situation where the pain threshold had to be reached in order to carry out essential tasks, it is desired that the pressure can be kept below pain tolerance, which is the maximum pressure that can be applied to a human body before the pain becomes intolerable.

The pressure being applied to human body surface or human interface pressure as it is often referred to, is considered to be an important parameter in comfort evaluation. Ideally, the contact surface between end effector and the human body should be kept as large as possible in order to minimize pressure. Pressure at the human interface is unavoidable for various physical interactions; however it is also dangerous since too much of it may cause discomfort, pain, damage and injury and discomfort.

Human interface pressure is an important consideration in the development of robotic applications. It is essential to understand the pressure patterns that are appropriate of the human body which is the pattern that reduce discomfort or enhance comfort. The understanding of pressure pattern can make product design and usage very safe and satisfying.

Pain or discomfort occurs when the special nerve endings, nociceptors, detect an unpleasant stimulus. Human body consists of millions of nociceptors in skin, muscles, joints, bones and internal organs. These nociceptors use nerve impulse to relay pain messages to networks of nearby nerve cells. The detected impulses are transmitted instantaneously from cell to cell and are facilitated by neurotransmitters. Transmitted messages travel along nerve system to the spinal cord and central nervous system, brain. There are some predictions that the pain signals must reach a threshold before they are relay.

As aforementioned, there are two important pressure threshold values, namely the pain threshold and pain tolerance. Here are their more precise definitions. The term pain threshold is defined as the minimum pressure that induces pain and involves measurement of stimulus intensity, whereas pain tolerance is the maximum pressure a person can tolerate without excessive effort. In other words the pain threshold is the point at which stimulus is reported by a person as pain and the pain tolerance is the amount of pain a person can withstand before breaking down either emotionally or physically [13,14,15,16,17,18].

In this study, it is decided to use pain threshold as a guideline pressure limit. This means avoiding causing pain to human at all. A number of investigations have been done to study the pain response to pressure applied onto human limbs. In human body, some regions with a thicker layer of soft tissue had lower pain thresholds than those with a thinner tissue layer. The patellar tendon and distal end of the fibula were the best and the worst pressure tolerant regions. The pain threshold could be age dependent. One of the methods researchers have applied to find pain threshold was to apply pressure to the test regions through a circular, flat ended indenter of 10mm diameter connected to a force transducer until the subjects claimed not being able to stand the pressure. In some methods indenters of 5mm and 13mm diameters were used. The peak threshold pressure for 10mm diameter indenter is 0.81 MPa [17].

Given the pain threshold values, and the definition of pressure (Eq. 1)

$$p = \Delta F / \Delta A \qquad (1)$$

Where:
p is the pressure (in Pascal)
F is the normal force (in Newton)
A is area (in Meter square)

The maximum force which can be safely applied to human is now only decided by the size of contact area between the gripper/end effector and human body. The introduction of a second end effector, which essentially increases contact area by supporting the segment being handled, will consequently decrease the force required to perform these handling tasks and thus lessen the possibility of inflicting pain.

6. Conclusion

After analysing all the human body shapes and sizes, it became apparent that most likely and frequent contact areas where gripping action would take on the human body segments are of cylindrical shapes and the sizes varied from 16mm diameter to 235mm in diameter/breadth/thickness. From these results, it can be concluded that an end effector capable of handling (especially gripping) cylindrical-shaped objects within a diameter range of 16mm to 235mm will be sufficient to perform the majority of human body manipulation, considering the possible deformation (indentation mostly) of soft tissue, this number could be further reduced. As far as pain threshold is concerned, in order to maintain a certain degree of dexterity which is essential to complex and possibly delicate human-robot interaction, the pressure reduction problem cannot be simply remedied by increasing the contact surface between the end effector and human body. Other possible options, such as the introduction of additional end effector(s) and the application of deformable material layers on contact surface, remains to be evaluated in-depth in the future.

References

[1] Paolo Fiorini, Khaled Ali and Homayoun Seraji, 1997 IEEE 'Health Care Robotics: A Progress Report', Proceedings of the 1997 IEEE International Conference on Robotics and Automation, Albuquerque, New Mexico- April 1997

[2] Q.Meng and M H. Lee, 2001 IEEE 'Behavior-Based Assistive Robotics for the home' Center for intelligent systems, Department of Computer Science, University of Wales, Aberystwyth, Ceredigion, SY23 3DB, UK

[3] Dipl.-Ing. C. Schaeffer, Dipl.-Ing T. May, 1999'Care-O-bot: A System for Assisting Elderly or Disabled Persons in Home Environments'. Institut Produktionstechnik and Automatisierung (IPA) Nobelstrabe 12, 70569 Stuttgart, Germany

[4] Jorgen Bjornstrup, October 1995, 'Estimation of Human Body Segment Parameters, History Background, the history of human body segment parameter estimation'. ISSN 0906-6233, Internat Tec-Report Not submitted anywhere

[5] Hanavan, E.P., 1964, 'A Mathematical Model of the Human Body' (AMRL-TR-64-102, AD-608-463) (Ohio: Aerospace Medical Research Laboratories, Wright Patterson Air Force Base)

[6] Dempster, 1955, 'Space requirements of the seated operator' (WADC-55-159, AD-087-892) (Ohio: Aerospace Medical Research Laboratories, Wright Patterson Air Force Base)

[7] I. Leandry, C. Breque, P. Lacouture and V. Valle, "3D shape reconstruction from relief calculation: application to human body". Axe PEM/PPRIME, UPR 3346, CNRS, Universite' de Poitiers, ENSMA, BP 30179 86962 Futuroscope Chasseneuil Cedex, France. EPJ Web of conference 6, 12003 (2010)

[8] Ergonomic Models of Anthropometry, Human Biomechanics and operator-Equipment Interfaces: Proceedings of a Workshop (1988)

[9] Young Hoo Kwon, 1998, ' Modified Hanavan Model'

[10] Hanavan, Technical Report, 64-102 (2008)

[11] Advances in Human Factors/Ergonomics 4, K.H.E. Kroemer, H.J. Kroemer, and Kroemer-Elbert.

[12] National Aeronautics and Space Administration Volume 1, Section 3 Anthropometry and Biomechanics,May2008 (http://msis.jsc.nasa.gov/sections/section03.htm)

[13] Dr Mandy Stirling on hehalf of CACFOA, National Anthropometry Survey of Firefighters, Designing for safety, performance and comfort

[14] Wilson C.C, Lee, Ming Zhang, Arthur F.T. Mak, How well can Different Regions of Residual Limb tolerate pressure, Jockey Club Rehabilitation Engineering Center, the Hong Kong Polytechnic University, Hong Kong China

[15] Wilson C.C, Lee, Ming Zhang, Arthur F.T. Mak, 2005, Regional Differences in Pain Threshold and Tolerance of the Transtibial Residual Limb: Including the Effects of Age and Interface Material. , Jockey Club Rehabilitation Engineering Center, the Hong Kong Polytechnic University, Hong Kong China

[16] David H Jones, Robert D Kilgour and Alain S Comtois, 2007, Test Retest Reliability of Pressure Pain Threshold Measurements of the Upper Limb and Torso in Young Healthy Women. Department of Exercise Science, Concordia University, Montreal, The journal of Pain, vol 8, No 8 (August 2007)

[17] Waldemar Karwowski 2006, International Encyclopedia of Ergonomics and Human Factors, Second Edition – 3 Volume Set, ISNBN:041530430X, CRC Press, Inc. Boca Raton, FL, USA

[18] "definition of Pain tolerance in the Medical dictionary". Mosby's Dental Dictionary, 2nd edition. Elsevier, Inc. 2008. Retrieved 6 October 2010.

Innovative Production Machines and Systems
D.T. Pham, E.E. Eldukhri and A.J. Soroka (eds)

An Integrated Manufacturing Approach Towards Improving Business Performance in Manufacturing SMEs

A.J Thomas [a], R A Barton [b]

a. *Faculty of Engineering, Coleg Sir Gar, South Wales, SA15 4DN*
b. *Orthoclinical Diagnotics GB, Pencoed, Bridgend, South Wales*

Abstract

For companies to remain competitive in an increasingly demanding global market, they must discover innovative ways in which to work effectively whilst ensuring that new markets are penetrated and increased business performance is achieved. Improving business performance not only requires companies to provide competitively priced products but it must develop innovative strategies to enhance market share through new product development. Therefore an integrated manufacturing strategy must be developed in order to ensure that the factory of the future is able to meet this new demand. This paper proposes a integrated manufacturing approach which merges the manufacturing efficiencies achieved through Lean and Agility, with the need to utilize new technologies, respond to changing market demands and employ effective marketing and product innovation strategies.

1 Introduction

This paper will explore and analyse the methods and potential benefits for a specialist manufacturing company seeking to leverage learnings gained in the fields of Business Process Improvement (BPI) and new technology implementation in order to achieve increased manufacturing performance. The following introductory sections will discuss the concepts of BPI, Lean and new technology implementation in the context of the specialist manufacturing company chosen for this case study.

The interactions between these aspects are then considered in order to propose an Integrated Manufacturing Framework (IMF) that is able to react quickly to market conditions and changes in customer demand without significantly increasing operating cost. Therefore, it can be surmised that the aim of IMF is to understand and control the sales and costs drivers, as well as considering the interaction between them. The paper then goes on to consider the implementation of such a model. The mechanism for delivery must overcome technical and methodological barriers and enable effective Business Process Improvement by means of identification of the most appropriate aspects of the manufacturing practices to be employed in the prescribed business environment.

1.1 Business and Manufacturing Performance

In basic terms, successful businesses achieve growth through ensuring that their operational costs are driven as low as possible whilst developing their sales volumes through innovative product development and capturing new high value markets [1]. Any company that runs with high manufacturing costs will always be in danger of running at a loss (or certainly at reduced profit margins). This can also occur if the sales volumes anticipated by a company fail to increase or even decline whilst operational costs continue to rise. An ideal company would have a successful plan in which to reduce waste and costs whilst ensuring that increased sales are achieved through the development of new and innovative products linked to an effective sales and marketing strategy.

If tight control of a company's operating costs and sales growth is not maintained then there is an imminent danger that increased operational costs accompanied by a fall in total sales could render the company insolvent quickly (see Figure 1). Alternatively, when product sales and operating costs are moving upwards then the company may be able to achieve sustainability although increased business performance is less likely since the profit margins will in essence, remain constant, as shown in Figure 2. This is acceptable but not ideal in competitive markets. A company looking to increase business performance and hence growth would be ideally engineer its business to ensure that its total costs and sales output trajectories diverge thus creating an area for growth, (see Figure 3). In realistic terms however, a company is likely to achieve optimum business performance whereby sales output is maintained to the desired trajectory whilst the organisation systematically removes waste from its product/process and business in order to achieve increased business performance.

Figure 1, Sales and costs moving in opposite directions, negative profit margin

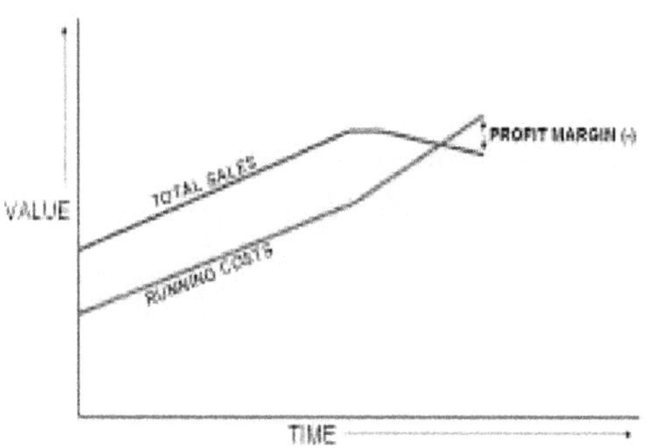

Figure 2, Sales and costs rising steadily to give a constant profit margin, no growth

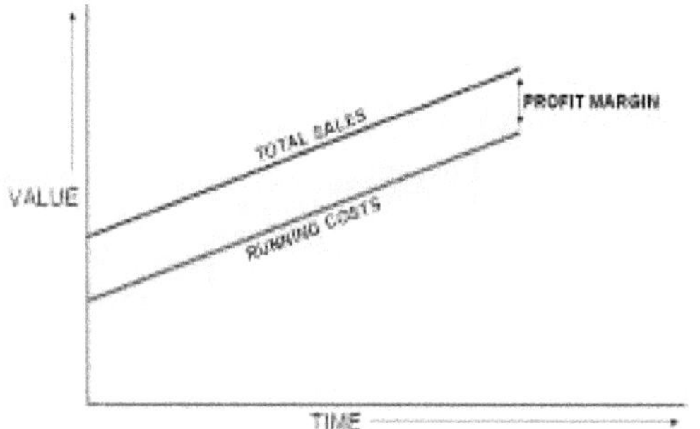

Figure 3, Sales and costs moving in opposite directions, increasing profit margin

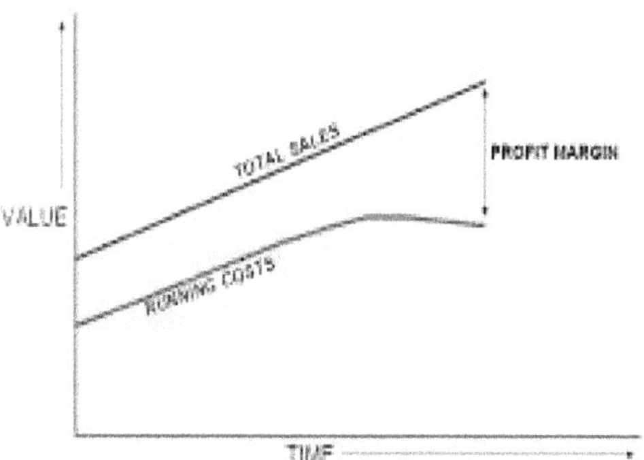

By drilling down into the cost profiles and business drivers of a company, it is clear that for a company to systematically grow through increased sales (however, acquisition strategies and other forms of business growth and enhanced business performance are equally relevant and effective), it must achieve increased customer satisfaction as well as improved internal business performance. Therefore, a business must ensure that the standard Quality, Cost, Delivery, and Flexibility (QCDF) are features are present and developed correctly to achieve growth. This offers a clear structure for continuous improvement, raising levels of customer satisfaction and greatly improving the management of production. QCDF is not, however, sector-specific. These key measures can be applied to improve production performance throughout manufacturing industry from the auto-industry to semi-conductors, electronics, aerospace, telecommunications, textiles, building products, food and chemicals processing.

QCD is often thought of as an acronym in its own right, but for the purposes of this paper the authors have included flexibility 'F' as an added factor. This is particularly relevant in consideration of a integrated strategic approach since the pursuit for increased Quality, minimal Cost and cost effective yet reliable Delivery naturally limits the company's ability to provide Flexibility in a cost effective way. This can be exemplified when considering the development of a product which has a new capability or functionality. Whilst this product is likely to enhance sales demand it is likely to increase the requirement for greater manufacturing flexibility through changing the manufacturing systems reconfiguration and coping with demand fluctuations as the product is introduced [2]. Each of the four elements are linked to different aspects of the business, and depending upon the requirements of the process or the market place, it is essential that an effective approach towards achieving the correct balance between the elements is achieved in order to maintain business performance.

1.2 Business Process Improvement Strategies

Over the years, a number of business improvement strategies, methodologies, tools and techniques have been proposed and developed aimed at enhancing the productivity of organisations. Total Quality Management, Business Process Re-engineering, Just-In-Time and "Lean", "Agility" and Six Sigma are just a few examples. Despite the reported success of these initiatives at increasing profits and market share for manufacturers, there are still a significant number of companies battling to keep their heads above the turbulent waters [3]. The issue for many organisations is that these proposed solutions, although purporting economic benefits in the short term, fail as long-term business improvement strategies since they rarely become the explicit or even implicit focus of the change initiative in companies [4]. The result of failing to embed the strategy correctly into these companies often is that any improvements are lost as initiatives are abandoned and shop floor employees regress to previous methods of working [5].

"Lean" is a set of management principles and techniques geared towards eliminating waste in the manufacturing process and increasing the flow of activities that, from the customers' perspective, add value to the product [6], [7]. It represents one of the most popular business improvement models of the current day with almost 50% of UK manufacturers applying "Lean" principles in at least part of their plant [8]. However, following many accounts outlining the success of "Lean", concerns about its ability to deliver realistic improvements in manufacturing competitiveness in organisations have grown over recent years [9]. The success of "Lean" in companies often mirrors the classic change curve - improvements in productivity after an intervention are soon followed by a steady decline to baseline, and sometimes even below baseline levels [10].

The implementation of "Lean" is mainly process-orientated and concentrates on the tools and techniques aimed at reducing waste in the system. Although "Lean" programmes consider strategy and alignment factors in the form of Policy Deployment, this has usually been poorly implemented and rarely sustained [10]. In most cases, there is very limited attention paid to different leadership styles and behaviour of the employees working within the organisational system and the need for cultivating a culture and employee mindset that will welcome and nurture "Lean" behaviours. "Lean" programmes also rarely give any significant consideration to the role of technology, especially Information Technology, in supporting an economically sustainable system. The lack of a integrated approach to "Lean" implementation suggests that "Lean" is one of the many business improvement models that are piecemeal driven and as such is open to implementation in a variety of different ways with the resulting success being equally varied. In essence, "Lean" focuses on only a small number of conditions necessary for increased manufacturing competitiveness.

Unlike previous research which tends to adopt a static approach to increased manufacturing competitiveness, this paper considers the need to move away from the traditional "Lean" approach and suggests and enhanced development of the Integrated Manufacturing Framework as the approach for future development. The paper suggests that the IMF links a number of BPI tools and techniques into a cohesive framework to deliver a 'competitiveness' solution for industry.

A review of the literature suggests that most organisations only concentrate upon one or two of these tools / techniques as a means of delivering improved manufacturing competitiveness [11], [12] but lack a framework for a more integerated and dynamic perspective on sustaining change. This paper attempts to address this area and proposes an integrated manufacturing approach which assists companies to decide how it might achieve enhances business performance and increased manufacturing competitiveness by taking account of each factor and the complex interaction between them. The paper closes with the development of an integrated manufacturing approach.

2 Proposing an Integrated Manufacturing Approach

The proposed framework is an integrated approach to the use of key business paradigms to achieve distinct and significant levels of production operations performance that are unique to each company. There are a number of production improvement methodologies which outline their benefits and abilities to provide sustainable growth and increased competitiveness to companies. However, many of these approaches either relate specifically to certain industrial sectors such as the automotive sector [13] or, focus primarily on the manufacturing performance aspect of competitiveness rather than also considering the roles of sales and marketing and their associated performance and integration into the wider manufacturing system to form a dual strategy to achieve increased business performance [14].

The Integrated Manufacturing Framework produced in Figure 4 shows a mechanism by which the operational context of a production system is able to change depending upon the market conditions faced by a company. The figure shows two clear pathways indicating on the top route a focus upon New Product development, Voice of the Customer analysis and a range of tools and techniques aimed at meeting future customer demands.

The lower route highlights a range of internal development tools aimed at reducing production costs are returning improved operational excellence to a company. It is suggested therefore that through this framework that a company will adjust rapidly to one or both of these routes as demand necessitates. Therefore, when a company is aiming to attract new contracts or repeat work then it is advisable to mobilise activities towards the upper route through ensuring that the company responds to their customers requirements. Likewise, if the customer is an established one and a long terms contract has been awarded then it is wise for a company to mobilise efforts towards operational excellence and cost reduction strategies which eventually will be used to harness further work from the customer. In many cases though, a balance between the two routes is advised with efforts being tapped between the two routes for optimum operational effectiveness.

This is where the marketing and sales departments have a crucial role to play in providing the feedback to the operations managers to decide upon which strategy to employ for the greater good of the company

Figure 4 An Operational Model of the Integrated Manufacturing Framework

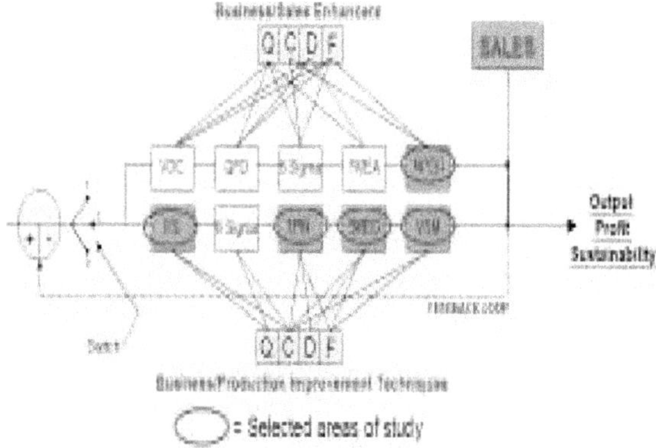

Case Study

The case company is an established pharmaceutical company. The company has a highly robust new product development and introduction system (NPD/I) which ensures products are delivered effectively into the marketplace whilst adhering to strict drugs agency requirements.

The company has also developed new Lean manufacturing methods and systems and has over time become increasingly proficient in various waste reduction strategies. This has enabled the company to achieve modest improvements in manufacturing performance.

The company however feels that the application and development of Lean often plays a secondary role to the NPD/I process and as a result the Lean implementation projects are often undertaken in a piecemeal way with no clear strategic approach being applied.

As a result of this approach, Lean initiatives are often seen to impede the NPD/I process in that it removes key staff from the NPD process with little or no major improvement being seen in manufacturing performance as a result of the lean initiatives.

In order to develop a more coherent approach towards Lean and NPD, the company decided to implement the IMF as a means of improving manufacturing performance. The initial stage of the company's work involved mapping each product line within the company and identifying where the products were on their product life cycle. In each case an analysis was undertaken as to whether for each product, the product was in its introduction stage, growth, maturity or decline stage. This analysis was undertaken from sales information and the due date for project completion with each customer.

Following this analysis, each product was than mapped against the IMF routes. Therefore, for products who were in their decline stage / introduction stage, the company would move towards increasing activity towards the NPD/I route in an attempt towards either developing new products to replace the products in the decline stage or to improve the product features of the products in the introduction stage.

Likewise, for products in their growth and maturity stages of the Product life Cycle, the company concentrated upon activities in the Lean improvement route in an attempt to improve manufacturing performance on these products and returning greater bottom line improvements and increased profit margins.

Where products were deemed to be on the cusp of moving between maturity / decline and introduction / growth, the company mobilized their workforce towards developing activities along the both routes on the IMF. For instance, for a product line moving from the maturity stage to the decline stage on the PLC, the company would initially allocate resources to concentrate upon the Lean cycle. As the product moves closer to the end of contract stage, the company would start to move its resources away from manufacturing /process improvement towards undertaking more Voice of the Customer and client analysis activities in order to start to develop new products which will eventually replace the existing contract provision

The opportunities this approach offered the company were many and included:

- Improved analysis of various product lines leading to applying the most appropriate strategic approach to a given product based upon its position on its PLC.
- More effective allocation and use of human and technical resources
- Improved clarity for project management purposes

3 Conclusions

The authors of this paper have proposed a possible method to achieve a balanced approach to the enhancement of business performance and growth. The proposed plan involves both business enhancers, and productivity improvement techniques. Some of these systems can be implemented with low investment where as others need a strong financial backing, which may be approved by cost-benefit analysis derived by the model. Whichever systems are chosen, they must have an objective that is followed from start to finish, with strong focus on particular aspects of customer satisfaction.

The key to success will be the management of implementation of whichever systems is chosen. As stated earlier in this paper, a company can find itself in a worse position by changing systems but missing the target by a small but critical margin. Lean manufacturing has many advantages and possibilities within any industry, but it will only be successful if implemented correctly. It is essential that the individuals driving the changes have the sufficient knowledge to train and educate the whole workforce. Lean is a change of culture within a company, and although many savings are made from improvements at a shop floor level it needs to be supported from a higher level of management. Another aspect to consider is that not all companies will be compatible with an entirely lean process. Some companies will need to maintain a degree of flexibility and agility.

Hence the model shown in figure 4 is proposed as a tool to identify the appropriate scope and scale of implementation of lean, agility and control techniques to improve customer satisfaction, hence increase sales, meet the demands of new markets and still achieve reduction of operating cost.

3.1 Further work

Further work will be published as the model is promoted to senior management for a more integrated use and subsequent results are seen in the company. Both Cardiff and Newport universities are then in a strong position to further roll-out the model to a wider audience, potentially through KTP projects, and add more detail to the tool-set and techniques generically appropriate to the businesses studied, with particular attention to the interaction of activities across the three streams of cost reduction, sales enhancement and control.

References

[1] Olsen, E.G., (2006) "Not bt technology Alone: sustaining winning strategies", Journal of Business strategy, 27, 4, pp33-42

[2] Wiendahla, H.-P., ElMaraghyb H.A., Nyhuisa, P., Zähc, M.F., Wiendahld, H.-H., Duffiee N., and Briekea M., (2007) "Changeable Manufacturing - Classification, Design and Operation", CIRP Annals - Manufacturing Technology56,2,pp 783-809

[3] Hines, P., 2004, Manufacturing in London: Where Should Development Effort be Focused, Theme Paper No. 5, London Development Agency, London.

[4] Bateman, N., 2001, Sustainability: A Guide to Process Improvement, Lean Enterprise Research Centre, Cardiff University, Cardiff.

[5] Thomas A. J., Rowlands, H., Byard, P., Rowland-Jones, R. 2009, """ Lean" Six Sigma: An Integrated Strategy for Manufacturing Sustainability", International Journal of Six Sigma and Competitive Advantage, ISSN 1479-2494, February

[6] Rich, N., 2001, Turning Japanese?, PhD. Thesis, Cardiff Business School, University of Wales, Cardiff, 2001.

[7] Womack, J., Jones, D., (1996) "Lean Thinking: Banish Waste and Create Wealth in your corporation", Simon and Schuster, New York, NY.

[8] Engineering Employers Federation, 2001, Catching Up with Uncle Sam: The EEF Final Report on US and UK Manufacturing Productivity, EEF, London.

[9] www.Lean.org.uk Accessed 21/07/2010

[10] Hines, P., Francis, M., Found, P., (2006) "Towards lean product lifecycle management", Journal of Manufacturing Technology Management, 17, 7, pp866-887.

[11] Small, M.H 2006, "Justifying investment in advanced manufacturing technology: a portfolio analysis", Industrial Management and Data Systems, 106, 4, pp 485-508

[12] Liyanage J. P., 2007,"Operations and maintenance performance in production and manufacturing assets: The sustainability perspective", Journal of Manufacturing Technology Management, 18, 3, pp 304-314

[13] Herron C, Braiden P.M. 2006, "A methodology for developing sustainable quantifiable productivity improvement in manufacturing companies" International Journal of Production Economics. Vol 104, Iss 1, pp 143-153

[14] Childerhouse, P., Towill, D. R., 2004,"Reducing Uncertainty in European Supply Chains", Journal of Manufacturing Technology Management, 15, 7, pp585-598.

Innovative Production Machines and Systems
D.T. Pham, E.E. Eldukhri and A.J. Soroka (eds)

Formulation of a Themed Response to Diagnosed System Complexity

Richard Barton [a] & Andrew Thomas [b]

[a] *Newport Business School, Newport University, Allt-yr-yn Campus, Allt-yr-yn Avenue, Newport, NP20 5DA*

[b] *Faculty of Engineering, Coleg Sir Gar, South Wales, SA15 4DN*

Abstract

As throughput and productivity become more critical to ensure the UK Manufacturing sector remains cost-competitive in the global marketplace, simplified material flow has become a key theme for business improvement. Aspects of supply, demand and internal process chain have associated contributions to simplified flow, hence there are associated improvement / management practices dedicated to increasing throughput in these functional areas. However, what is not always clear is the route to progress across the board. To achieve simplicity, how does the company frame improvement activities? To identify particular gaps and areas for improvement, the Quickscan Audit methodology (QSAM) has been developed to pick up on weaknesses, but it has been seen that there are few common recommendation for improvement between quickscanned companies. This paper outlines how one company has triangulated the QSAM uncertainty assessment with simplicity requirements and is developing a business complexity theme to close the gap.

Keywords; Simplicity, complexity, six sigma, lean, quickscan

1 Introduction

The increase in variety, customization and number of tiers within the supply chain has increased the number of activities requiring greater interaction amongst different supply chain players which has introduced different forms of complexity within organizations and supply chains. Today, it is not the sole organizations competing, but the supply chain as a whole and the performance of every part of the supply chain affects the performance of the others and ultimately the customer (Perona & Miragliotta, [1]). Hence, the interactions between functions, and interfaces between both internal and external stakeholders are becoming more critical as process times are optimised and sources of waste

2 QuickScan Audit Methodology (QSAM)

QSAM is considered to be a well-rounded methodology, with the apparent benefits of bringing together the best features of horizontal surveys (i.e. statistical significance) and also vertical case studies (i.e. good understanding of the real world scenario). While, at the same time, QSAM avoids the worst pitfalls of the traditional research approaches by utilising a range of data gathering techniques and analysis; interview, observation, secondary data and qualitative analysis in order to triangulate the various data and establish conclusions with a high level of confidence in the academic rigour.

In addition, QSAM also endeavors to analyse recorded information in time series format to associate previous and future activities with trends in performance. This is supported by the historical records of other 'quickscanned' companies and value chains which now number close to one hundred.

Taking these factors into account, the QSAM methodology is strong in three levels of the business, relating to the benefits of undertaking this research

1. In the short term, the performance of the company itself must be understood and benchmarked with a viewpoint of reaching a strategic level of improvement, hence internal operations must be well quantified by some detailed analysis, which is relatively well catered for by the process champion being based in the company.
2. In the medium term, it is possible to establish the context of uncertainties throughout the supply chain, and how particular bottleneck activities can be selected and investigated in terms of their disproportionately large impact, hence targeted for improvement of the overall system.
3. activities throughout the supply chain to bring benefit to the host company. This can be compared with the wider data set which is available from previous audits, with a view to emulating best-in-sector performance.

The case study company (referenced Company A for confidentiality reasons) undertook the QSAM study in conjunction with Cardiff University in order to understand areas of potential improvement. Particular attention was paid to the response to customer demand and the de-coupling point and the response to this is described by Jenkins et al [2]. However the wider findings of the study revealed weaknesses around data processing and metrics availability, isolation of specialist knowledge in functional silos and discord between data sources regarding material flow and process performance. The following sections describe the symptoms of systems complexity, the simplicity solution, and go on to discuss the example of Company A who have explored the practical implications of systems complexity reduction to meet the simplicity requirements.

3 Simplicity

The underlying causes of sub-optimal performance, especially around the operational management aspects mentioned above, were referred to in terms of system complexity, but without a collective understanding or true characterisation of what the system complexity entailed either in the generic or this specific case. The first point of reference to be considered was work undertaken by other experienced QSAM practitioners, and the most apparently relevant output included the simplicity rules established by Towill [3], shown below in table 1.

These rules are an excellent demonstration of critical characteristics which, if put in place in a supply chain, would minimise end-to-end lead time and optimise material movement. However, it was difficult to consider the manufacturing system in Company A in these terms without including the existing constraints and regulatory requirements. So the question is... if the simplicity rules are the answer, what are the barriers to overcome and how can the change management process be administered to meet the simplicity requirements without compromising existing capabilities?

The most popular operations management paradigm in recent years, which shares many of the philosophical intentions of the simplicity rules, is lean manufacturing which is intended to focus on customer value, optimise material flow and offer improved customer value by removing waste and usually cost.

Table 1. The 12 Simplicity Rules. Source: Towill [3]

Rule 1	Only make products that can be quickly despatched and invoiced to customers
Rule 2	Only make in one time bucket those components needed for assembly in the next.
Rule 3	Streamline material flow and minimize throughput time
Rule 4	Reduce information lead times via the use of the shortest planning periods
Rule 5	Only take deliveries from suppliers in small batches as and when needed for processing and assembly
Rule 6	Synchronization of time buckets throughout the chain
Rule 7	Form natural clusters of products and designing processes appropriate to each value stream so that the requirements of diverse customer requirements can be best served.
Rule 8	Eliminate all uncertainties in all processes
Rule 9	Develop a structured approach to change. Understand, Document, Simply, Optimize (UDSO)
Rule 10	Highly visible and streamlined information flows
Rule 11	Use proven and robust decision support systems in the management of the supply chain
Rule 12	The operational target of the seamless supply chain needs to be commonly accepted and shared by all members of the change team

4 Lean

At one time lean manufacturing was confined to the automotive industry, but now many lean principles are becoming a standard procedure in a wide range of industries. The reason for this is simple: Where they are implemented with a good performance management system, lean principles (Womack and Jones, [4]) have a proven track record of operational and strategic success, which results in increased value to the customer.

Through tackling the issue of waste reduction, and successfully implementing many of the principles of lean manufacturing, significant improvements can be derived in the areas of:

- Overhead operating costs
- Lead time
- Process queues
- Profit
- Improved working environment
- Speed of defect detection

Ashkenas [5] supports this, stating that 'reducing business complexity drives long-term competitive advantage as by increasing the capacity to address issues fast and efficiently'

However, what is not fully appreciated is the potential downside to lean thinking; that rigidity in the production schedule, a lack of variability in shop floor control and subsequently reduced flexibility in the throughput capacity may serve to reduce agility (not increase it as many lean practitioners have come to suggest in light of this proposition). While the authors do not wish to be caught up in a discussion around terms and semantics, there are a variety of different ways to consider these implications. One generic theme is to consider that a level of complexity/variability/uncertainty may prove to be very healthy in terms of minimising the customer response time and being able to service unexpected but potentially important changes i customer demand; be that volume or product configuration. So, again, what is the degree to which simplicity should be pursued, and what level of complexity is key to maintaining the edge over competition? The model for scoping out the detail behind uncertainty in the QSAM activities is the uncertainty circle.

5 Uncertainty

The uncertainty circle method is based on the control of a company's internal process in responding to the effects of customer demand and in turn the ability of the company's internal processes to place orders on to their suppliers (towill [6]). Uncertainty within the supply system can be significantly affected by the company's own process, supplier interface, demand interface and by poor system controls, and can be represented by [bad] characteristics such as;

- inability to cope with demand patterns
- irregular frequency and quantity of order profiles
- lack of accurate tracking of customer requirements
- players in the system generating 'rogue' orders for price leverage
- wrong decision rules
- stale, noisy or incomplete information.
- erratic, frequent and problematic downtime of its machinery and equipment
- changing customer schedules

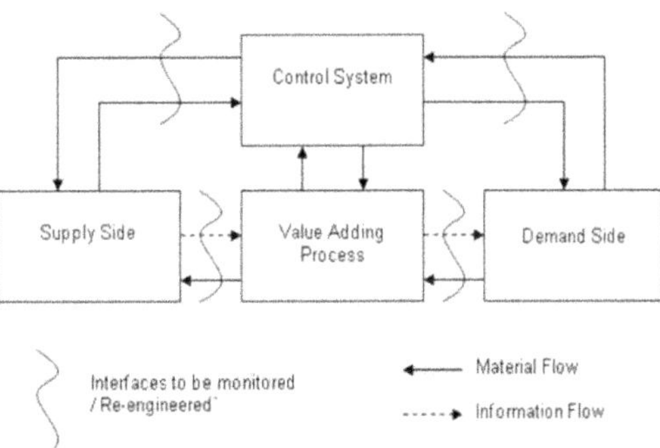

Fig 1. Uncertainty circle principle (Source: Mason-Jones and Towill [3])

However, as discussed in the previous section, some degree of these characteristics could easily be seen as advantageous to an agile and/or responsive company. Wilding [7], argues that complexity has a positive role to play as it encourages organizations to be innovative and grow by restructuring their work patterns. In order to understand the 'breakeven' point at which these characteristics change from good to bad, the case study company, Company A, has taken the approach of characterising all system complexity and defining what is necessary, what is unnecessary but essential and what is purely system driven waste.

6 Complexity

The first stage was to examine the existing literature and information in the public domain for complexity causes, effects and symptoms and collating them into a format against which the company could be audited.

A complex system can be defined as one that made up of a number of different parts interacting in a non-simple way, such that it becomes difficult to predict the system's behaviour without detailed analysis or computation (Bozarth et al [8], Deshmukh et al, [9]). According to Waldrop [10], complexity is a dynamism that makes systems "qualitatively different from static matter like computer-chips or snowflakes which are only complicated". Dynamic and Detailed complexity is displayed by processes, products and relationships in a supply chain (Bozarth, [8]). Another definition explains complexity as a systemic characteristic that integrates key dimensions of the supply chain environment like size, variety, objectives, information, uncertainty and cost (Pham et al, [11]).

Miller and Vollman stated that every plant has two factories; one involved in production and the other in generating transactions using control systems (cited from New,[12]).The latter increases waste and expenses, thus organizations must adopt technologically simpler management methods to improve their efficiency and gain competitive advantage. Gunasekaran [13] further broadens the scope of the area by stressing that the measurement system used must represent a balance between both the financial and non financial measures and the organization's strategic goals for effective decision making and control.

Perona and Miragliotta [1] describe complexity sources and their impact on the performance, in terms of variety in functional silos such as production, engineering and sales processes. This seems superficial at first but proves to be a good framework to break down the higher level complexity definitions.

Wilding [7] emphasized uncertainty and the 'far-from-equilibrium' behaviour within supply chains that causes complexity. Three independent yet interactive factors were identified; deterministic chaos, amplification and parallel interactions. While Milgate [14] identified an alternative take on the three complexity dimensions; uncertainty, technological intricacy and organisational systems. Ashkenas [5] identifies four factors; structural mitosis, product proliferation, process evolution, managerial behaviour.

Vachon and Klassen [15] developed a 2D matrix for supply-chain complexity, describing the technological, managerial and product variables and studied their effect on delivery performance. Technology and information processing are introduced as the two dimensions to complexity. Uncertainty and complicatedness were seen to cause dynamic and static complexities respectively. This proves to be most useful as it goes on to combine many of the thought processes described by the authors listed above, and is shown in the table below.

Table 1. 2D Matrix for supply chain complexity (Source: Vachon and Klassen, [15])

TECHNOLOGY

INFORMATION PROCESSING

	Process/Product Structure	Management System Infrastructure
Complicatedness	•Production skills •No. of sub processes •No. of parts/components •Level of interaction between parts •Level of process decomposability	•Customization and variety •Supply network (extent) •No. of echelons •Geographical span of suppliers/customers
Uncertainty	•Production capability •Supplier capabilities •Throuput and setup time variations	•Production scheduling changes •Late supply deliveries •Demand side volatility

With these outlines, descriptions and descriptors in mind, Company A undertook voice of the customer style interviews and data gathering discussions to support the findings of the QSAM and populate a more detailed cause and effect diagram. Ironically the result was far from simple and too expansive to show here, but established four main themes shown in figure 2 below.

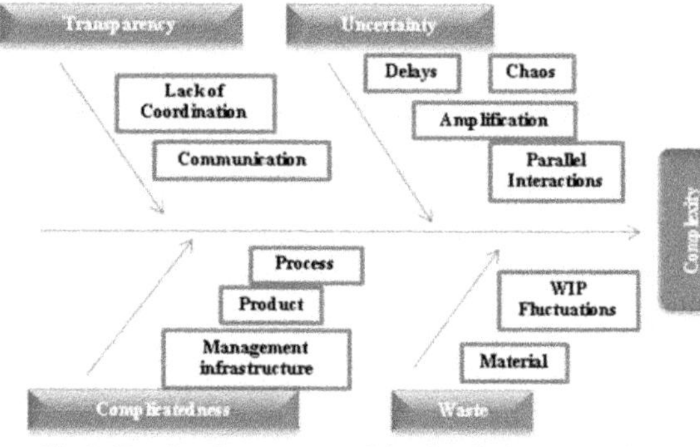

Fig 2. Top level summary of the detailed cause and effect determined by Company A.

In terms of practical aspects, one significant compromise, contrary to the recommendation of Gunasekaran [13] was the philosophical approach to be taken in terms of common understanding of complexity. In Company A, the tangible manifestation of complexity was considered to be contribution to fixed overhead of the particular business function in question, i.e. cost.

Table 2.
Reduction and Management levers for elements of complexity.

	Complexity Cause	Control lever	Complexity cause	Control Lever	
UNCERTAINTY	Delays	•Flexibility •Pricing Models for Suppliers & Shippers •JIT	Product	•Info-tool for component reutilization •Quality Circles	COMPLICATEDNESS
UNCERTAINTY	Chaos	•Control System Flexibility •Risk Assessments •Lateral Managerial Relationships	Process	•Automation (Simulation models) •Info system for PP&C •Outsourced Production •Operations Flexibility •Process Mapping •Streamlining •Process Re-engineering •Six Sigma KANBAN DMAIC	COMPLICATEDNESS
UNCERTAINTY	Amplification	•Capacity Management •Hybrid (SIS) Systems •FCS systems	Regulations	•Vendor Rating System	COMPLICATEDNESS
UNCERTAINTY	Parallel Interactions	•Flexibility	Management Infrastructure	•Outsourced Delivery & Warehousing •Balanced Scorecard •Strategy Maps •Assessment of Management Actions •Use of Direct Display methods •Random Surveys	COMPLICATEDNESS
TRANSPARENCY	Communication	•Supply Chain Relationships •Vendor 'Open-door policy' •Integrated Info-systems •Quality Circles •'Work-Out' •Pricing Models •Strategy maps	Material	•Lean	WASTE
TRANSPARENCY	Coordination	•Team-work (work centres) •Cross-training •Cross-functional Interactions •Pricing models •Supply chain Collaborations	WIP	•Lean •'Work-Out'	

With this measure in mind, Company A then classified each theme in terms of its contributory factors and divided these between the reduction and management levers as described by Perona and Miragliotta, [1] and Hoole [16]. The results are shown above in table 2.

These formed the basis upon which improvement activities could take place. Currently, projects are ongoing and results are to be collated on a project scorecard.

7 Conclusions

Simplicity is not the panacea to business competitiveness, it must be tempered by the need to retain service and functional individuality, often causing system complexity.

By identifying the symptoms of complexity using the quickscan audit, the company could identify causes from the literature, identify their components in the business and go on to identify improvement activities for benefit without compromising the core business, hence competitive edge.

The focus on cost is a significant limitation, and as the complexity culture matures in Company A, a more rounded set of measurement data should be established. However, this single vision allowed the functional managers to identify particular improvement activities to reduce complexity without compromising customer response of product functionality. The results should be shard in the near future with increased detail around the improvement activities.

References

[1] Perona, M. and Miragliotta, G. (2004). Complexity Management and Supply Chain Performance assessment. A field study and a conceptual framework. International Journal of Production Economics. 90 (-), 103-115.

[2] Jenkins L., Wong G. and Barton R (2010) "Understanding the Decoupling Point to Optimise Manufacturing Strategy", Proceedings of the 8th International Conference on Manufacturing Research (ICMR2010), 14th - 16th September 2010, V. I. Vitanov and D. Harrison (eds).

[3] Towill, D. R. (1999) "Simplicity wins: Twelve Rules for Designing Fffective Supply Chains". Control the Journal of the Institute of Operations Management 25 , pp. 9-13.

[4] Womack, J., Jones, D., (1996) "Lean Thinking: Banish Waste and Create Wealth In Your Corporation", Simon And Schuster, New York.

[5] Ashkenas,R (2010) "Simply Effective: How to cut through complexity in your organization and get things done" Boston: Harvard Business Press. pp 82, 93-107

[6] Mason-Jones, R. And Towill, D.R. (1998), "Shrinking the supply chain uncertainty circle", Control, The Institute of Operations Management, Vol 24 No 7, pp 17-22.

[7] Wilding, R. (1998). The Supply Chain Complexity Triangle. International Journal of Physical Distribution and Logistics. 28 (8), 599-616.

[8] Bozarth,C.,C. , Warsing,D.,D., Flynn,B.B. and Flynn,E,J. (2009). The Impact of Supply Chain Complexity on Manufacturing Plant Performance [Online]. International Journal of operations management, Vol 27, No 1.

[9] Deshmukh, A.V. , Talavage J.J., and Barash,M.M. (1998). Complexity in manufacturing systems part 1 : Analysis of static complexity . IEEE Transactions. 30 (-), 645-655.

[10] Waldrop , M. ,1992. Complexity. Simon & Schuster, New York.

[11] Pham, D. T., Dimov, S. S. and Hagan, V.O'. (2001). Advances in Manufacturing Technology XV. London: Professional Engineering Publishing Limited. 49-60, 351-356.

[12] New, S.J. (1996). A Framework for analysing Supply Chain Improvement. International Journal of Operations and Production Management. 16 (4), 19-34.

[13] Gunasekaran , A. , Patel, C. and McGaughey , R.E.. (2004). A Framework for Supply chain performance measurement. International Journal of Production Economics. 87 (-), 333-347.

[14] Milgate, M. (2001). Supply Chain Complexity and Delivery Performance: An International Exploratory study. Supply Chain Management: An International Journal. 6 (3), 106-118.

[15] Vachon, S. and Klassen, R.D. (2002). An Exploratory Investigation on the effects of Supply Chain complexity on Delivery performance. IEEE Transactions of Engineering Management. 49 (3), 218-230.

[16] Hoole, R. (2005). Five ways to simplify your supply chain. Supply Chain Management : An International Journal. 10 (1), 3-6.

Innovative Production Machines and Systems
D.T. Pham, E.E. Eldukhri and A.J. Soroka (eds)

RFID-enabled quality controlled logistics as a tool to improve the sustainability of chilled produce distribution and supply chains: a critical analysis based on the features of actual international supply chains

P T Kidd[1] and Xuan-Tien Doan[2]

[1]Cheshire Henbury, PO Box 103, Macclesfield, SK11 8UW, UK, Email: paulkidd@cheshirehenbury.com
[2]134 CMT8 Street, Thu Dau Mot, Binh Duong,Vietnam, Email: xuantien@hotmail.com

Abstract

Considered is the sustainability of chilled produce distribution and supply chains. A technique known as *Quality Controlled Logistics*, which aims to control logistics based on the current and predicated quality of fresh produce, is critically analysed with respect to the actual features of modern international distribution and supply chains. Several features of these distribution and supply chains render RFID-enabled *Quality Controlled Logistics* unsuitable as a tool for improving sustainability. Hence it is proposed that other, non Information and Communication Technology (ICT) solutions should be pursued to improve sustainability, and that ICT could be better deployed in support of improving the quality of produce that is fed into these distribution and supply chains.

Keywords: Sustainable food distribution and supply chains; quality controlled logistics; RFID; sustainable agriculture; ICT for sustainable growth; sustainable intensification.

1. Introduction

Anecdotal evidence exists which suggests that losses in chilled, fresh produce, distribution and supply chains, are significant. In reality however, hard factual information about the scale of such losses is extremely difficult to find. One of the reasons for this is that most interest in food waste has centred on the consumer side [1, 2]. However, given the growing importance of reducing food losses, efforts are currently underway to establish exactly what losses do occur in chilled fresh produce distribution and supply chains [3].

Even though the scale of the losses is still unclear, it is evident that the food industry is interested in addressing supply chain sustainability issues [4]. Moreover, several techniques exist which may improve the sustainability of chilled fresh produce distribution and supply chains, which initially might be seen as adding some weight to the contention that the technologies already exist that will enable the solution of climate change problems [5]. One of these existing technology sets that could contribute towards this objective are Information and Communication Technologies (ICTs).

One particular ICT of interest is Radio Frequency Identification (RFID) technologies. Of interest in particular are RFID-enabled systems that seek to predict quality of fresh produce, based on measurements taken (using RFID sensors) while the produce is in the distribution and supply chain. Such systems use this use this quality information to control logistics. This approach is known as *Quality Controlled Logistics* [6].

This paper arises from a large scale study, begun in 2010, into the sustainability of agriculture, which has included an in-depth examination of the meaning of sustainability, and the investigation of issues surrounding food production, such as food distribution and supply, and food processing, as well as the biological and agronomy issues that are central to any efforts to apply ICTs in this area [7]. Within this paper the idea of using RFID-enabled *Quality Controlled Logistics*, as a tool to improve the sustainability of chilled fresh produce distribution and supply chains is considered. Using characteristics of actual international distribution and supply chains, issues relevant to the application and use of *Quality Control Logistics* are critically analysed.

2. Quality controlled logistics

The idea of using ICTs for chilled fresh produce distribution and supply chain applications is not new. Morris *et. al.* [8] addressed this matter at shipping container level, by fitting these containers with sensing devices that provided real-time monitoring capabilities. This enabled actual storage conditions to be monitored. By interfacing the remote real time measurement and tracking system to prediction software, such as a *remaining shelf life* prediction model, the system provided a decision support tool. Suggested by Morris *et. al.* [8] as possible outcomes of this decision support tool, were recommendations to cancel the shipment owing to a predicted loss of quality, or recommendations to adjust storage conditions to preserve storage life. Here can be seen the potential interest in such technologies as sustainability improvement tools, either by reducing losses, or avoiding the transportation of produce that is unlikely to reach its destination with the desired quality.

A similar idea has been proposed by Jedermann *et. al.* [9] only this time the sensing is based on RFID devices. These sensors are then linked to quality prediction models, with the functionality directed at initiating an alarm when produce goes beyond a specified temperature limit.

The concept of *Quality Controlled Logistics* [6] is essentially the same as that proposed by Morris *et. al.* [8], i.e. (typically) using the sensor information and quality predictions to take logistics decisions. For example, it is suggested that quality predictions (again based on a prediction model) could be used to take decisions about redirection of produce, with produce with a lower quality being directed to retailers willing to accept such produce. Another suggestion is that the quality predictions could be used to implement *first-to-expire-first-out* logistics, rather than *first-in-first-out* logistics. A similar proposal has also been made by Dada and Thiesse [10], who suggest using RFID-based sensors and a quality prediction model for quality based issuing of produce. Simulation studies have also been undertaken [6] that purport to show that by steering logistics flows using quality-oriented data, it is possible to deliver to retailers, produce with longer remaining shelf life, and to reduce, by 25%, waste at importers depots and retail outlets. Another simulation study of a supply chain [11] explores this concept of diverting lower quality produce to other outlets (discount stores). The results show economic benefits for the distribution channel in terms of being able to sell to discount retailers, produce that is unacceptable to supermarkets, while at the same time providing supermarkets with produce with a longer remaining shelf life. Elsewhere it has been suggested that any specific produce showing unacceptable deterioration could be identified and removed from the shipment [12]. These studies again demonstrate the potential interest in using the concept as a tool to improve the sustainability of chilled fresh produce distribution and supply chains.

3. Characteristics of international chilled fresh produce distribution and supply chains

This section describes the characteristics of international chilled produce distribution and supply chains relating to the shipment of short shelf-life produce, with specific reference to cut fresh flowers, sent by air freight from Equatorial and Tropical Regions into Europe. The information presented in this section is common knowledge to those working in the sector, but may not be widely known to those outside the sector. No specific supply chain is considered here, just general characteristics based on knowledge of the sector and modern supply chain practices. What is described here also relates to other types of fresh produce with a short shelf life sent to Europe by air freight, mostly from developing countries.

What follows therefore is a typical description of the process, from cutting of the flowers, to arrival at a distribution centre in Europe, at a level of detail necessary for the critical analysis that follows this section.

The first step in the process is harvesting. This is

typically done in the early morning (7-9am), the coolest part of the day, thus helping to keep *field heat* temperature of the crop to a minimum. Later in the process this *field heat* will need to be removed, so it makes sense to ensure that it is low as possible at the point of harvest. Sometimes however, it is also necessary to harvest later in the day, in the afternoon, depending upon demand.

Immediately after being harvested, the flowers are placed in buckets which contain post harvest solutions to help preserve the produce. Using open trolleys, the buckets are then carried to a temporary storage area that provides shade for the cut flowers while they a waiting to be transported to a pack house. These shaded temporary storage points typically lie very close to the place (e.g. greenhouses) where the crop is grown. Depending upon the sophistication of the grower, these shaded areas may be temperature controlled, but not always. Ideally the time spent in these shaded areas should be kept as short as possible. If the pack house in located on the farm, then regular collections (every 15 minutes) can be arranged. On smaller farms the pack house is likely to be located off-farm, serving a number of small growers. While being kept in the temporary shaded areas, the flowers may also be sprayed with a post harvest fungicide to help protect the flowers from botrytis infections during the transportation phase; botrytis infections being a major potential for loss of quality, especially for flowers such as roses. Because of the botrytis problem, flowers such as cut roses will need this preventative spraying.

Once flowers arrive at the pack house, cut flowers typically pass thought six stages, including: defoliation; 1st cold storage; grading; 2nd cold storage; wrapping; and blast chilling.

At the defoliation stage, flower stems are defoliated up to a third of the stem length. However, sensitive flowers might not go through this stage. The temperature within the defoliation room is maintained between 12 and 15 °C which is not optimal for the flowers, but is necessary for the comfort of people working on the flowers. During the 1st cold storage, the flowers are stored at 4-8 °C for at least 4 hours to help remove *field heat* before being graded. The purpose of this chilling is to ensure a longer vase life.

In the next stage, grading, the flowers are sorted according to their stem lengths and are bunched together as per customer specifications. Those that are bruised, or have shorter stem lengths than specified, or have pests or diseases, are rejected. Overall, this stage takes approximately 40 minutes. From the grading hall, the flowers are conveyed to a cold room for the 2nd cold storage. This period should be at least 4 hours. The storage temperature depends upon the variety of flowers. At the packaging stage, the bunches are wrapped in customer branded plastic sleeves, a packet of flower food is inserted, and bar codes and *display until* labels (as well as price/special offer labels) are attached to the flower bunches, which are subsequently packed into large cardboard boxes for dry shipping. Finally these boxes are sent to the blast chiller room, where they ideally remain for 1.5 hours, to remove the remaining *field heat*, with the aim of reducing the temperature to 3-5 °C.

When the shipment is complete, corresponding to the size of the customer order for the day, the boxes are loaded onto chilled trucks (on aircraft pallets), and transported to the airport, where they are off loaded into chilled storage conditions, to await loading onto an aircraft.

The whole process from harvest to dispatch can take an average of 11 h. Because of this, flowers cut one morning, would not be dispatched to the airport, till lunchtime on the following day, but in that time the flowers spend most of the time in chilled storage conditions. Journey time to the airport depends upon the distance, but typically in developing countries the flowers may be in the truck for several hours. Typically, each truck would be carrying 750 large boxes of flowers (stacked on aircraft pallets, ready for loading onto the aircraft).

At an appropriate time, the shipment is taken from cold storage at the airport, loaded onto the aircraft, and then transported to Europe. Typically flights tend to depart in the evenings, and arrive in Europe in the early morning, with flight times varying from 7 to 12 plus hours, depending on distance. On arrival, after customs clearance, the pallets are loaded into chilled trucks and transported to the importer's distribution centre, where they are usually kept in cold storage for a few days before being unpacked and then repacked into the supermarket's in-store display system, and then sent to the supermarket's own distribution centres. Sometimes however, flowers might not be stored, but used as soon as they arrive. It depends upon the supply-demand situation.

Overall, the duration from harvest to arrival at the importer's distribution centre can be between 2 to 3 days, and for most of this time the flowers are kept under varying storage temperature conditions.

While the above refers to cut flowers, the description of the process also corresponds to the handling of fresh food produce (green beans, etc.), with slight variations determined by produce features.

4. Critical analysis of quality controlled logistics applications

RFID can be used for track and trace applications [13], which in Europe are primarily driven by regulation, specifically by food traceability requirements. This is an application that can be expected to grow in use, simply for these regulatory reasons. What is addressed here, are the more sophisticated applications that propose to use quality predictions to control logistics (distribution and supply) chains. Specifically of interest are: the use of *first-to-expire-first-out* issuing policies; the idea of cancelling shipments; removing specific parts of the shipment; and produce diversion. These are considered in the context of the general characteristics of the distribution and supply chains used to ship, by air freight, short shelf life produce, such as cut flowers, into Europe, from developing countries.

With regard to cancelling the shipment, this would be extremely unlikely to be practicable in many cases. There are several reasons for this. The first is because of the short time from cutting to arrival at the shipping airport, which mostly occurs within a typical (approximate) timeframe of 24 hours. Provided that the flowers (and other produce), are properly handled, it is unlikely there would be significant stress issues that would render such a drastic step viable (although there may be some exceptions to this when handling produce that has temperature sensitive flavour characteristics). In some cases however, the sophistication at the growers' site might cause temperature stress problems, as they may not have appropriate on-farm chilling facilities. This however, is a problem of lack of availability of the right technologies at the growers' site. It would not be useful to apply RFID just to predict quality problems in such circumstances, as this would only penalise the growers as they would not get paid for produce that was rejected in this way, and it does not make commercial sense for growers to become involved in such a process. A more appropriate investment would be to make improvements to on-farm facilities. There are also measures that can be taken to reduce heat stress in such circumstances, such as providing shaded short-term storage areas for the cut produce, close to growing houses, as described in the previous section.

Once produce is sent from the farm or growing areas to the pack house, produce enters the cold chain. Provided that chilled trucks are used for transportation to the airport, the cold chain will typically be maintained at least up to the point where produce is taken from the cold store at the airport, out to the aircraft, to be loaded.

At this point it might be argued that it would be worthwhile to forward predict the condition of the flowers when they arrive in Europe. But here there are some difficulties. Generally, with good supply chain practices, and apart from the period between cutting and arrival at the pack house, cold chain continuity will have been maintained. If there are any issues, then they might lie in the area of *field heat* removed, which, if this has not been done properly, might cause problems later on.

In the case of cut flowers, one of the difficulties is that models that predict the vase life of cut flowers are not well developed, and also because the senescence of cut flowers is still not well understood. Another difficulty is that there are several factors at work that can be harmful to cut flowers (and other produce) while in transit; primarily, temperature, humidity, and physical (mechanical) damage. In the case of the latter, this is not something that can be predicted, and reliable models for the former two do not exist. Similar points can also be made in general about other types of fresh produce. To be of use a model for each specific cultivar is needed, and this creates tremendous problems given the large number of cultivars in commercial use.

It should also be noted that there are a few known temperature trouble spots in the air freight distribution and supply chain. One occurs between the flowers being taken out of cold storage at the airport, being loaded onto the plane, and the plane departing. Here the issue of temperature stress can arise if there are any delays in loading the flowers on the plane, or if the plane itself is delayed. Another trouble spot occurs if planes are diverted, on route, because of aircraft technical problems or air traffic control problems. These are essentially unpredictable occurrences. To cancel the shipment at this stage would not be practicable, as it would mean unloading one part of the plane's cargo, thus adding extra costs, and delaying the delivery of other cargo on the plane, which may also be other items of fresh produce, which could suffer damage by being delayed in this way. There would also be additional local disposal costs.

Even if it would be possible to forward predict quality problems, and then cancel the shipment at last minute before the cargo plane departed, the shipper will have booked space on the cargo plane and would still have to pay for this, so in effect they would be paying out for the transportation of nothing. Moreover, once the plane is outbound from the airport there is no choice but to continue with the shipment. This is why the idea of produce diversion is not very practicable, since the plane will not make a detour from its intended destination just because the quality of one part of the cargo is predicted to fall below a defined level.

Another factor that stands in the way of produce diversion, is that the flowers are typically grown for specific customers, and are packed into customer (supermarket) branded wrapping at the farm (an increasing common practice in the fresh produce sector), and consequently these flowers cannot be simply diverted to another market. This would involve significant costs in removing the flowers from the boxes, removing the customer specific packaging, and then repackaging them, or sending them onto a flower auction, for which there are a limited number of options. Thus it is simply not cost effective to consider activities that add costs in a sector where stakeholders generally operate with low profit margins and rely on sheer volume of turnover to make a profit.

The fact that cut flowers are packaged in customer specific packing at source also raises issues concerning the practicability of using *first-to-expire-first-out* issuing policies. Customer specific packaging also comes with specific *display until dates*, which are used by retailers so that they know when to removed unsold items from display. To practice *first-to-expire-first-out* issuing, it would be necessary for a distributor to be able to put together a shipment to a retailer, of the right quantity, with the same or very similar *display until dates*. This is easier said than done, for such an issuing policy implies creating one single shipment to a retailer with what is available, and leaving for later items, with shorter remaining life. Even if this is were feasible, it is very unlikely that retailers would be easily convinced that this was the right thing to do. In effect, one would have to be able to convince them that older produce did in fact have a longer vase life than more recently delivered produce. It could also imply sending produce to retailers with expired *display until* dates. All this has implications.

To implement the above it would be necessary to be able to make very reliable, repeatable and accurate predictions, which is unlikely to be achievable in the short-to-medium term timeframe. There is of course, also the cost implication for the supply chain stakeholders, who would probably have to invest in modifying software and retraining staff, and would also encounter additional costs as they would need to add *display until* labels in Europe, where wages are much higher than they are in developing countries.

Another factor that should also be considered is the size of the shipment, which consists of several hundred boxes, typically with 50 to100 flower bunches in each. These boxes are stacked onto aircraft pallets, which are in themselves large, and are then covered in netting to prevent the boxes moving while in transit. This is another aspect of real distribution and supply chains that does not fit well with *Quality Controlled Logistics*. This method of shipping flowers means that it would be necessary to use several hundred (expensive) RFID based sensors, and what is more, it would be impossible, with the type of boxes currently used and the means of stacking, to extract signals from these, as these signals would be blocked by a mass of biological material (the cut flowers). What is more, to obtain a reliable prediction of quality, internal box level tracking would be needed, as it is what is happening within the boxes which need to be recorded, as this can be quite different to ambient conditions, owing to biological factors such as respiration, among other issues. Conditions within the individual boxes can also vary. This is why container level sensing for quality prediction is not practicable. It is also easy to see that the removal of one box from the shipment would be impracticable.

5. Conclusions

Analysis of the characteristics of international fresh produce distribution and supply chains, the specific features of fresh produce, the technical challenges involved, packaging practices, and the enormous volume of produce involved, suggests that many of aspects of *Quality Controlled Logistics* are impracticable. Some of these issues have been addressed in the preceding section. What then are the implications?

One conclusion that can be drawn is that the concept of *Quality Controlled Logistics* is not well matched to the characteristic of modern chilled fresh produce distribution and supply chains. For this reason, its usefulness as a tool to support improvements in the sustainability of these chains is very doubtful.

This suggests that improvements in sustainability may need to be achieved through other solutions that significantly reduce or eliminate the need for RFID-enabled *Quality Controlled Logistics*. This would then allow the use of low-cost RFID devices in much simpler track and trace applications (for regulatory compliance), which would be beneficial for supply chain stakeholders, as they would be able to use general systems deployed in other logistics applications, thus enabling some economies of scale to be achieved. Applying RFID-enabled *Quality Controlled Logistics* is an additional cost burden, given that it is a specialised application for the fresh produce sector. It is also an application that is probably not needed, as other solutions can be considered.

These other solutions include: (i) plant breeding to develop cultivars more robust to the vagaries of chilled supply chains; (ii) further development and application of existing technologies that will help to reduce loss of quality (e.g. modified atmosphere packing); (iii) improved supply chain practices that reduce exposure to stresses that cause loss of quality; and (iv) attention to developing improved growing practices to ensure that higher quality produce is fed into food distribution and supply chains.

All the above, in effect, should reduce the need for ICT-based solutions and thus the need for *Quality Controlled Logistics*. Instead, ICT-based applications may be better deployed in support of, for example, assisting growers to develop better growing practices that deliver produce of higher quality.

It can also be argued that these other solutions might in fact be better as these reduce the need for the investment in more sophisticated ICTs in the supply chain, which can be difficult to deal with given that there are many supply chain actors involved, no well established mechanisms for sharing costs and benefits, and also often inequality in terms of the power of one actor to influence another or to determine supply chain practices.

In effect, what is being proposed here is actually a systems level approach to sustainability [14], within the context a bio and agronomy centric perspective.

References

[1] Kantor, L.S., Lipton, K., Manchester, A., Oliveria, V. Estimating and addressing America's food losses, *Food Review*, Jan-April, (1997) 2-12.

[2] WRAP. Food Waste Report: The Food We Waste. 2008. (http://www.wrap.org.uk/retail/case_studies_research/report_the_food_we.html)

[3] DEFRA. Food Matters: One Year On. 2009. http://www.defra.gov.uk/foodfarm/food/pdf/food-matters-oneyearon090806.pdf

[4] Smith, B.G. Developing sustainable food supply chains. *Phil. Trans. R. Soc.* B, 363, (2008) 849-861.

[5] Pacala, S. and Socolow, R. Stabilization Wedges: Solving the Climate Problem for the Next 50 Years with Current Technologies, *Science*, 305, (2004) 968-972.

[6] van der Vorst, J.G.A.J., Tromp, S., van der Zee, D-J. A simulation environment for the redesign of food supply chain networks: modelling quality controlled logistics. In Kuhl, M.E., Steiger, N.M., Armstrong, F.B. Joines, J.A. (Eds.), Proc. 37th conference on winter simulation, 1658-1667, Orlando, Florida, 2005.

[7] Kidd, P.T. *Sustainable Intensification of European Agriculture: Bio and Agronomy Centric ICT as an Enabling Component*. Cheshire Henbury Publications, Macclesfield, 2011 (to be published).

[8] Morris, S.C., Jobling, J.J., Tanner, D.J., Forbes-Smith, M.R. Prediction of Storage or Shelf Life for Cool Stored Fresh Produce Transported by Reefers. *Acta. Hort.* 604, (2003) 305-311.

[9] Jedermann, R. Ruiz-Garcia, L., Lang, W. Spatial temperature profiling by semi-passive RFID loggers for perishable food transportation. *Comput. Electron. Agric.*, 65(2), (2009) 145-154.

[10] Dada, A., Thiesse, F. Sensor applications in the supply chain: the example of quality-based issuing of perishables. In: Floerkemeier, C., Langheinrich, M., Fleisch, E., Mattern, F., Sarma, S.E. (eds.) The Internet of Things, Springer-Verlag, Berlin, 140-154, 2008.

[11] Wienk, I., Canters, R., Kreft, F., Tromp, S. *SustainPack project deliverable 6.14: report on first simulation model that predicts the feasibility of communicative packaging concepts.* 2005 (http://www.sustainpack.com).

[12] StePac, Xsense System, 2010 (http://www.stepac.com/Xsense-QA-System.asp)

[13] Smith, I.G. and Furness A. (eds.) *Food Traceability Around the World.* Vicarage Publications, Halifax, England, 2008.

[14] Kidd, P.T., Systems level approaches to addressing the energy and climate change objectives of the European Union: the need for agility to adapt R&D management approaches to achieve the sustainable intensification of manufacturing. IPROMS 2010.

Innovative Production Machines and Systems
D.T. Pham, E.E. Eldukhri and A.J. Soroka (eds)

Systems level approaches to addressing the energy and climate change objectives of the European Union: the need for agility to adapt R&D management approaches to achieve the sustainable intensification of manufacturing

P T Kidd

Cheshire Henbury, Tamworth House, PO Box 103, Macclesfield, SK11 8UW, UK
Email: paulkidd@cheshirehenbury.com

Abstract

Considered are system level issues in relation to addressing the energy and climate change objectives of the European Union. System level issues, such as the concept of *Energy Positive Manufacturing Neighbourhoods*, *Embodied Energy*, and *Energy Return on Investment*, point towards a new approach to managing manufacturing R&D. The barriers to this include existing, *taken for granted assumptions* about what to fund, how to manage manufacturing research, and also how research funding agencies design research programmes. It is argued that research into R&D management is needed to develop new methods of research better suited to systems-based considerations, moving beyond the paradigm trap whereby research tends to be focused upon *operational energy* savings, even though there is no guarantee that such work will provide a basis for global reductions in energy consumption and greenhouse gas emissions. It is also argued that adaptation of R&D management approaches to meet the new circumstances, highlights the strategic importance of agility, which has in the past being too much identified with operational changes that are in fact just matters of flexibility.

Keywords: Sustainable manufacturing; energy efficiency in manufacturing; carbon abatement in manufacturing; operational energy efficiency; embodied energy; climate change; greenhouse gas emissions, agility; R&D management; energy return on investment; energy positive manufacturing neighbourhoods; sustainable intensification of manufacturing.

1. Introduction

Within the Strategic Research Agenda (released into the public domain in the latter part of 2006) of the European *Manufuture* Technology Platform [1] there is not a single reference to the rising costs of energy, to the depletion of oil reserves, to climate change and global warming, or to greenhouse gas emissions, all of which have been key strategic issues for European manufacturing since at least the year 2000, and in reality probably much earlier. This is a fact that really does raise some serious credibility issues with respect to Europe's *Manufacturing Futures* thinking, especially given the claimed strategic nature of the *Manufuture* Research Agenda and its focus on *Assuring the Future of Manufacturing in Europe*.

Only fairly recently, and following the political decision in 2007 by EU heads of Government [2], has *Manufuture* begun to address the matter of climate change related issues, in what is in effect, an afterthought [3, 4]. Business as Usual, or BAU as

Pacala and Socolow [5] label it, seems to have been the overriding preoccupation of *Manufuture*.

This paper addresses Europe's energy and climate change objectives from the perspective of manufacturing industry. Explored are some of the systems level issues which lead to the proposal that, to effectively address these objectives, a new approach to R&D Management is needed for European Manufacturing research. This need for a new approach is relevant, not only to researchers themselves and the way that they operate, but also to research funding agencies. Pointers towards the research needs relating to the development of such a new approach to R&D management are considered. In other words, what is advocated is research into the way manufacturing research is undertaken, with the aim of modifying these methods to fit better the new circumstances brought about by the need to address climate change objectives. The central importance of agility [6, 7] is highlighted.

2. European manufacturing at a crossroads

The main energy and climate change objectives of the EU are: a 20% increase in the use of renewable energy (wind, solar power, etc.); a 20% cut in energy consumption; and a 20% reduction in greenhouse gas emissions; all by the year 2020.

Further intensification of the production of manufactured goods in Europe (either directly in Europe, or indirectly through offshoring), and the associated consumption that this creates, upstream and downstream, needs to be done in a sustainable way. While sustainability can be viewed as a multi-dimension concept, in the context of energy and climate change objectives, sustainability means achieving a *global* reduction in energy use and greenhouse gas emissions, while continuing to seek economic growth in the manufacturing sector. Moreover, the requirement is to address multiple objectives, simultaneously, with the objectives of: achieving security of energy supply, reducing dependency on fossil fuels such as oil and natural gas, stabilising and then lowering greenhouse gas emissions, and reducing energy consumption [8, 9]. Collectively this can be termed, *sustainable intensification*, with a note that other environmental factors relating to pollution, reduction in the demand for increasingly scarce resources (such as water, copper, etc.), etc. must be included as well. The key phrase in these considerations is: *global reductions*.

The above is not a trivial problem, and it will not be solved through incremental changes and improvements, of the type for example, that might arise by improving the energy efficiency, by a few percent, of a machine tool. What are needed are radical transformations.

The radical nature of what is implied by the pursuit of sustainable intensification (in the context defined above) can be better understood by examining a few systems level issues.

The first of these system level issues is a solution that is beginning to be contemplated elsewhere (mostly in the context of urban residential areas). This solution, which is increasing being seen as a viable approach for managing energy, is that of *Energy Efficient Neighbourhoods* [10], which are sometimes also referred to as *Energy Positive Neighbourhoods*.

An *Energy Positive Neighbourhood* is a systems level concept where the neighbourhood generates more energy than it consumes, with the surplus energy being either stored locally or exported. This definition is neutral in terms of the specific details about the energy generated and used, which makes the concept a holistic one, as it encompasses all feasible energy production and use within the neighbourhood, and its transfer around the neighbourhood, in response to changing needs. This holistic aspect is one of the key advantages of the concept, as it enables those within the neighbourhood to manage and reduce their total energy use. The concept is also generic as it can be tailored to specific circumstances and priorities that prevail in neighbourhoods.

Although the definition is formulated in terms of energy use, it is clearly linked to other strategic objectives, since to achieve an *Energy Positive* circumstance requires a reduction in energy consumption within the neighbourhood. One of the consequences of this is that greenhouse gas emissions resulting from energy consumption should also be reduced. This reduction could be further enhanced by the use of mirco-generation within the neighbourhood, for example by using various small-scale generating capabilities and the use of waste streams for energy production.

The proposed concept is also an integrating one, as it brings together many technologies and techniques to deliver an *Energy Positive* circumstance. One of these technologies are Information and Communication Technologies (ICTs) which have one unique feature, in that these can be used in an integrating fashion, both in terms of providing design tools, and for the purpose of operational monitoring and control, as well as delivering specific solutions in their own right.

Presently, the concept of an *Energy Efficient* or an *Energy Positive Neighbourhood* is being addressed with the context of urban areas and cities (mostly residential or with a mixture of residential, retail, office and light commercial premises), but in principle the concept can also be applied in industrial areas dominated by manufacturing facilities of various kinds.

One of the key aspects of the concept is that of cooperative behaviour. This certainly includes cooperation among energy users and suppliers to achieve a flatting of demand curves, which will help energy suppliers to avoid the use of rapid response generating capacity, which is both expensive to run and which commonly involves the use of natural gas, with the associated greenhouse gas emissions. In the case of *Energy Positive Manufacturing Neighbourhoods* there would also be a need for cooperation among the manufacturing businesses within the neighbourhood, to manage their energy production and use to achieve an *Energy Positive* circumstance.

The second system level issue that should be considered, and one that poses a barrier (resulting from past behaviour) to innovative thinking, is that the traditional focus of energy reduction in manufacturing has been on the reduction in *operational energy* use (e.g. see [11, 12]). But there is another aspect to energy issues in manufacturing (and elsewhere), in that to reduce *operational energy* use, one first has to invest energy in replacing old equipment with new, or by adding new technologies, all of which implies increased energy consumption (e.g. in transportation of raw materials, manufacturing, installation, disposal of the old, etc.). Hence there are *Energy Investment Criteria* to consider. In other words, does the *operational energy* saving that will be delivered through the new technologies, etc., exceed the investment in energy that has to be made to make the proposed changes, and what is the *energy payback time*?

The above is referring to what in the construction sector is called *Embodied Energy* [13, 14] The point is that there is no value to society at large, and to the objective of achieving global reductions in energy consumption, if manufacturing companies invest in energy saving measures, which may provide a financial return on investment for specific manufacturing companies, but which only serve to increase global energy consumption. What is needed from any energy saving project within a company, is a good return on the *Energy Investment* in the global context, within a timescale appropriate to the lifecycle of the technologies used. This is a systems level issue, and serves to reinforce the need for consideration of concepts such as *Energy Positive Manufacturing Neighbourhoods*. The key issue is to deliver global energy savings at neighbourhood level, and if this concept can be successfully addressed, then this provides the basis for expanding the boundaries of the neighbourhood, or linking together these neighbourhoods, so that global energy savings can be achieved at increasing (higher) levels of system complexity.

This then is just a glimpse into the complexity of the issues involved in this domain. Further insights are provided in [8, 9, 14]. The main point here is that solutions for addressing the challenging energy and climate change related problems faced by manufacturing lie in systems level approaches and with behavioural changes. This has several implications that need to be addressed, one of which relates to R&D Management, but before exploring such matters, a significant barrier to progress needs to be highlighted; that of paradigm shifts.

All those involved in manufacturing have a set of core beliefs and assumptions specific and relevant to the sector, which are shared in common with others [6]. This is the essence of a paradigm: it is taken for granted and not seen as problematic by those who adhere to it. The paradigm defines how things are done and what is acceptable [15].

A paradigm, however, is a double edged sword, because it is both helpful and unhelpful, depending on the circumstances. Associated with a paradigm is a mind-set, a way of thinking which means that people become over-sensitised to some particular aspect of their domain (in this case manufacturing) or available information, at the expense of other parts. This over-sensitising is useful, because it helps people to become sensitised to important things, and to patterns that remind people of problems successfully solved, and this often serves very well.

The problem with paradigms however, is that they blind people to discontinuities that render past approaches and solutions inappropriate. Discontinuities are defined as non-linearities that either render aspects of prevailing practices inappropriate, or which provide new opportunities, or open up entirely new ways of working. This makes extrapolating into the future based on the past, an exercise of little value.

When discontinuities occur, the success stories of yesterday can have little relevance to the problems of tomorrow. In fact, according to Handy [16] these success stories might even be damaging since the world, at every level, has to be reinvented to some extent.

The failure to recognise discontinuities often leads to the adoption of incorrect change strategies. The assumption is that whatever needs to change is just a matter of evolution in approaches. The evolutionary view is based on the belief that there is nothing fundamentally wrong with the way things are being done, and that all that is required are some slight adjustments to compensate for the changed circumstances.

The belief that only slight adjustments are needed to a paradigm in order to bring the way things are done into line with new requirements has a technical name. In terms of change this view corresponds to model known as *morphostatic change* [6]. This simply means that the type of change people face is incremental. The established order is maintained by treating disturbances as external noise requiring minor adjustments. This leads, for example, to efforts to improve the sustainability of manufacturing by focusing on low-level energy efficiency issues, while maintaining the fundamentals of the existing system. Yet the complexity and nature of the challenges point to more than just *morphostatic change*.

What is being advocated in this paper is the presence of discontinuities that render *morphostatic change* inappropriate. In effect, the challenges posed by energy and climate change objectives are so demanding that just working in the way that has been the norm in the past, is no longer a viable option. In these circumstances what is needed is a major change to the existing paradigm. In terms of the theory of change what is advocated is called *morphogenic change*. This simply means a type of change that produces a different order to that which existed previously. Disturbances are treated as information about the inappropriateness of established practices, and this leads to fundamental changes in methods and principles [6].

The central argument here is that the presence of discontinuities means that a new approach to Manufacturing R&D management is needed, when it comes to the matter of addressing energy and climate change issues. This is one of the most important ramifications of the systems level issues highlighted above. Essentially, to address these objectives requires both a systems level approach to Manufacturing R&D management as well as to manufacturing operations.

3. Towards a new approach to manufacturing R&D management

Typically Manufacturing R&D is addressed in a bottom up way, with researchers pursuing topics of interest, and with research funding agencies willing to support such interests. Sometimes this bottom-up research is undertaken within the context of a strategy (defined by a specific research programme which defines topics of interest, which (it is hoped) have some strategic significance). This provides a strategic dimension.

However, the strategy itself may be flawed owing to the paradigm issues discussed above. For example, the research funding agency may believe, or may even have been advised, that funding low-level energy efficiency projects, e.g. improving the efficiency of machine tools, is appropriate. However underlying this decision is a set of taken for granted assumptions, e.g. researching the energy efficiency of machine tools, logistics chains, etc. is appropriate because it will reduce *operational energy* use, and hence in principle save money for those who invest in more energy efficient operations, which is an important issue in a high-energy cost environment. But there is no guarantee that such research will make any contribution at all to the reduction in global energy consumption.

Importantly this has implications for the research funding agencies. They need, in fact, a better way of deciding upon a research strategy and taking decisions about what to fund. This provides an example of why research into Manufacturing R&D Management is needed, for there is a need to investigate and define Research Strategy development processes that capture the complexity of the new circumstances, which represent a discontinuity with the past.

Importantly however, it should be mentioned here, that the above does not imply an *either or* circumstance [7]; a binary choice between lower level (more bottom-up research), and top-down systems level approaches to research. This would be a simplistic interpretation, and one that typifies the mind-set of industrial era thinking, that is still prevalent today, even though it is claimed that society has entered the knowledge era [17]. Between the two choices exists a continuum of possibilities, largely unexplored, where positioning is determined by technological feasibility, production economics, lifecycle carbon and energy matters, safety issues, product characteristics, and so forth.

What is actually needed is an approach that delivers both dimensions. There may in fact be an important principle here, and it is the principle of combining opposites. This involves moving beyond *either/or* constructs and considering oxymorons, that is to say, seemingly contradictory notions achieved simultaneously [18]. Combination of opposites is, for example, what *mass customisation* (cost efficient manufacturing of individually customised products) is about [19].

Similar remarks also apply to the way that Manufacturing R&D is undertaken. Common to both research strategy development and it implementation, and the actual execution of the research, is a need for taking into consideration the system level issues, capturing for example, *Embodied Energy* considerations, and *Energy Return on Investment*, along with other matters such as Carbon Lifecycle emissions, energy security of supply, etc. These considerations point towards a research agenda in R&D Management, directed at developing new research methods along with supporting ICT tools. This, it is suggested, is part of the basis for the creation of a circumstance where *sustainable intensification* of manufacturing will be possible. In effect, what is being proposed here is research driven by what is called *double loop learning* [20] whereby the problems with the way R&D is currently managed, are considered, and then new methods are proposed that better match the changed circumstances. This also illustrates the real importance and meaning of Agile Manufacturing [6, 7], namely an ability to deal with structural changes occurring in the business environment, as a result, for example, of discontinuities (non-linearities), and then to adapt method and processes as needed to meet the requirements of these new circumstances.

4. Conclusions

Ahead lies some significant challenges for Europe's manufacturing companies (and for manufacturing businesses located in other regions). In the past, when other fundamental challenges and issues have arisen (e.g. lean production concepts; the importance of manufacturing strategy; the need to adopt new organisational models; fragmentation of markets; new competitive conditions; new technologies; and so on), manufacturing, as a sector, has often been very slow to recognise such changes, or has not fully appreciated their significance and importance, which is one of the core problems (perhaps) inherent to a sector with a long history, and with (too) many taken for granted assumptions.

There are also issues such as levels of awareness and understandings of complex matters not typically addressed by the manufacturing R&D community, and also issues such as over emphasis on certain topical matters (e.g. carbon foot-printing) which fail to adequately capture the full complexity of what needs to be addressed.

This paper has highlighted that Europe's climate change objectives represent yet another fundamental challenge, whereby, taken for granted assumptions and *business as usual* could prevent Europe from doing the right things at the right time, or put another way, could result in Europe doing the wrong things now, and the right things too late.

It is not the intention of this paper to propose what a new system of R&D Management should be that reflects the new circumstances and future needs, but only to point out the existence of a discontinuity, and the need for R&D Management Research to address this. This is but one step in a process of radical change that will be necessary to address challenging current climate change related objectives, and the even more challenging objectives that will no doubt follow in the run up to 2020.

What is addressed in this paper also illustrates the real importance and meaning of Agile Manufacturing [6, 7], namely an ability to deal with structural changes occurring in the business environment, as a result, for example, of discontinuities (non-linearities). This interpretation of agility has more strategic importance that presenting Agile Manufacturing as a means of dealing with operational changes, which are in fact matters related to Flexible Manufacturing [7].

It should also be mentioned that within the issues raised in this paper lies the potential for development of real and sustainable competitive advantage, as well as for European manufacturing to become a role model, helping to demonstrate to emerging economies that a less resource intensive route to economic development exists; a route that leaves for future generations resources such as oil, which is already becoming a scarce and expensive resource.

References

[1] *Manufuture* Strategic Research Agenda, Assuring the Future of Manufacturing in Europe, Report of the High-Level Group, September 2006. (http://www.manufuture.org/manufacturing/?page_id=10).

[2] BBC 2007, EU agrees renewable energy target, 2007, BBC News Web Site, (http://news.bbc.co.uk/1/hi/world/europe/6433503.stm).

[3] Westerkamp, E. (2007), Sustainable Manufacturing, *Manufuture* 2007 Conference, Dec 2-4, Porto, Portugal, (http://manufuture2007.inescporto.pt/documents/morning-day-1).

[4] Jovane, F., Westkämper, E., and Williams, D. (eds.). The ManuFuture Road: Towards Competitive and Sustainable High-Adding-Value Manufacturing. Springer-Verlag, Berlin/Heidelberg, 2009.

[5] Pacala, S. and Socolow, R. Stabilization Wedges: Solving the Climate Problem for the Next 50 Years with Current Technologies, *Science*, 305, (2004), 968-972.

[6] Kidd, P.T. *Agile Manufacturing: Forging New Frontiers*. Addison-Wesley, Wokingham, 1994.

[7] Kidd, P.T. Agile Holonic Network Organizations. In *Encyclopedia of Networked and Virtual Organizations*. Putnik G.D. and Cunha M.M (eds.). IGI Global Publications, Hershey PA, 2008.

[8] Kidd, P.T. The Future of European Manufacturing: Driven by Globalisation or Global Warming? In: *European Visions for the Knowledge Age: A Quest for New Horizons in the Information Society*, Kidd P.T. (Ed.). Cheshire Henbury Publications, Macclesfield, 2007.

[9] Kidd P.T., Sustainable energy consumption in European manufacturing: the driver for a new research agenda addressing climate change, energy supply, and global competitiveness. In: Innovative Production Machines and Systems: 4th I*PROMS Virtual International Conference, D.T. Pham, E.E. Eldukhri & A.J. Soroka (Eds.). Whittles Publishing, Dunbeath, 2009.

[10] European Commission, Towards an ICT Infrastructure for Energy Positive Neighbourhoods, 2009. (http://ec.europa.eu/information_society/activities/sustainable_growth/docs/elsa/elsa_report/ELSA-EnergyPositive-Report1.pdf).

[11] Ross M. Improving the Efficiency of Electricity Use in Manufacturing, *Science*, 244, (1989) 311–317.

[12] Thollander P., Danestig M. & Rohdin P. Energy Policies for Increased Industrial Energy Efficiency: Evaluation of a Local Energy Programme for Manufacturing SMEs. *Energy Policy*, 35, (2007) 5774-5783.

[13] Yohanis, Y.G. & Norton, B. Life-cycle operational and embodied energy for a generic single-storey office building in the UK, *Energy*, 27(1), (2002) 77-92.

[14] Kidd P.T., Energy efficiency improvements and carbon emission abatement in European manufacturing: a strategic decision framework in response to climate change. In: Innovative Production Machines and Systems: 5th I*PROMS Virtual International Conference, D.T. Pham, E.E. Eldukhri & A.J. Soroka (Eds.). Whittles Publishing, Dunbeath, 2010, pp. 363-368.

[15] Johnson G. Process of Managing Strategic Change. *Management Research News*, 11, (1988) 43-46.

[16] Handy, C. *Beyond Certainty: The Changing Worlds of Organisations*. Arrow Business Books, Reading, 1996.

[17] Kidd, P.T. *European Visions for the Knowledge Age: A Quest for New Horizons in the Information Society*. Cheshire Henbury Publications, Macclesfield, 2007.

[18] Davis, S. *Future Perfect*. Addison-Wesley, Reading MA, 1987.

[19] Pine, B.J. *Mass Customization: The New Frontier in Business Competition*. Harvard Business School Press, Boston MA, 1993.

[20] Argyris, C. and Schon, D. *Organizational Learning: A theory of action perspective*. Addison-Wesley, Reading MA, 1978.

www.ingramcontent.com/pod-product-compliance
Ingram Content Group UK Ltd.
Pitfield, Milton Keynes, MK11 3LW, UK
UKHW051127260726
13967UKWH00010B/2916